IMPA Monographs

Volume 4

This series, jointly established by IMPA and Springer, publishes advanced monographs giving authoritative accounts of current research in any field of mathematics, with emphasis on those fields that are closer to the areas currently supported at IMPA. The series gives well-written presentations of the "state-of-the-art" in fields of mathematical research and pointers to future directions of research.

More information about this series at http://www.springer.com/series/13501

Letterio Gatto • Parham Salehyan

Hasse-Schmidt Derivations on Grassmann Algebras

With Applications to Vertex Operators

Letterio Gatto
Politecnico di Torino
Torino, Italy

Parham Salehyan
São Paulo State University
São José do Rio Preto Campus
São José do Rio Preto, Brazil

IMPA Monographs
ISBN 978-3-319-81134-5 ISBN 978-3-319-31842-4 (eBook)
DOI 10.1007/978-3-319-31842-4

Printed on acid-free paper

This Springer imprint is published by Springer Nature
The registered company is Springer International Publishing AG Switzerland

A Sheila e ai nostri 10 *figli,*
Giuseppe e Giuliano

be mādar va pedar-e azizam

Preface

This book is a revised and expanded version of the material that the first author presented in a mini-course given at IMPA, about the same topics, on the occasion of the *30° Colóquio Brasileiro de Matemática* (July 2015), and is largely based on the collaboration with the second author and Inna Scherbak. It aims at introducing and advertising the notion of *Hasse–Schmidt derivation on a Grassmann algebra*, which we propose as a natural language to put many seemingly unrelated subjects into a unified framework.

The broadened perspective it offers is able to capture, for example, a generalization of the Cayley–Hamilton theorem, the theory of linear ODEs and their generalized Wronskians, the exponential of a matrix with indeterminate entries (another reading of Putzer's method dating back 1966), the characterization of decomposable tensors in an exterior power of a free abelian group and the bosonic expressions of the vertex operators generating the so-called free fermionic vertex superalgebra.

The latter comes from the representation theory of the *oscillator Heisenberg algebra*, which is one of the fancy terms scattered throughout the text, other such being *bosonic and fermionic Fock spaces*, *boson–fermion correspondence*, *vertex operators*, *vertex algebra* and KP *hierarchy*, to name a few. We believe that the common formalism underpinning the aforementioned subjects proves interesting for its own sake. At the same time we wish to stress, once and for all, that this book is about none of them. Readers are not assumed to have any prior familiarity with such a language, for which we point out throughout the text many existing and excellent references.

The book is suited for Ph.D. students wishing to learn additional features of exterior algebras in connection with more advanced topics. It is our hope, however, that diverse scholars, both algebraic geometers and not, will profit from it:we tried to illustrate material that may not belong to everybody's field of research – or taste – but at the same time has the subtle potential of appealing to many, due to its far-reaching breadth.

Torino, Italy Letterio Gatto

São José do Rio Preto, Brazil Parham Salehyan
January 2016

Acknowledgments

First and foremost, we acknowledge the organizing committee of the *30° Colóquio Brasileiro de Matemática* (IMPA, 27–31 July 2015) for hosting the short course that led to these notes, notably professor *Carolina Araujo* to whom our deepest feelings of gratitude go: her support was invaluable and eventually instilled in us the necessary enthusiasm to take on this project.

We are also deeply indebted to professor *Paulo Sad*, for having encouraged us to submit the present text for publication within this monograph series, and to Simon G. Chiossi, who generously and patiently read all the preliminary drafts, making abundant remarks here and there and enabling us to substantially improve the exposition, not to speak of his truly invigorating support. Extra special thanks are also due to professors Marcos Jardim and Paolo Piccione for unreserved support.

The first-named author (LG) is indebted to Inna Scherbak, for her enduring and friendly collaboration and for her many remarks on the first version of this work; to Maxim Kazarian, who first suggested the relationship between derivations on an exterior algebra and the boson–fermion correspondence; to Ilka Agricola, Simon Chiossi, Martin Gulbrandsen and Andrea Ricolfi for hosting preliminary talks on the subject in Marburg and Stavanger; and especially to Jorge Cordovez for having been very close to us during the redaction process and to Maria Jack for language consultancy. Moreover, LG is deeply thankful to Anders Thorup and Gary Kennedy for long-standing friendship; to Alex Kasman, Igor Mencattini and Carlos Rito for stimulating discussions on PDEs; to Alessandro Ardizzoni and Paolo Saracco for enlightening explanations on infinite exterior powers; and to professor Louis Rowen for precious suggestions, not to mention his tenacious backing.

Both authors are grateful to the head of the Scientific Office at the Italian Embassy in Brasilia, Professor Roberto Bruno, for his personal and institutional support.

For many diverse reasons, we are also thankful to Roberto Alvarenga, Dan Avritzer, Claudio Bartocci, Roberto Bedregal, Pablo Braz e Silva, Ugo Bruzzo, Frediano Checchinato, Valeria Chiadò-Piat, Marco Codegone, Giovanni Colombo, Marc Coppens, Fernando Cukierman, Caterina Cumino, Eduardo Esteves, Fabio

Favaro, Umberto Ferri, Anna Fino, Paolo Giaquinta, Flavia Girardi, Sabrina Gonzatto, Antonella Grassi, Abramo Hefez, Joachim Kock, Daniel Levcovitz, Angelo Felice Lopez, Angelo Luvison, Simone Marchesi, Marina Marchisio, Luisa Mazzi, Nivaldo Medeiros, Paolo Mulassano, Barbara Mussio, Antonio Nigro, Karl–Otto Stöhr, Marco Pacini, Diego Parentela, Elisa Pilot, Pietro Pirola, Piotr Pragacz, Francesco Raffa, Gaetana Restuccia, Paulo Ribenboim, Andrea Ricolfi, Clara Silvia Roero, Lucio Sartori, Boris Shapiro, Aron Simis, Fernando Xavier Souza, Carlos Tomei, Mario Trigiante, Emanuela Truzzi, Israel Vainsencher and, especially, *Laura Tonini*, who helped LG to look at the world with different eyes.

Moreover, both authors are deeply grateful to Robinson Nelson dos Santos of the Springer Editorial Staff for careful editorial assistance and Mr. Santhamurthy Ramamoorthy of the Spi Global, for his very precious help to improve the very final version of the notes, as well as the Production Editor Mr. Anthony Reagan Chinappan.

Last, but certainly not least, we owe a lot to the precious and friendly help of the entire staff at IMPA, notably *Maria Celano Maia, Priscilla Fernandes Pomateli, Rogério Dias Trindade, Rafael Simão Rodriguez, Noni Geiger, Ana Paula da Fonseca Rodrigues and Sergio Ricardo Vaz*, as well as *Suely Lima, Pedro Luis Darrigue de Faro, Sonia M. Alves, Juliana C. Bressan, Leticia R. Nascimento, Jurandira F. R. Nascimento, Paula Cristina R. Dugin, Rosana de Souza, Vanda Silvestre and Roseni Victoriano.*

This work was partially supported by: Instituto Nacional de Matemática Pura e Aplicada (IMPA); Italy's 'National Group for Algebraic and Geometric Structures, and their Applications' (GNSAGA–INdAM); FAPESP-Brazil, Proc. 2012/02869-1 and 2015/04513-8, Ufficio Scientifico dell'Ambasciata d'Italia a Brasilia; PRIN 'Geometria delle varietà algebriche'; UNESP – Campus de São José do Rio Preto; Dipartimento di Scienze Matematiche 'G. L. Lagrange' del Politecnico di Torino; Istituto Superiore Mario Boella and Madi; Medas Solutions and Filters srl. All these institutions are warmly acknowledged.

Partial List of Notation

We adopt general standard conventions: $\mathbb{Z}, \mathbb{N}, \mathbb{Q}, \mathbb{C}$ denote, respectively, the ring of the integers, the monoid of non-negative integers and the field of the rationals and of the complex numbers. We also use the notation $\mathbb{Z}_+ = \mathbb{Z}_{\geq 0} = \mathbb{N}$. The positive integers are denoted by $\mathbb{N}^*$. The unit of a ring A is denoted 1_A or simply 1 if clear from the context. The identity endomorphism of an A-module M is denoted by $\mathbb{1}_M$.

$M[[t]]$	Formal power series with M-coefficients	24
$A[X]$	The ring of polynomials in an indeterminate X with coefficients in A	24
$M((t))$	Formal Laurent series $M[t^{-1}, t]]$	25
$\det(\cdot)$	Determinant of a square matrix or of an endomorphism	25
B_r	The polynomial ring $\mathbb{Z}[e_1, \ldots, e_r]$ in r weighted indeterminates	26
$(B_r)_w$	The submodule of B_r of elements of weight $w \geq 0$	27
$\mathfrak{p}_r(X)$	Generic monic polynomial of degree r	26
$\mathcal{P}$ and $\mathcal{P}_r$	The set of all partitions and of the partition of length at most r	27
$\mathcal{P}_{r,n}$	The set of partitions of $\mathcal{P}_r$ bounded by n-r	27
D^j		26
$\Delta_{\boldsymbol{\lambda}}(\mathbf{a})$	The Schur polynomial associated to a partition $\boldsymbol{\lambda}$ and to a sequence $\mathbf{a}$ in a commutative ring	28
$E_r(t)$	The polynomial $1 - e_1 t + \cdots + (-1)^r e_r t^r \in B_r[t]$	28
$H_r(t)$	The inverse $\sum_{n \in \mathbb{Z}} h_n t^n$ of $E_r(t)$ in the ring $B_r[[t]]$	29
H_r	The sequence $(h_j)_{i \in \mathbb{Z}}$ of the coefficients of $H_r(t)$	29
$(u_i)_{i \in \mathbb{Z}}$	The fundamental sequence of $B_r[[t]]$	30
U_i		31

Introduction

As the title gives away, this book is concerned with *Hasse–Schmidt (HS) derivations on Grassmann (or exterior) algebras*. These were proposed in [37] as an alternative formalism for dealing with Schubert calculus on complex Grassmannians. They are a straightforward extension of a corresponding notion [60] arising in commutative algebra. The guiding expository strategy with these notes is to exploit HS derivations to dig up a number of lands in the realm of elementary linear algebra to uncover the roots of more advanced topics.

- **HS-derivations, in general.** To begin with, let A be a commutative ring with unit and $(\mathfrak{M}, \star)$ an associative A-algebra. An *$\mathfrak{M}$-valued formal power series* in an indeterminate z over $\mathfrak{M}$ is an infinite sum $\sum_{i\geq 0} m_i z^i$ encoding the data of an $\mathfrak{M}$-valued sequence $(m_0, m_1, \ldots)$. Denote by $\mathfrak{M}[[z]]$ the set of all $\mathfrak{M}$-valued formal power series, endowed with its standard A-algebra structure. Extending the terminology of [112, p. 207], by an *HS*-derivation on $(\mathfrak{M}, \star)$ we shall understand an algebra homomorphism $\mathcal{D}(z) : \mathfrak{M} \to \mathfrak{M}[[z]]$, i.e. an A-linear map such that $\mathcal{D}(z)(m_1 \star m_2) = \mathcal{D}(z)m_1 \star \mathcal{D}(z)m_2$ for all $m_1, m_2 \in \mathfrak{M}$. Equivalently, $\mathcal{D}(z)$ is a formal power series whose underlying sequence $(D_0, D_1, \ldots)$ of endomorphisms of $\mathfrak{M}$ obeys the following *higher-order* Leibniz-like rule

$$D_n(m_1 \star m_2) = \sum_{j=0}^{i} D_{n-j}m_1 \star D_j m_2, \qquad (i \geq 0). \tag{1}$$

- **HS-derivations: à la Hasse and Schmidt.** If $(\mathfrak{M}, \star)$ happens to be a commutative A-algebra, the definition sketched above is the same that Hasse and Schmidt introduced back in 1937, under the name of *higher derivation* [60]. The purpose of those authors, mainly interested in arithmetic questions, was to build a characteristic-free algebraic analogue of the Taylor series of a smooth function. Two years later, Schmidt constructed *Wronskian determinants* associated to higher derivations in order to study Weierstrass points on curves in positive characteristic [137]. Ever since, the literature concerning *HS*-derivations in the commutative-algebra setting has grown relentlessly. Besides the erstwhile aforementioned work of Hasse and Schmidt and the standard reference [112, Ch. 4], it is dutiful to cite the papers by Ribenboim [127–130], Saymeh [135], Laksov-Thorup [95, Section 4], Vojta [146] and the more recent works by Esteves and Skjelnes [27, 142] as an indication of the vitality of the subject.

- **HS-derivations: on a Grassmann algebra.** Surprisingly, the literature appears to miss, if not ignore completely, the case where $(\mathfrak{M}, \star)$ is the *exterior algebra* $(\bigwedge M, \wedge)$ of a free A-module M. The main purpose of this book is to fill such gap. We shall do so by defining an *HS*-derivation on $\bigwedge M$ as an algebra homomorphism

$$\mathcal{D}(z) : \bigwedge M \to \bigwedge M[[z]], \tag{2}$$

where $\bigwedge M[[z]] := (\bigwedge M)[[z]]$ denotes the algebra of $\bigwedge M$-valued formal power series. That is nothing fancy, of course, as it merely amounts to clone p. 207 of Matsumura's book [112]. It turns out that the unique sequence $(D_n)_{n\geq 0}$ of A-endomorphisms of $\bigwedge M$ defined by $D(z)\eta = \sum_{n\geq 0} D_n\eta \cdot z^n$, for all $\eta \in \bigwedge M$, satisfies the relation

$$D_n(\eta \wedge \theta) = \sum_{j=0}^{n} D_{n-j}\eta \wedge D_j\theta, \tag{3}$$

which is exactly formula (1) up to replacing '$\star$' with the wedge product '$\wedge$'. Formula (3) is obtained by comparing the coefficients of z^n on either side of the equality $\mathcal{D}(z)(\eta \wedge \theta) = \mathcal{D}(z)\eta \wedge \mathcal{D}(z)\theta$, which holds by definition, as in the commutative case. It follows that D_0 is a homomorphism of A-algebras ($D_0(\eta \wedge \theta) = D_0\eta \wedge D_0\theta$), while D_1 is an A-derivation of $\bigwedge M$, that is, besides being A-linear, it satisfies the ordinary Leibniz rule $D_1(\eta\wedge\theta) = D_1\eta\wedge\theta+\eta\wedge D_1\theta$. This passing comparison with the classical commutative case had a purpose: to convince the reader that calling *HS-derivation* an algebra homomorphism $\bigwedge M \to \bigwedge M[[z]]$ rightly answers the need for a truly appropriate terminology.

- **Integration by parts.** If an *HS*-derivation is additionally invertible, thought of as an $\mathrm{End}_A(\bigwedge M)$-valued formal power series, its inverse $\overline{\mathcal{D}}(z) \in \mathrm{End}_A(\bigwedge M)[[z]]$ turns out to be an *HS*-derivation too. This seemingly harmless remark provides us with the *integration by parts* formula

$$\boxed{\mathcal{D}(z)\eta \wedge \theta = \mathcal{D}(z)(\eta \wedge \overline{\mathcal{D}}(z)\theta)}, \tag{4}$$

a rather obvious equality that, on the other hand, is probably the most powerful tool of the theory. Formula (4) will be applied throughout the book to a wide variety of situations: to name a few, a useful and major generalization [51] of the known Cayley–Hamilton theorem, whereby any endomorphism is a root of its own characteristic polynomial, Giambelli's formula for the *Schubert derivations* treated in Chapter 5 and the approximate expressions of the vertex operators occurring in the representation theory of the oscillator Heisenberg algebra. To sense more intensely the flavour of the kind of results we shall depict within our framework, an especially relevant situation is described below.

- **Decomposable tensors in an exterior power**. Adopting the same convention of Chapter 5, let us denote by M_0 the free abelian group of infinite countable rank with basis $(b_0, b_1, \ldots)$. An example of such a module is the additive group $\mathbb{Z}[X]$ of polynomials with integral coefficients. The reader who feels a little unfamiliar with the notion of exterior algebra and exterior power of M_0 is referred to Section 3.2. However, recall that by definition $\bigwedge^0 M_0 = \mathbb{Z}$ and that, for $r \geq 1$, the rth exterior power $\bigwedge^r M_0$ is the free abelian group generated by $([\mathbf{b}]^r_{\boldsymbol{\lambda}} \mid \boldsymbol{\lambda} \in \mathcal{P}_r)$, where $\mathcal{P}_r$ is the set of finite sequences $(\lambda_1, \ldots, \lambda_r) \in \mathbb{N}^r$ such that $\lambda_1 \geq \ldots \geq \lambda_r$, and

$$[\mathbf{b}]^r_{\boldsymbol{\lambda}} := b_{0+\lambda_r} \wedge b_{1+\lambda_{r-1}} \wedge \cdots \wedge b_{r-1+\lambda_1}.$$

In particular, each $\eta \in \bigwedge^r M_0$ can be uniquely expressed as $\sum_{\boldsymbol{\lambda} \in \mathcal{P}_{r,n}} a_{\boldsymbol{\lambda}} [\mathbf{b}]^r_{\boldsymbol{\lambda}}$, where $\mathcal{P}_{r,n} := \{\boldsymbol{\lambda} \in \mathcal{P}_r \mid \lambda_1 \leq n - r\}$ is the set of *partitions* of length at most r bounded by $n - r$. An element $\eta \in \bigwedge^r M_0$ is said to be *decomposable* if $\eta = m_1 \wedge \cdots \wedge m_r$ for some $m_1, \ldots, m_r \in M_0$. The issue is to determine equations for the locus of *decomposable tensors* in $\bigwedge^r M_0$. The solution to this classical problem is very well known (see, e.g. [55, 58, 64]), but we shall revisit it here as follows. First recall that if $\mathbf{a} := (a_i)_{i \in \mathbb{Z}}$ is a bilateral sequence of elements in a commutative ring, the *Schur determinant* associated to '**a**' and $\boldsymbol{\lambda} \in \mathcal{P}_r$ is

$$\Delta_{\boldsymbol{\lambda}}(\mathbf{a}) := \det(a_{\lambda_j - j + i})_{1 \leq i,j \leq r}.$$

Let $B_0 := \mathbb{Z}$ and, for $r \geq 1$, define $B_r := \mathbb{Z}[e_1, \ldots, e_r]$ to be the polynomial ring in the r indeterminates $(e_1, \ldots, e_r)$. The reason for the notation is that the complexification $B := B_\infty \otimes_{\mathbb{Z}} \mathbb{C}$ coincides with the *bosonic* (whence the 'B') Fock representation of the Heisenberg oscillator algebra, i.e. the *Weyl affinization* [73, p. 51] of the trivial Lie algebra $\mathbb{C}$ (cf. Remark 1.2.3).

Accordingly, for any $r \geq 0$, we define polynomials $E_r(z) \in B_r[z]$ as follows: $E_0 = 1$ and $E_r(z) := 1 - e_1 z + \cdots + (-1)^r e_r z^r$ for $r > 0$, in order to construct the unique B_r-valued sequence $H_r := (h_i)_{\in \mathbb{Z}}$ such that

$$\sum_{i \in \mathbb{Z}} h_i z^i = \frac{1}{E_r(z)},$$

where the equality is understood in the abelian group of formal Laurent series $B_r[[z^{-1}, z]]$. In particular, $h_0 = 1$ and $h_j - 0$ if $j < 0$. Toe each $h_n \subset B_r$ wc associatc the following two Laurent polynomials

$$\overline{\sigma}_-(z)h_n := h_n - h_{n-1}z^{-1} \qquad \text{and} \qquad \sigma_-(z)h_n := \sum_{i \geq 0} h_{n-i} z^{-i}.$$

Denote by $\overline{\sigma}_-(z)H_r$ (resp. $\sigma_-(z)H_r$) the sequence $(\overline{\sigma}_-(z)h_i)_{i\in\mathbb{Z}}$ (resp. $(\sigma_-(z)h_i)_{i\in\mathbb{Z}}$). Then, the main result of Section 6, Theorem 6.4.3, guarantees that the linear combination

$$\sum_{\lambda\in\mathcal{P}_{r,n}} a_\lambda [\mathbf{b}]^r_\lambda \tag{5}$$

is decomposable in $\bigwedge^r M_0$ if and only if the integers a_λ satisfy the following system of bilinear equations:

$$\boxed{\mathrm{Res}_z \sum_{\lambda,\mu\in\mathcal{P}_r} a_\lambda a_\mu E_{r-1}(z)\Delta_\lambda(\sigma_-(z)H_{r-1}) \otimes \frac{1}{E_{r+1}(z)}\Delta_\mu(\overline{\sigma}_-(z)H_{r+1}) = 0,} \tag{6}$$

where 'Res_z' denotes the residue of a Laurent series, i.e. the coefficient of the monomial z^{-1}. It is dutiful to remark that, in spite of its elegance, formula (6) is not very efficient from a computational point of view. Examples 6.4.6 and 6.4.8 show that detecting decomposable tensors already in the second wedge power of a module of rank 4 or 5 is quite painful computationally, while the condition $\eta\wedge\eta = 0$ for $\eta \in \bigwedge^2 M_0$ gives the result in a few instants (Exercise 5.11.1).

There are a number of features, however, that make formula (6) interesting, shortly listed below:

1) It implicitly involves, through the formal power series $1/E_{r+1}(z)$ and the polynomial $E_{r-1}(z)$, the data of two distinguished *HS*-derivations on $\bigwedge M_0$. They are denoted by $\sigma_+(z)$ and $\overline{\sigma}_+(z)$, respectively, and called *Schubert derivations* in compliance with the terminology introduced in [49]. The former is characterized by the equality $\sigma_+(z)b_j = \sum_{i\geq 0} b_{i+j}z^i \in M_0[[z]]$, for all $j \geq 0$, while the latter is its inverse in $\bigwedge M_0[[z]]$.
2) For each $r > 0$, the Schubert derivations enable to turn the basis $(b_0, b_1, \ldots)$ of M_0 into a generic linear recurrence sequence of order r (Chapter 2), which endows M_0 with a structure of free B_r-module of rank r, denoted by M_r. Decomposability in $\bigwedge^r M_0$ is then read in $\bigwedge^r M_r$, a free B_r-module of rank one generated by $b_0 \wedge \ldots \wedge b_{r-1}$. This is crucial for simplifying the algebraic manipulations leading to (6).
3) Formula (6) renders explicit the fact that detecting decomposable tensors on $\bigwedge^r M_r$ amounts to exploit the interaction between the B_{r-1}-module $\bigwedge^{r-1} M_{r-1}$ and the B_{r+1}-module $\bigwedge^{r+1} M_{r+1}$. For $r = \infty$, they become infinite exterior powers coinciding over B_∞, and the relationship turns into an interaction between fermionic Fock spaces of different charges.
4) Chapter 5 will show that formula (6) is manifestly related with Schubert calculus, as it is built out of four fundamental blocks only: the Schubert derivations $\sigma_+(z)$, $\overline{\sigma}_+(z)$ and two novel *HS*-derivations, denoted by $\sigma_-(z)$ and $\overline{\sigma}_-(z)$, respectively. The latter appear as a mirror of sorts of $\sigma_+(z)$ and $\overline{\sigma}_+(z)$ but, in contrast to them,

enjoy a nice stability property enabling to describe their action directly on the ring B_r, as explained in Section 6.

5) Classical Plücker equations of the Grassmann cone of $\bigwedge^r M_0$ are uniformly encoded in just one formula. Last, but not least, taking the limit for r going to infinity, and extending the coefficient ring to $B := B_\infty \otimes \mathbb{Q}$, equation (6) recovers precisely the celebrated KP hierarchy, so named after Kadomtsev and Petviashvili, from the theory of infinite-dimensional integrable systems. Or, as Sato's Japanese School puts it, it provides the Plücker embedding of the universal Grassmannian manifold (UGM) parametrizing infinite-dimensional subspaces of an infinite-dimensional vector space.

Having said all that, it is perhaps time to pass to the contents of the single chapters, as a teaser of what is to come.

- **Chapter 1** is a prelude of sorts, mainly to furnish motivations and not strictly essential for grasping the rest. It may be read casually, glossing over details that we made no attempt to clarify at this juncture. It begins by describing a few hidden algebraic peculiarities, unveiled by the work of many authors, of two renowned non-linear PDEs, the Korteweg–deVries equation (KdV) and the Kadomtsev–Petviashvili equation (KP). The so-called boson–fermion correspondence is also anticipated in this chapter, and Vertex operators generating the free fermionic vertex superalgebra, in the sense of [32, Ch. 5], are outlined in Section 1.3.3. They act on a polynomial ring in infinitely many indeterminates via the boson–fermion correspondence and encode the full KP hierarchy, a certain infinite system of non-linear PDEs, which in turn includes the KP equation itself.

Due to remarkable work by Sato [133] (see [134] for an illuminating survey) and Date–Jimbo–Kashiwara–Miwa [18, 19, 70], the KP hierarchy can be amazingly interpreted as the infinite set of Plücker equations of the Grassmann cone of decomposable tensors in an infinite wedge power. This closes the circle on one hand and paves the way to Chapter 6 on the other.

- **Chapter 2** presents generic *linear recurrence sequences* (LRS) with values in a module over a B_r-algebra. Section 2.1 deals with LRSs viewed as formal power series, while Section 2.2 describes the *weight graduation* of B_r using partitions. Section 2.5 discusses the formal Laplace transform and initial-value problems for linear ODEs. We work out two examples taken from [42] to explain how the method works. The reward for developing all this is the universal expression for solutions to Cauchy problems of linear ODEs with constant coefficients and analytic forcing terms [49, 50]. The formula arises from a distinguished sequence $(u_i)_{i\in\mathbb{Z}}$ of B_r-valued generic LRSs, for $i \geq -r+1$, introduced in [49] and generalized in [44]. They will resurface in Chapter 7, when their elementary properties will be used to determine the explicit expression of the vertex operators.
- The first two sections of **Chapter 3** collect basic algebraic facts that belong to any scholar's background. In view of having a self-contained exposition, Section 3.1 opens with the notion of tensor algebra of a module, of which the exterior algebra

is a quotient. The purpose of these preliminaries is to prepare the reader with the construction of the less common Clifford algebra of a module, as we do in Section 3.3. The infinite exterior power of an infinite dimensional vector space can be seen as a representation of a suitable canonical Clifford algebra. The concept of derivation on a Grassmann algebra, in the standard sense of derivations on any algebra, is reviewed in Section 3.4 to emphasize that iterated derivations yield higher-order Leibniz rules which can be themselves organized into an *HS*-derivation.

- The pivotal notion of *Hasse–Schmidt derivation on an exterior algebra* is finally introduced in **Chapter 4**. Its basic formalism is developed, and the key role of the *integration by parts* formula (4) is discussed at length. *HS*-derivations on exterior algebras also provide a natural framework to state and prove, as in Section 4.2, a generalization, based on [51], of the classical celebrated Cayley–Hamilton theorem: *any endomorphism is a root of its characteristic polynomial.* It will be employed in Chapter 5 to equip a free abelian group of rank $n \in \mathbb{N} \cup \{\infty\}$ with a structure of free B_r-module of rank $r \leq n$. Section 4.3 offers a few simple, but noteworthy, applications to matrix exponentials and first integrals of linear ODEs with constant coefficients, adopting the same language of Section 2.5.
- **Chapter 5** is concerned with the Schubert derivations $\sigma_+(z), \overline{\sigma}_+(z)$, introduced with a different notation in the papers [14, 37, 47] to deal with Schubert calculus for complex Grassmannians and recently used in [15] for application to the existence of certain indecomposable vector bundles. The σ in the notation is reminiscent of its relationship with Schubert calculus, while the subscript '+' keeps into account its kinship with vertex operators.

The bridge connecting the *Schubert derivation* to classical Schubert calculus for Grassmannians, as summarized in Section 5.4, rests on a Pieri-like formula (Theorem 5.2.2) and a Giambelli formula satisfied by the coefficients of $\sigma_+(z)$. The latter is proven by invoking an elegant and flexible determinantal formula due to Laksov and Thorup [96, 97] that will be also used, along the way, to construct approximations of the vertex operators generating the fermionic vertex superalgebra.

Section 5.5 is entirely devoted to examples of manipulations with Schubert derivations carrying enumerative geometrical interpretations, such as computing degrees of Schubert varieties.

- The main point of **Chapter 6** is that the Schubert derivation $\sigma_+(z)$, dealt with in Chapter 5, actually tells us one half of the story only. In fact, as anticipated in the first part of this introduction, there exists a sort of 'mirror' Schubert derivation, $\sigma_-(z) := \sum_{i\geq 0} \sigma_{-i} z^{-i} : \bigwedge M_0 \to \bigwedge M_0[[z^{-1}]]$, which is the unique *HS*-derivation such that $\sigma_{-j} b_i \mapsto b_{i-j}$ if $i \geq j$ and 0 otherwise (Section 6.2). The definition of $\sigma_-(z)$ and its inverse $\overline{\sigma}_-(z)$ may seem rather artificial at first sight. However, they arise spontaneously from the theory. The very definition of $\sigma_+(z)$ and *integration by parts* (4) imply the equality

$$\sum_{i\geq 0} b_i z^i \wedge \eta = \sigma_+(z) b_0 \wedge \eta = \sigma_+(z)(b_0 \wedge \overline{\sigma}_+(z)\eta). \tag{7}$$

The fact that, by Theorem 6.2.6, the leftmost term of (7) satisfies the equality

$$\sigma_+(z)(b_0 \wedge \overline{\sigma}_+(z)\eta) = z^r\sigma_+(z)\overline{\sigma}_-(z)(\overline{\sigma}_r\eta \wedge b_0),$$

in $\bigwedge^{r+1} M_0((z))$, shows at a glance how $\sigma_+(z)$ naturally carries $\sigma_-(z)$ with it and the kinship of both with the vertex operators. The stability properties enjoyed by $\sigma_-(z)$ and $\overline{\sigma}_-(z)$ enable to define them directly in terms of $\mathbb{Z}$-module homomorphisms $B_r \to B_r[z^{-1}]$ as in (5), which in turn allow to describe the decomposable locus in exterior powers as we explained at the beginning of this introduction.

- **Chapter 7** is another take on the same material of Chapter 6, albeit from a more concrete point of view, due to the identification of $M_0 := \mathbb{Z}[X]$ with the $\mathbb{Z}$-module spanned by generic LRSs of finite order. A basis of the latter is given by the fundamental sequence of formal power series (u_i) introduced and discussed in Chapter 2. A formalism involving infinite exterior powers is explicitly used in this chapter, which is mainly devoted to the deduction of the bosonic expression of vertex operators arising in the fermionic vertex superalgebra, mentioned in the Prologue (Chapter 1). Among other things, we shall touch upon the bosonic and fermionic representation of the finite dimensional Heisenberg algebra, as well as the extension of the *HS*-derivation to the fermionic spaces of order r, in the sense of Section 7.6.

Every chapter ends with one section about the related literature and another with exercises meant both as a complement to the subject and to highlight the connections with other topics that are not addressed directly in the text. Exercises range from routine ones, or examples that the authors worked out early on, to more involved that might offer novel insights into the theory.

We should add, to conclude, that space and time constraints obliged us to leave out of the general picture many interesting and deserving aspects. In particular, we would have liked to include the equivariant cohomology of Grassmannians [48] and revisit Laksov's two papers [93, 94] as well. We disregarded Chern classes of vector bundles defined in terms of *trace polynomial operators* (though they appear in an exercise) and left out the description of the notable Kempf–Laksov formula [84] in terms of the exterior algebra of a module over the cohomology ring of the base of the bundle. Nor have we touched the important and fascinating relationship with the celebrated Jacobi triple product identity, of which the boson–fermion is a categorification in the sense of [31].

This book was written in the attempt to be as interdisciplinary as possible. This is reflected in the assortment of areas touched upon: from algebra to differential equations, from geometry to mathematical physics. Since the level is more elementary than that of the aforementioned general references, no prerequisites beyond standard linear/multilinear algebra are needed. In fact, starting from Chapter 2, the text is entirely self-contained. An effort was made to render the presentation seamless and avoid sudden jumps, to the point that even well-known concepts such as tensor algebras are defined from scratch.

Corrections, additional material, solutions of selected exercises and updates will be posted at the `url`:

http://calvino.polito.it/~gatto/impa_monograph.htm

Please, visit it now and consult it frequently.

Contents

Chapter 1
Prologue

This chapter lays down a non-technical and expository pathway to two non-linear PDEs: the Korteweg–de Vries (KdV) equation[1], modelling solitary waves, and the Kadomtsev–Petviashvili (KP) equation[2], a generalization of the KdV originally motivated by applications to plasma physics. Although the chapter's contents might not seem related to the rest of the book in a straightforward manner, the KdV and KP equations are, surprisingly, linked to a number of subjects that any mathematician has had early close encounters with. To name but a few, cubic plane curves, linear recurrence sequences, linear ODEs and (generalized) Wronskians[3] associated with fundamental systems of solutions, exterior algebras of free modules, Plücker[4] embeddings of finite-dimensional Grassmannians[5] and Schubert[6] calculus. In sum, these two equations provide us with an excuse to return to the early stages of our scholarly education and shed a new light on it. The chapter culminates with the appearance, as a kind of deus ex machina, of the bosonic expression of the vertex operators occurring in the so-called vertex algebra of free charged fermions. These operators act on a polynomial ring in infinitely many indeterminates and encode the full system of PDEs known under the name of KP hierarchy, which arises as

---

[1]Diederik J. Korteweg (Den Bosch, 1848–Amsterdam, 1941) and Gustav de Vries (Amsterdam, 1866–Harlem, 1934) were Dutch mathematicians, after whom the Institute of Mathematics of the University of Amsterdam is currently named.

[2]Boris B. Kadomtsev (1928–1998) and Vladimir I. Petviashvili (1936–1993), Russian physicists

[3]Józef Hoene-Wroński (Wolsztyn, 1776–Neuilly-sur-Seine, 1853). A Polish mathematician with multifaceted interests. He was also an inventor, a lawyer and an economist. See [122].

[4]Julius Plücker (1801–1868), German mathematician and physicist. His formulas computing the geometric genus of a plane curves with singularities are particularly renowned.

[5]After Hermann Günther Grassmann, great German mathematician born in Stettin in the year 1809, when the town was part of the Kingdom of Prussia. He died in 1877 in the same town, by which time it was part of the German Empire.

[6]Hermann Cäsar Hannibal Schubert (Potsdam, 1848–Hamburg, 1911), the creator of the celebrated calculus which bears his name

L. Gatto, P. Salehyan, *Hasse-Schmidt Derivations on Grassmann Algebras*,
IMPA Monographs, DOI 10.1007/978-3-319-31842-4_1

compatibility conditions for another system of infinitely many PDEs (expressed in Lax form, see [101, 140] or [78, p. 73]).

What in this chapter we refer to as boson–fermion correspondence essentially boils down to an isomorphism between a free polynomial algebra in infinitely many indeterminates and the infinite exterior power of an infinite-dimensional vector space. When approximated using finitely generated polynomial algebras and finite-dimensional vector spaces, the correspondence proves to be formally the same as the Poincaré[7] duality between cohomology and homology of complex Grassmannians, as ruled by the famous Giambelli[8] formula. The latter can, in turn, be understood as a generalization of the classical theorem by Abel[9] and Liouville[10] whereby the derivative of the Wronskian associated with a fundamental system of solutions is proportional to the Wronskian itself. In sum, just the beginning of a long story.

Some of the seemingly sophisticated terminology, mentioned in the prologue without an explanation, will be introduced and thoroughly explained in due course. This reflects the book's distinctive cyclic structure: the first chapter can be read a second time after the last one; hence it functions as both a prologue and a natural epilogue.

1.1 The KdV and the KP Equations

1.1.1. The KdV equation is a non-linear partial differential equation (PDE) that was deduced [89] to model the dynamics of *solitary waves*, or *solitons*. These had been noticed for the first time in 1834 by John Scott Russell [131], while observing two galloping horses pulling a boat through a narrow channel. For more on this picturesque story, see [81, p. 45]. Although the most general form of the KdV equation is $af_t + bff_x + cf_{xxx} = 0$, with $a, b, c \in \mathbb{C}$, mainly for pedagogical reasons, and to adhere more closely to the many excellent expositions of the subject, such as [5, 117] or [78, p. 75], we shall write it as follows:

$$4\frac{\partial f}{\partial t} - 12f\frac{\partial f}{\partial x} - \frac{\partial^3 f}{\partial x^3} = 0, \tag{1.1}$$

---

[7]Jules Henri Poincaré was one of the most famous French mathematicians, who left important contributions in nearly every field of mathematics, besides theoretical physics and philosophy of science. He was born in Nancy in 1854 and died in Paris in 1912.

[8]Born in Verona in 1879, Giovanni Zeno Giambelli was an Italian mathematician, one of the most brilliant students of Corrado Segre. The formula bearing his name is extremely important in Schubert calculus and its variants (quantum, equivariant or for other homogeneous spaces). He died in Messina in 1953, probably without suspecting the future influence of his work. See [91] for more details on his biography.

[9]Niels Henrik Abel (Finnøy, 1802–Froland, 1829) was a Norwegian mathematician. His name is linked to many different contributions, including the impossibility to solve the quintic equation by radicals. He died a couple of days before being appointed professor at the University of Berlin.

[10]Joseph Liouville (Saint-Omer, 24 marzo 1809–Parigi, gave outstanding contributions in nearly all the branches of mathematics, from number theory (he provided a proof of the existence of transcendental numbers) to theoretical mechanics (Liouville tori)).

where f is a map of class C^3 defined locally on the real plane (x,t). In spite of its experimental origin, equation (1.1) reveals several interesting features, not to speak of its striking relationships with many topics in algebraic geometry. It can be written in a number of equivalent ways, all geometrically flavoured, and finding exact explicit solutions is not that difficult even without a PDE expertise.

1.1.2. A natural and standard way to tackle (1.1) is to seek solutions of the form

$$f := -\mathbf{p}(x-ct)+K,$$

viewing the parameter c as the speed of the 'travelling wave' and K as an arbitrary constant, free to vary according to the convenience. Substituting the above ansatz in (1.1) gives

$$\mathbf{p}''' - 12\mathbf{p}\cdot\mathbf{p}' - 4c\mathbf{p}' + 12K\mathbf{p}' = 0. \tag{1.2}$$

Chosing $3K = c$ transforms equation (1.2) into the simpler $\mathbf{p}''' = 12\mathbf{p}\cdot\mathbf{p}'$, which begs to be integrated to

$$\mathbf{p}'' = 6(\mathbf{p})^2 - \frac{1}{2}g_2, \tag{1.3}$$

where $-g_2/2$ is a constant. Multiplying either side of (1.3) by $\mathbf{p}'$ gives

$$\mathbf{p}''\mathbf{p}' = 6(p)^2\mathbf{p}' + \frac{1}{2}g_2\mathbf{p}'.$$

One more integration eventually yields

$$(\mathbf{p}')^2 = 4\mathbf{p}^3 - g_2\mathbf{p} - g_3. \tag{1.4}$$

At the cost of looking for solutions among complex functions of one complex variable, equation (1.4) is satisfied by the celebrated *Weierstrass*[11] *$\wp$-function*

$$\wp_\Lambda(z) = \frac{1}{z^2} + \sum_{\lambda\in\Lambda\setminus\{0\}} \frac{1}{(z-\lambda)^2} - \frac{1}{\lambda^2}, \tag{1.5}$$

where $\Lambda := \mathbb{Z}\lambda_1 \oplus \mathbb{Z}\lambda_2$ is a *lattice*[12] in $\mathbb{C}$. The coefficients g_2 and g_3 depend on the choice of Λ and are *modular forms*[13] of weights 4 and 6, respectively:

$$g_2 := g_2(\Lambda) = 60G_4 \quad\text{and}\quad g_3 := g_3(\Lambda) = 140G_6,$$

---

[11] Karl Theodor Wilhelm Weierstrass (1815–1897), German mathematician often cited as the "father of modern analysis". However, his contributions to mathematics are countless. For instance *Weierstrass points* are very important in the geometry of algebraic curves.

[12] A discrete subgroup Λ of $\mathbb{C}$ such that $\dim_{\mathbb{R}} \Lambda\otimes_{\mathbb{Z}}\mathbb{R} = 2$.

[13] A *modular form* is a complex-valued function f, defined on the complex upper half-plane $\mathbb{H}$, such that $f(gz) = (cz+d)^k f(z)$, where for any $g = \begin{pmatrix} a & b \\ c & d \end{pmatrix} \in Sl_2(\mathbb{Z})$ one defines $gz := \frac{az+b}{cz+d}$. See on this the exciting survey [152].

where for $k > 1$

$$G_{2k}(\Lambda) = \sum_{\lambda \in \Lambda \setminus \{0\}} \lambda^{-2k}$$

is the renowned *Eisenstein*[14] modular form of *weight* $2k$ [141, p.157]. Conversely, for general values of $g_2, g_3 \in \mathbb{C}$, there exists a *lattice* Λ_{g_2,g_3} in the complex plane, such that the function (1.5) satisfies equation (1.4) for $\Lambda = \Lambda_{g_2,g_3}$. In fact, an equivalence class of lattices under the action of the group $Sl_2(\mathbb{Z})$ is parametrized by a point τ of the Poincaré half-plane $\mathbb{H} := \{z \in \mathbb{C} \mid \Im(z) > 0\}$, and the *j-invariant* $j : \mathbb{H} \to \mathbb{C}$

$$j(\tau) = \frac{1728 g_2(\tau)^3}{g_2(\tau)^3 - 27 g_3(\tau)^2}$$

is surjective, i.e. for each pair $g_2, g_3 \in \mathbb{C}$, there exists $\tau \in \mathbb{H}$ such that the lattice $\mathbb{Z} \oplus \mathbb{Z}\tau$ does provide, via the $\wp$-function, a parametrization of (1.4). We have thus found solutions to (1.1) of the form

$$f := -\wp_\Lambda(x - ct) + \frac{c}{3},$$

called *periodic solutions*. Actually, the use of elliptic functions to solve (1.1) suggests another kind of substitution; see below.

1.1.3. The g-dimensional *Siegel*[15] *generalized domain* is the set of $g \times g$ Hermitian[16] matrices with positive definite imaginary part:

$$\mathbb{H}^g := \{\Omega \in \mathbb{C}^{g \times g} \mid \Omega^T = \overline{\Omega}, \Im(\Omega) > 0\},$$

where ^{T} denotes transposition and $^-$ complex conjugation. If $g = 1$, then $\mathbb{H} := \mathbb{H}^1$ consists of complex numbers τ with positive imaginary part. Clearly, the matrix $\Lambda_\Omega := (\mathbb{1}_{g \times g}, \Omega) \in \mathbb{C}^{g \times 2g}$, where $\mathbb{1}_{g \times g}$ is the $g \times g$ identity matrix, defines a g-dimensional lattice in $\mathbb{C}^g$. Recall that a *principally polarized abelian variety* is a pair (X, Θ), where $X = \mathbb{C}^g / \Lambda_\Omega$ and Θ is an ample divisor class, which in turn is the fundamental class of the zero locus of the *theta function* on $\mathbb{C}^g$:

$$\theta_\Omega(\mathbf{z}) = \sum_{\mathbf{n} \in \mathbb{Z}^g} \exp \pi \sqrt{-1}(\mathbf{n}^T \cdot \Omega \cdot \mathbf{n} + 2\mathbf{n}^T \cdot \mathbf{z}).$$

[14] Ferdinand Gotthold Max Eisenstein (Berlin, 1823–1852), German mathematician who gave outstanding contributions to number theory.

[15] Carl Ludwig Siegel (Berlin, 1896 – Gottingen, 1981), German mathematician known for his deep contributions to analytic number theory.

[16] After Charles Hermite (Dieuze, 1822–Paris, 1901), French mathematician well known for his studies on complex-valued quadratic forms bearing his name.

The above map is not invariant under the action of the lattice Λ_Ω on $\mathbb{C}^g$. However, the relation

$$\theta_\Omega(\mathbf{z} + \mathbf{n} + \Omega \cdot \mathbf{m}) = \exp 2\pi\sqrt{-1}\left(-\frac{1}{2}\mathbf{m}^T \cdot \Omega \cdot \mathbf{m} - \mathbf{m}^T \cdot \mathbf{z}\right)\theta_\Omega(\mathbf{z})$$

shows that its zero locus is well-defined modulo Λ_Ω. If $g = 1$ we have

$$\theta_\tau(\mathbf{z}) = \sum_{n \in \mathbb{Z}} \exp \pi\sqrt{-1}(n^2\tau + 2nz), \tag{1.6}$$

where $\Im(\tau) > 0$. The theta function (1.6) is related to the Weierstrass $\wp$-function associated with the lattice $\Lambda_\tau := (1, \tau)$, according to [141, pp. 155–156]:

$$\wp_{\Lambda_\tau}(z) = -\frac{d^2}{dz^2}\log\theta_\tau\left(z + \frac{1}{2}(1+\tau)\right) + k.$$

This observation suggests, first of all, that solutions to (1.1) can be sought in terms of theta functions on an elliptic curve rather than in terms of the Weierstrass $\wp$-function.

Secondly, it is perhaps the origin of a computational trick due to Hirota [62, 63]. This consists in looking for KdV solutions of the form

$$f = \frac{\partial^2}{\partial x^2}\log v(x,t), \tag{1.7}$$

where $v = v(x,t)$ is a sufficiently regular map with no zero in the domain of concern. The substitution of (1.7) into (1.1) gives

$$\begin{aligned}
0 &= 4\frac{\partial^3 \log(v)}{\partial x^2 \partial t} - 12\frac{\partial^2 \log(v)}{\partial x^2} \cdot \frac{\partial^3 \log(v)}{\partial x^3} - \frac{\partial^5 \log(v)}{\partial x^5} \\
&= \frac{\partial}{\partial x}\left(4\frac{\partial}{\partial t}\left(\frac{\partial \log(v)}{\partial x}\right) - 6\left(\frac{\partial}{\partial x}\left(\frac{\partial \log v}{\partial x}\right)\right)^2 - \frac{\partial^3}{\partial x^3}\left(\frac{\partial \log v}{\partial x}\right)\right) \\
&= \frac{\partial}{\partial x}\left(4\frac{\partial}{\partial t}\left(\frac{v_x}{v}\right) - 6\left(\frac{\partial}{\partial x}\left(\frac{v_x}{v}\right)\right)^2 - \frac{\partial^3}{\partial x^3}\left(\frac{v_x}{v}\right)\right),
\end{aligned}$$

from which

$$4\frac{\partial}{\partial t}\left(\frac{v_x}{v}\right) - 6\left(\frac{\partial}{\partial x}\left(\frac{v_x}{v}\right)\right)^2 - \frac{\partial^3}{\partial x^3}\left(\frac{v_x}{v}\right) - \gamma = 0, \tag{1.8}$$

where $\gamma = \gamma(t)$ is an arbitrary function that does not depend on x. Expanding (1.8) produces

$$\frac{\gamma v^2 + 4v_x v_t + 3v_{xx}^2 - 4v_t v_{xxx} - 4v v_{xt} + v v_{xxxx}}{v^2} = 0,$$

and by clearing the denominators, one eventually obtains *Hirota's bilinear form* of the KdV equation:

$$\gamma v^2 + 4v_x v_t + 3v_{xx}^2 - 4v_t v_{xxx} - 4v v_{xt} + v v_{xxxx} = 0. \tag{1.9}$$

The Example of the Simple Pendulum 1.1.4. There are many classical and elementary problems of mathematical physics whose solutions call for the use of elliptic functions. One such is provided by the equations of motion of a rigid body with a fixed point, due to Euler[17]. The reader interested in explicit solutions may consult [41] for the full details. The quintessential instance, though, is perhaps the *simple pendulum*. The linearized dynamics (in a neighbourhood of the stable equilibrium point) is controlled by the *harmonic oscillator*'s classical equation $\ddot{\theta} + \omega^2\theta = 0$ (Figs. 1.1).

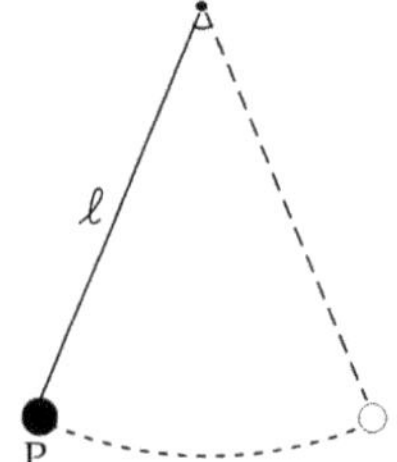

Fig. 1.1 Simple pendulum: its linearized equation is that of the *one*-dimensional harmonic oscillator

Such linear ODE possesses the *first integral*[18] $\dot{\theta}^2 + \omega^2\theta^2 = 2E$, where E is a constant called the *total energy* [19]. The non-linear dynamics of the simple pendulum, instead, is governed by the equation

$$\ddot{\theta} + \omega^2 \sin\theta = 0. \tag{1.10}$$

It, too, possesses a first integral

$$\dot{\theta}^2 - 2\omega^2 \cos\theta = 2E, \tag{1.11}$$

obtained by multiplying (1.10) by $\dot{\theta}$ and integrating. Reparametrizing using $u = \tan(\theta/2)$, we get

$$\cos\theta = \frac{1-u^2}{1+u^2} \quad \text{and} \quad \dot{\theta} = \frac{2\dot{u}}{1+u^2}.$$

[17]Leonhard Euler (Basel, 1707–Saint Petersburg, 1783) was a Swiss polymath mathematician; famous also for discovering the T-shirt slogan $e^{\pi\sqrt{-1}} + 1 = 0$.

[18]A function that is constant along the integral curves of a vector field.

[19] More precisely it is the total energy divided by the mass of the point P and by the length ℓ of the supposedly massless string.

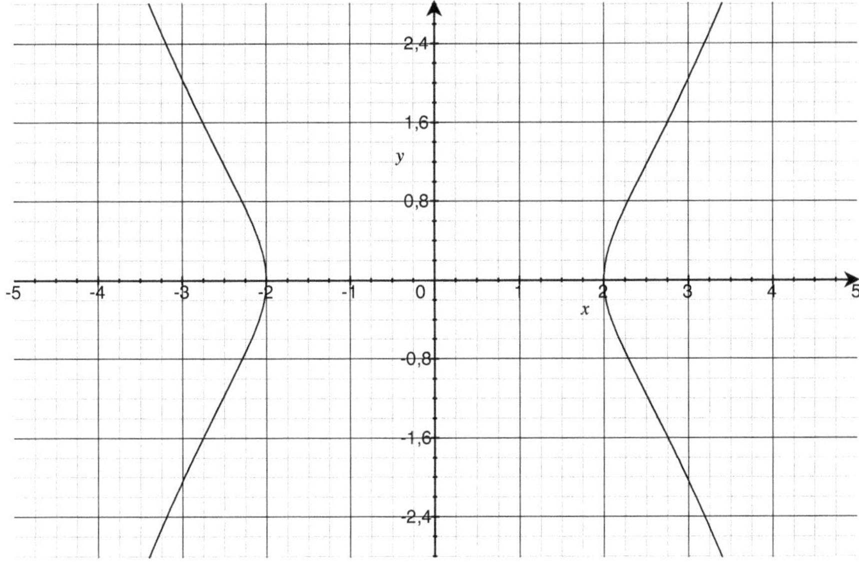

Fig. 1.2 The graph of the real part of the curve $2v^2 - \beta(u^2+1)(u^2-\alpha^2) = 0$ in the (u, v) plane (here $\beta = 1/6, \alpha = 2$)

Substituted into (1.11) and simplifying:

$$2\dot{u}^2 - \omega^2(1-u^2)(1+u^2) - E(1+u^2)^2 = 0, \tag{1.12}$$

which we rewrite in the form

$$2\dot{u}^2 = \beta(u^2+1)(u^2-\alpha^2), \tag{1.13}$$

where for generic values of E and ω^2, we have put $\beta := E - \omega^2$ and denoted by α a square root of

$$\frac{E+\omega^2}{\omega^2 - E}.$$

Writing $v =: \dot{u}$, the equation $2v^2 - \beta(u^2+1)(u^2-\alpha^2) = 0$ describes (the affine part of a) double covering of the projective line with the four ramification points ($\pm\sqrt{-1}$ and $\pm\alpha$), i.e. it defines an *elliptic curve* (Fig. 1.2).

Equation (1.12) can be solved through a standard sequence of steps. First we change the variable u, by acting on $\mathbb{P}^1$ via some element of $Sl_2(\mathbb{C})$, to send one ramification point to ∞ and another one to 0. We can make a further variable change to kill the term of degree 2 and so obtain the elliptic curve in *Weierstrass canonical form* (1.4). The desired canonical form is finally attained via the Möbius[20] transformation

$$X := \frac{\beta}{24}\frac{(5\alpha^2-1)u+5\alpha+\alpha^3}{u-\alpha}$$

which maps $\alpha \mapsto \infty$. Solving for u returns

$$u = \alpha \cdot \frac{24X+5\beta+\alpha^2\beta}{24X-\beta-5\alpha^2\beta},$$

which substituted into (1.12) gives

$$\dot{X}^2 = 4X^3 - g_2X^2 - g_3 \tag{1.14}$$

after clearing the denominator on a neighbourhood where it is non-zero, having set

$$g_2 := \frac{\beta^2}{48}(1-14\alpha^2+\alpha^4) \tag{1.15}$$

and

$$g_3 := \frac{\beta^3(\alpha^2-1)}{1728}\left(\alpha^4+34\alpha^2+1\right). \tag{1.16}$$

In the variable X, the solution to (1.13) then reads

$$X = \wp_{\Lambda_g}(z(t)),$$

where Λ_g is a lattice such that $(\wp_{\Lambda_g}(t), \wp'_{\Lambda_g}(t))$ parametrizes (1.14) with g_2 and g_3 given, respectively, by (1.15) and (1.16). In other words

$$\theta(t) = 2\arctan\left(\alpha\cdot\frac{24\wp_{\Lambda_g}(z(t))+5\beta+\alpha^2\beta}{24\wp_{\Lambda_g}(z(t))-\beta-5\alpha^2\beta}\right)$$

is the explicit solution of (1.10), which has thus been linearized. The parameter $z = z(t)$ can be viewed as the equation of the linear flow with unit speed: $z(t) = t+a$ with a constant.

[20] August Ferdinand Möbius (Bad Kösen, 1790–Leipzig, 1868), German mathematician and astronomer, well known to the general public for an important non-orientable surface (the Möbius strip)

1.1.5. In the 1970s Kadomtsev and Petviashvili [80] generalized the KdV equation motivated by applications to plasma physics. They turned it into an ostensibly more complicated equation:

$$3\frac{\partial^2 f}{\partial^2 y} - \frac{\partial}{\partial x}\left(4\frac{\partial f}{\partial t} - 12 f \cdot \frac{\partial f}{\partial x} - \frac{\partial^3 f}{\partial^3 x}\right) = 0, \tag{1.17}$$

whose solutions are functions of three variables (x, y, t). It is apparent that each KdV solution (constant along y) also solves (1.17). If Kadomtsev and Petviashvili had been algebraic geometers, their generalization of (1.1) might – with hindsight – have seemed an attempt to promote the many algebraic beauties their equation discloses. Once (1.17) is turned into bilinear form using the same Hirota computational trick (1.7) employed for the KdV equation, it is very easy to find whole families of exact solutions, if one seeks them in the form

$$f := \frac{\partial^2}{\partial x^2}(\log w),$$

where $w := w(x, y, t)$. In that case

$$\begin{aligned} 0 &= 3\frac{\partial^2}{\partial^2 x}\frac{\partial^2 \log w}{\partial^2 y} - \frac{\partial^2}{\partial x^2}\left(4\frac{\partial}{\partial t}\left(\frac{w_x}{w}\right) - 6\left(\frac{\partial}{\partial x}\left(\frac{w_x}{w}\right)\right)^2 - \frac{\partial^3}{\partial x^3}\left(\frac{w_x}{w}\right)\right) \\ &= \frac{\partial^2 \log w}{\partial^2 y} - \left(4\frac{\partial}{\partial t}\left(\frac{w_x}{w}\right) - 6\left(\frac{\partial}{\partial x}\left(\frac{w_x}{w}\right)\right)^2 - \frac{\partial^3}{\partial x^3}\left(\frac{w_x}{w}\right)\right), \end{aligned}$$

from which

$$3\frac{\partial^2 \log w}{\partial^2 y} - 4\frac{\partial}{\partial t}\left(\frac{w_x}{w}\right) - 6\left(\frac{\partial}{\partial x}\left(\frac{w_x}{w}\right)\right)^2 - \frac{\partial^3}{\partial x^3}\left(\frac{w_x}{w}\right) + \gamma_1 x + \gamma_2 = 0.$$

Here $\gamma_i := \gamma_i(y, t)$ $(i = 1, 2)$ are arbitrary functions that do not depend on the variable x. Consequently

$$\begin{aligned} &\frac{1}{w^2}\Big(- 3w_y^2 + 4w_t w_x + 3w_{xx}^2 - 4w_x w_{xxx} + 3ww_{yy} - 4ww_{xt} + ww_{xxxx}\Big) \\ &\quad + \gamma_1 x + \gamma_2 = 0, \end{aligned}$$

which can be written as

$$\begin{aligned} &ww_{xxxx} + 3w_{yy}w - 4ww_{xt} + 3w_{xx}^2 - 3w_y^2 + 4w_t w_x - 4w_x w_{xxx} + \\ &\quad + (\gamma_1 x + \gamma_2)w^2 = 0. \end{aligned}$$

This is the most general expression for the *Hirota bilinear form* of the KP equation.

1.1.6. Let us restrict ourselves to the case $\gamma_1 = \gamma_2 = 0$, which we rewrite below due to its exceptional importance:

$$ww_{xxxx} + 3w_{yy}w - 4ww_{xt} + 3w_{xx}^2 - 3w_y^2 + 4w_t w_x - 4w_x w_{xxx} = 0. \tag{1.18}$$

A fundamental peculiarity of (1.18) is that the sum of the coefficients is zero, a fact that enables one to find many solutions in a simple manner. Following Kasman [81] we give the

Definition 1.1.7. A function $w := w(x, y, t)$ is *nicely weighted* if $w_{xx} = w_y$ and $w_{xxx} = w_t$.

Proposition 1.1.8. *Nicely weighted functions solve (1.18).*

Proof. If w is nicely weighted, we have

$$w_{xxxx} = \frac{\partial^2 w_{xx}}{\partial x^2} = \frac{\partial^2 w_y}{\partial x^2} = \frac{\partial w_{xx}}{\partial y} = \frac{\partial w_y}{\partial y} = w_{yy}.$$

Similarly $w_{xx}^2 = w_y^2$; hence a nicely weighted w is a solution of (1.18). □

Producing nicely weighted functions is extremely easy. The most obvious one is $w = \exp(x\lambda + y\lambda^2 + t\lambda^3)$, for

$$\frac{\partial^n w}{\partial x^n} = \lambda^n w, \qquad \frac{\partial^n w}{\partial y^n} = \lambda^{2n} w, \qquad \frac{\partial^n w}{\partial t^n} = \lambda^{3n} w.$$

This case is not that interesting, however, as $f = (\log w)_{xx} = (x\lambda + y\lambda^2 + t\lambda^3)_{xx} = 0$ only gives the trivial solution of the original KP equation. To produce a host of nontrivial solutions, one may consider the family of functions g_i defined by

$$1 + \sum_{i\geq 1} g_i(x, y, t)\lambda^i = \exp(x\lambda + y\lambda^2 + t\lambda^3).$$

Notice that

$$\sum_{i\geq 0} \frac{\partial g_i}{\partial x}\lambda^i = \lambda \exp(x\lambda + y\lambda^2 + t\lambda^3) = \sum_{i\geq 0} g_i \lambda^{i+1}$$

from which $\dfrac{\partial g_i}{\partial x} = g_{i-1}$. Similarly

$$\frac{\partial g_i}{\partial y} = g_{i-2} = \frac{d^2 g_i}{\partial x^2} \qquad \text{and} \qquad \frac{\partial g_i}{\partial t} = g_{i-3} = \frac{\partial^3 g_i}{\partial x^3}.$$

In other words each g_i is nicely weighted, and as such it solves the Hirota equation (1.18). It is not difficult to see, for instance, that

$$g_3 = \frac{x^3}{3!} + xy + t$$

is nicely weighted. It follows that

$$f := \frac{\partial^2}{\partial x^2} \log \left(\frac{x^3}{3!} + xy + t \right) = \frac{3(12tx - x^4 - 12y^2)}{(6t + 6x + x^3)^2}$$

is an exact solution to the KP equation (1.17), as the brave reader can check with a little patience.

1.2 Vertex Operators and Affine Lie Algebras

1.2.1. Let $B := \mathbb{Q}[x_1, x_2, \dots]$ denote the polynomial algebra in the infinitely many indeterminates $(x_1, x_2, \dots)$ with rational coefficients. For each $i > 0$, the multiplication by x_i and the partial derivative ∂_j with respect to x_j are $\mathbb{Q}$-endomorphisms of B and generate a Lie[21] algebra, since $[\partial_i, x_j] = \delta_{ij}$. Let

$$B[[z^{-1}, z]] := \left\{ \sum_{i \in \mathbb{Z}} a_i z^i \mid a_i \in B \right\}$$

denotes *formal Laurent series* [22] with B-coefficients in some other indeterminate z. The maps

$$\Gamma(z) := \exp\left(\sum_{i \geq 1} x_i z^i \right) \exp\left(-\sum_{i \geq 1} \frac{1}{iz^i} \frac{\partial}{\partial x_i} \right) : B \to B[[z^{-1}, z]] \tag{1.19}$$

and

$$\Gamma^*(z) := \exp\left(-\sum_{i \geq 1} x_i z^i \right) \exp\left(\sum_{i \geq 1} \frac{1}{iz^i} \frac{\partial}{\partial x_i} \right) : B \to B[[z^{-1}, z]] \tag{1.20}$$

[21] Marius Sophus Lie (Nordfjordeid, 1842–Oslo, 1899), the Norwegian mathematician who introduced the notion of the homonymous algebra, which is still one of the building blocks of modern mathematics

[22] Pierre Alphonse Laurent (Paris, 1813–1854), French mathematician. His paper on the generalization of Taylor series was published posthumously.

are known in the literature[23] as *vertex operators* [78]. Let us explain why $\Gamma(z)$ and $\Gamma^*(z)$ take values in $B((z)) := B[z^{-1}, z]]$ only. As each element of B is – de facto – a polynomial in finitely many variables $x_{i_1}, \ldots, x_{i_r}$, it suffices to check the claim on monomials. To this end we first observe

$$\Gamma(z)x_j = \exp\left(\sum_{i\geq 1} x_i z^i\right)\left(x_j - \frac{1}{jz^j}\right) \in B((z))$$

and

$$\Gamma^*(z)x_j = \exp\left(\sum_{i\geq 1} -x_i z^i\right)\left(x_j + \frac{1}{jz^j}\right) \in B((z)).$$

Secondly, notice that

$$G(z) := \exp\left(-\sum_{i\geq 1} \frac{1}{iz^i}\frac{\partial}{\partial x_i}\right) \quad \text{and} \quad G^*(z) := \exp\left(\sum_{i\geq 1} \frac{1}{iz^i}\frac{\partial}{\partial x_i}\right)$$

are ring homomorphisms $B \to B[z^{-1}]$, because both are exponentials of derivations (cf. Lemma 3.4.2). Therefore $\Gamma(z)$ and $\Gamma^*(z)$ do send polynomials to $B((z))$.

Example 1.2.2. The image of $x_1^2 + x_1x_2 + x_2^2$ under $\Gamma(z)$ is

$$\begin{aligned}
&\Gamma(z)(x_1^2 + x_1x_2 + x_2^2)\\
&= \exp\Big(\sum_{i\geq 1} x_i z^i\Big)\left(\left(x_1 - \frac{1}{z}\right)^2 + \left(x_1 - \frac{1}{z}\right)\left(x_2 - \frac{1}{2z}\right) + \left(x_2 - \frac{1}{2z}\right)^2\right)\\
&= \exp\Big(\sum_{i\geq 1} x_i z^i\Big)\left(x_1^2 + x_1x_2 + x_2^2 - \frac{5}{2}\frac{x_1}{z} - 2\frac{x_2}{z} + \frac{7}{4z^2}\right)\\
&= \exp\Big(\sum_{i\geq 1} x_i z^i\Big)\left(x_1^2 + x_1x_2 + x_2^2 - \frac{1}{z}\left(\frac{5}{2}x_1 - 2x_2\right) + \frac{7}{4z^2}\right).
\end{aligned}$$

Historical Remark 1.2.3. It is often convenient to consider not just polynomials, but elements of the completion $\widehat{B} := \mathbb{Q}[[x_1, x_2, \ldots]]$ as well. In fact, differential operators on B can be identified with linear maps $B \to \widehat{B}$ (cf. [75, p. 97]). Prime examples of such operators are the multiplication by x_n, or $(T_a f)(x_1, x_2, \ldots) = f(x_1 + a_1, x_2 + a_2, \ldots)$, which by Taylor's formula is nothing but

$$(T_a f)(x_1, x_2, \ldots) = \left(\exp\sum_{i\geq 1} a_i \frac{\partial}{\partial x_i}\right) f.$$

[23] The definition we adopt of $\Gamma^*(z)$ is $1/z$ times the one used, for instance, in [78].

Vertex operators of the forms (1.19) and (1.20) arise naturally in the representation theory of affine *Kac–Moody algebras*. We shall not do any attempt to say what a Kac–Moody algebra is. However, its simplest instance is the so-called Weyl affinization $\widehat{\mathfrak{sl}_2(\mathbb{C})}$ of the simple Lie algebra $\mathfrak{sl}_2(\mathbb{C})$ of traceless 2×2 complex matrices. The underlying vector space is $\big(\mathfrak{sl}_2(\mathbb{C}) \otimes \mathbb{C}[t^{-1}, t]\big) \oplus \mathbb{C}\mathbf{k}$, and the Lie bracket is defined by

$$[a \otimes t^m, b \otimes t^n] = [a, b] \otimes t^{m+n} + m \cdot \mathrm{tr}(ab)\delta_{m,-n}\mathbf{k}$$

for $a, b \in \mathfrak{sl}_2(\mathbb{C})$, $m, n \in \mathbb{Z}$, while $\mathbf{k}$ is a central element, i.e. $\mathbf{k}$ commutes with every element of $\widehat{\mathfrak{sl}_2(\mathbb{C})}$. The affinization procedure applied to the trivial one-dimensional Lie algebra $\mathbb{C}$ gives rise to the oscillator Heisenberg[24] algebra $\mathcal{H} := \mathbb{C}[t^{-1}, t]$ with commutation relations $[t^m, t^n] = m\delta_{m,-n}\mathbf{k}$. In 1978 Lepowsky and Wilson [104] found a concrete description of $\widehat{\mathfrak{sl}_2(\mathbb{C})}$ by representing it on $\mathbb{C}[x_{\frac{1}{2}}, x_{\frac{3}{2}}, \ldots]$. Let Y_j be the coefficient of z^j ($j \in \frac{1}{2}\mathbb{Z}$) in the expansion of the *vertex operator*

$$Y := \exp\left(\sum_{n \in \frac{1}{2}\mathbb{N}^*} \frac{x_n}{n} z^n\right) \exp\left(-2 \sum_{n \in \frac{1}{2}\mathbb{N}^*} \frac{1}{z^n} \frac{\partial}{\partial x_n}\right).$$

The main theorem in [104], by Lepowsky and Wilson, says that

$$\mathbb{C} \cdot 1 \oplus \bigoplus_{n \in \frac{1}{2}\mathbb{N}^*} \mathbb{C} \cdot x_n \oplus \bigoplus_{n \in \frac{1}{2}\mathbb{N}^*} \mathbb{C} \cdot \frac{\partial}{\partial x_n} \oplus \bigoplus_{j \in \frac{1}{2}\mathbb{Z}} \mathbb{C} Y_j$$

is a Lie algebra of operators on $\mathbb{C}[y_j]_{j \in \frac{1}{2}\mathbb{N}^*}$, with respect to the usual commutator, isomorphic to the affine Kac–Moody algebra $\widehat{\mathfrak{sl}_2(\mathbb{C})}$. A few years later, this result was generalized in the milestone paper [75], which describes the so-called basic representation of a 'Euclidean Lie algebra' $\mathfrak{g}$ as a ring of operators on $B \otimes \mathbb{C} := \mathbb{C}[x_1, x_2, \ldots]$, spanned by the identity, the *annihilation* and *creation* operators (the multiplication by x_n and the derivative with respect to x_n, for all $n \in \mathbb{N}^*$) and the homogeneous components of operators of the form

$$\exp\left(\sum_{j \geq 1} \mu_{ij} x_j\right) \exp\left(\sum_{j \geq 0} \nu_{ij} \frac{\partial}{\partial x_j}\right).$$

Kac et al. [75] recall that a differential operator on $B_{\mathbb{C}} := B \otimes_{\mathbb{Q}} \mathbb{C}$ can be interpreted as a linear map $D : B_{\mathbb{C}} \mapsto \widehat{B_{\mathbb{C}}}$ and go on to prove in Corollary 3.1 that the relations

[24] Werner Karl Heisenberg (Würzburg, 1901–Munich, Bavaria, 1976), German theoretical physicist and one of the pioneers of quantum mechanics, known for his celebrated *indeterminacy principle*

$[x_i, A] = a_i A$ and $[\partial/\partial x_i, A] = b_i A$ imply that A is constrained to be equal to $C\exp(\sum_i a_i x_i)\exp(-\sum_{i\geq 1} b_i(\partial/\partial x_i))$, where C is a constant. If one takes $A = T_a$ with

$$a = \left(\frac{1}{z}, \frac{1}{2z^2}, \frac{1}{3z^3}\cdots\right),$$

a straightforward exercise shows that

$$[x_j, T_a] = -\frac{1}{jz^j}T_a \qquad \text{and} \qquad \left[\frac{\partial}{\partial x_j}, T_a\right] = z^j T_a,$$

and thus T_a is precisely the operator Γ as in formula (1.19).

The strategy employed in Chapters 6–7 to derive (1.19) and (1.20) is different and is based on the integration by parts formula enjoyed by Hasse-Schmidt derivations on a Grassmann algebra. It amounts to computing a 'finite' approximation of $\Gamma(z)$ and $\Gamma^*(z)$, viewed as maps from $B_r \mapsto B_r((z))$, where $B_r \otimes \mathbb{Q}$ is a suitable quotient of B, as anticipated in the introduction.

1.3 Vertex Operators and Vertex Algebras

The goal of this section is to motivate the word *vertex* attached to the operators (1.19), (1.20), in connection with the notion of vertex algebra, whose definition we outline with almost no detail. The example of charged free fermions is mentioned here merely to contextualize the terminology used in Section 1.2. In particular, the present section is not crucial for the ensuing discussion, and the reader may skip this part entirely or return here after the last chapter.

1.3.1. As in the introduction (see also 2.2), let $\mathcal{P}$ denote the set of all partitions and $\mathcal{P}_r$ the set of partitions with at most r non-zero terms. It is well known [109] that the ring $B = \mathbb{Q}[x_1, x_2, \ldots]$ possesses a $\mathbb{Q}$-basis parametrized by $\mathcal{P}$:

$$B := \bigoplus_{\lambda\in\mathcal{P}} \mathbb{Q}\cdot S_\lambda(\mathbf{x}),$$

where $\mathbf{x} := (x_1, x_2, \ldots)$ and, by definition,

$$\begin{cases} \sum_{j\geq 0} S_j(\mathbf{x})z^j &= \exp\left(\sum_{n\geq 1} x_n z^n\right), \\ \\ S_\lambda(\mathbf{x}) &:= \det(S_{\lambda_j - j + i}(\mathbf{x}))_{1\leq i,j\leq r}. \end{cases}$$

The elements $S_{\boldsymbol{\lambda}}(\mathbf{x}) \in B$ are called *Schur polynomials*. For example,

$$S_0(\mathbf{x}) = 1, \quad S_1(\mathbf{x}) = x_1, \quad S_2(\mathbf{x}) = \frac{x_1^2}{2} + x_2, \quad S_{(1,1)}(\mathbf{x}) = \begin{vmatrix} S_1(\mathbf{x}) & S_0(\mathbf{x}) \\ S_2(\mathbf{x}) & S_1(\mathbf{x}) \end{vmatrix} = \frac{x_1^2}{2} - x_2.$$

The ring B is the *bosonic Fock space*, a representation space of the Heisenberg oscillator Lie algebra. Let now V be a $\mathbb{Q}$-vector space with basis $(b_i)_{i\in\mathbb{Z}}$, and denote by $(\beta_j)_{j\in\mathbb{Z}}$ the basis of the *restricted dual* V^* of V, i.e. $\beta_j(b_i) = 0$. Consider the canonical symmetric bilinear form on $V \oplus V^*$:

$$g : (V \oplus V^*) \times (V \oplus V^*) \to \mathbb{Q},$$

defined by $g(v_1 \oplus \alpha_1, v_2 \oplus \alpha_2) = \alpha_2(v_1) + \alpha_1(v_2)$. Identifying b_i with $b_i \oplus \{\mathbf{0}\}$ and β_j with $\{\mathbf{0}\} \oplus \beta_j$, an easy check shows that $g(b_i, b_j) = g(\beta_j, \beta_j) = 0$ and $g(\beta_i, b_j) = g(b_j, \beta_i) = \delta_{ij}$. The *fermionic Fock space* F is an irreducible representation of the Clifford algebra $\mathcal{C} := C_g(V \oplus V^*)$ associated with g (Section 3.3). It is generated over $\mathcal{C}$ by a distinguished vector $|0\rangle$, called the *vacuum vector*, such that $b_i|0\rangle = 0$ for any $i \leq 0$ and $\beta_j|0\rangle = 0$ for any $j > 0$. Moreover, expressions of the kind

$$b_{i_1} \cdots b_{i_h} \cdot \beta_{j_1} \cdots \beta_{j_k}|0\rangle, \qquad h, k \in \mathbb{N}^*$$

where $i_1 > \cdots > i_h > 0$ and $0 \geq j_1 > \cdots > j_h$ form a basis of F over the rational numbers. The irreducible action of the Clifford algebra on F can be described with more familiar linear algebra tools, by resorting to the *infinite wedge representation* of Kac and Peterson. We set out to do so, and we call the expression

$$\Phi_i := b_i \wedge b_{i-1} \wedge b_{i-2} \wedge \cdots, \qquad i \in \mathbb{Z}.$$

the *vacuum vector* of *charge i*.

We indicate by $\Phi_{i+\boldsymbol{\lambda}}$ the 'excitation' of Φ_i by $\boldsymbol{\lambda} \in \mathcal{P}_r$:

$$\Phi_{i+\boldsymbol{\lambda}} = b_{i+\lambda_1} \wedge b_{i-1+\lambda_2} \wedge \cdots \wedge b_{i-r+1+\lambda_r} \wedge \Phi_{i-r}$$

and define

$$F_i = \bigoplus_{\boldsymbol{\lambda}\in\mathcal{P}} \mathbb{Q} \cdot \Phi_{i+\boldsymbol{\lambda}}.$$

Following [78], we call *fermionic Fock space* of total charge i the vector space F_i. This can be viewed as a *semi-infinite exterior power* of the infinite-dimensional vector space $\bigoplus_{j\in\mathbb{Z}} \mathbb{Q}b_j$. Given that $F_i \cap F_j = 0$ if $i \neq j$, it is customary to set $F := \bigwedge^{\infty/2} V := \bigoplus_{i\in\mathbb{Z}} F_i$. For practical purposes F_i works as the ordinary exterior power of a vector space of sufficiently high dimension. In particular, each monomial

$\Phi_{i+\lambda}$ changes sign whenever two of its factors are exchanged. Let $V^* := \bigoplus_{j\in\mathbb{Z}} \mathbb{Q}\cdot\beta_i$ be the *restricted dual* of V, i.e. the linear span of the forms satisfying $\beta_j(b_i) = 0$. Each b_i induces an element of $\mathrm{End}_{\mathbb{Q}}(F)$, namely, a 'wedge' homomorphism $b_i \wedge \cdot$, which is homogeneous of degree 1 with respect to the *charge* graduation:

$$F_i \ni \Phi_{i+\lambda} \stackrel{b_i\wedge}{\longmapsto} b_i \wedge \Phi_{i+\lambda} \in F_{i+1}.$$

Analogously, any $\beta_j \in V^*$ induces a homogeneous map $\beta_j \lrcorner : F \to F$ of degree -1, where '$\lrcorner$' denotes *contraction* against F (see Section 3.5.3). It is not hard to see that the operators $(\beta_i\lrcorner,\ b_i\wedge)_{i\in\mathbb{Z}}$ act on F precisely as the Clifford algebra defined abstractly above. Therefore, putting

$$|i\rangle := \Phi_i := b_i \wedge b_{i-1} \wedge \cdots$$

one sees indeed that, for $\boldsymbol{\lambda} = (\lambda_1, \ldots, \lambda_r)$,

$$\Phi_{i+\boldsymbol{\lambda}} = b_{i+\lambda_1} \wedge \cdots \wedge b_{i-r+1+\lambda_r} \wedge (\beta_i\lrcorner(\cdots\lrcorner\beta_{i-r}\lrcorner|i\rangle)).$$

For example,

$$\Phi_{-2+(4,2)} = b_2 \wedge b_{-1} \wedge b_{-4} \wedge b_{-5} \wedge \cdots = b_2 \wedge (b_{-1} \wedge (\beta_{-3}\lrcorner(\beta_{-2}\lrcorner|-2\rangle)).$$

This suggests that any *semi-infinite* exterior power should be expressible as a composite of *finitely* many endomorphisms of F. Speaking informally, the *boson–fermion correspondence* is the isomorphism of abelian groups

$$\varphi : B[\ell^{-1}, \ell] \to F \tag{1.21}$$

associating $\Phi_{i+\lambda}$ to $\ell^i S_{\boldsymbol{\lambda}}(\mathbf{x})$.

1.3.2. A *vertex algebra* $(\mathcal{F}, |0\rangle, T, Y)$ consists of the following data: a vector space $\mathcal{F}$ (it will be enough to consider a $\mathbb{Q}$-vector space for our purposes) called the *space of states*, a distinguished element $|0\rangle$ of $\mathcal{F}$ called *vacuum* vector, a *translation operator* $T \in \mathrm{End}_{\mathbb{Q}}(\mathcal{F})$ and the so-called *state–field correspondence* $Y : \mathcal{F} \to \mathrm{End}_{\mathbb{Q}}(\mathcal{F})[[z^{-1}, z]]$ that maps $v \in \mathcal{F}$ to a *vertex operator* (the *field*), i.e. to a formal Laurent series:

$$Y(v, z) = \sum_{n\in\mathbb{Z}} v_{(n)} z^{-n-1}, \quad (v_{(n)} \in \mathrm{End}_{\mathbb{Q}}(\mathcal{F})).$$

The above data is furthermore required to obey three axioms:

- $T|0\rangle = 0$, $Y(|0\rangle, z) = \mathbb{1}_V$ and $Y(v, z)|0\rangle \in \mathcal{F}[[z]]$; *(vacuum axiom)*

- $\big[T, Y(v,z)\big] = \dfrac{\partial Y(v,z)}{\partial z} = \sum_{n\in\mathbb{Z}} -na_{(n-1)}z^{-n-1}$; *(translation axiom)*
- for all pairs $v_1, v_2 \in \mathcal{F}$:
  $(z-w)^N[Y(v_1,z), Y(v_2,w)] = 0$, for some $N \in \mathbb{N}$. *(locality axiom)*

If the locality axiom holds already for $N = 0$, the vertex algebra is called *commutative*.

The Free-Fermionic Vertex Superalgebra 1.3.3. Despite the name of clear physical descent, which we borrowed from [32, Ch. V] and [29], whose origin we shall not go into, for us the vertex algebra of charged free fermions will be, by definition, that whose state space is precisely the fermionic space F seen earlier. The vacuum state is taken to be $\Phi_0 := v_0 \wedge v_{-1} \wedge \cdots$, and the translation operator T is a derivation of F which, by the translation axiom, needs only to be defined on $\Phi_{1+(i)}$ and $\Phi_{-1+(1^i)}$, where (i) denotes the partition with only one part i and (1^i) denotes the partition with i parts equal to 1:

$$T\Phi_{1+(i)} := (i+1)\Phi_{1+(i+1)} \qquad T\Phi_{-1+(1^i)} := -i\Phi_{-1+(1^{i+1})}.$$

The proof that these objects do fulfil the above axioms can be found in [32]. What is still missing is the vertex operator $Y(\Phi_{i+\lambda}, z)$. Although we shall retain to give here the general definition, it will suffice to know that

$$Y(\Phi_1, z) := \sum_{i\in\mathbb{Z}} z^i b_i \wedge \quad \text{and} \quad Y(\Phi_{-1}, z) := \sum_{j\in\mathbb{Z}} z^{-j-1}\beta_j \lrcorner$$

give, under the boson–fermion correspondence (1.21), two operators $B[\ell^{-1}, \ell] \to B[\ell^{-1}, \ell]((z))$. The image of a monomial $\ell^i S_\lambda(\mathbf{x})$ is $\varphi^{-1}(Y(\Phi_1, z)\Phi_{i+\lambda})$ under the former operator, while $\varphi^{-1}(Y(\Phi_{-1}, z)\Phi_{i+\lambda})$ is under the latter. These are precisely, up to a coefficient, the operators $\Gamma(z)$ and $\Gamma^*(z)$ of Section 1.2.1: as a matter of fact, one of our tasks will be to show how the *Schubert derivation*, a particular *HS*-derivation on the exterior algebra of a free abelian group, introduced in Section 5 leads, on one hand, to the (re)discovery of $\Gamma(z)$, $\Gamma^*(z)$, and allows to compute them on the other.

1.4 The KP Hierarchy via Vertex Operators

It belongs to an established tradition to denote by τ both points on the Poincaré half-plane as well as solutions to the *KP hierarchy*. Although the use of one symbol for different objects should be frowned upon, respecting this convention should hopefully cause no confusion.

Definition 1.4.1. A *tau function* for the *KP hierarchy* is an element τ of B satisfying the equation

$$\mathrm{Res}_z \Gamma^*(z)\tau \otimes \Gamma(z)\tau = 0,$$

where $\Gamma(z)$ and $\Gamma^*(z)$ are as in *(1.19)* and *(1.20)*, respectively.

1.4.2. Here is a reason why the above equation is geometrically significant. The group

$$Gl_\infty(\mathbb{Q}) := \{\mathcal{M} \in \mathrm{Aut}(V) \mid \mathcal{M}b_j = b_j \text{ for all but finitely many } j\}$$

can be identified [78] with the group of invertible matrices $\mathcal{M} := (a_{ij})_{i,j\in\mathbb{Z}}$ such that $a_{ij} - \delta_{ij} = 0$ for all but finitely many entries. Hence we may view an element in $Gl_\infty(\mathbb{Q})$ as an invertible matrix $(a_{ij}) \in Gl_n(\mathbb{Q})$ embedded in an infinite matrix where i) all off-diagonal entries are equal to 0, but finitely many, and ii) all diagonal entries are 1, but finitely many (Fig. 1.3).

Fig. 1.3 A matrix $(a_{ij}) \in Gl_3(\mathbb{Q})$ embedded in $Gl_\infty(\mathbb{Q})$

$$\begin{array}{cccccccccc}
\ddots & \vdots & \vdots & \vdots & \vdots & \vdots & \vdots & \vdots & \iddots \\
\dots & 1 & 0 & 0 & 0 & 0 & 0 & 0 & \dots \\
\dots & 0 & 1 & 0 & 0 & 0 & 0 & 0 & \dots \\
\dots & 0 & 0 & a_{i,i} & a_{i,i+1} & a_{i,i+2} & 0 & 0 & \dots \\
\dots & 0 & 0 & a_{i+1,i} & a_{i+1,i+1} & a_{i+1,i+2} & 0 & 0 & \dots \\
\dots & 0 & 0 & a_{i+2,i} & a_{i+2,i+1} & a_{i+2,i+2} & 0 & 0 & \dots \\
\dots & 0 & 0 & 0 & 0 & 0 & 1 & 0 & \dots \\
\dots & 0 & 0 & 0 & 0 & 0 & 0 & 1 & \dots \\
\iddots & \vdots & \vdots & \vdots & \vdots & \vdots & \vdots & \vdots & \ddots
\end{array}$$

1.4.3. The group $Gl_\infty(\mathbb{Q})$ acts on F_i under the *determinant representation* introduced by Kac and Peterson [76]:

$$\det(\mathcal{M})\Phi_{i+\boldsymbol{\lambda}} = \mathcal{M}b_{i+\lambda_1} \wedge \cdots \wedge \mathcal{M}b_{i-r+1+\lambda_r} \wedge \mathcal{M}b_{i-r} \wedge \mathcal{M}b_{i-r-1} \wedge \cdots$$

for $\mathcal{M} \in Gl_\infty(\mathbb{C})$. Via the boson–fermion correspondence (1.2), $Gl_\infty(\mathbb{Q})$ acts on B as well, via $\mathcal{M} \cdot S_{\boldsymbol{\lambda}}(\mathbf{x}) = \varphi_i(\det(\mathcal{M})\Phi_{i+\boldsymbol{\lambda}})$. A linear combination $\sum_{\boldsymbol{\lambda}\in\mathcal{P}} a_{\boldsymbol{\lambda}}\Phi_{i+\boldsymbol{\lambda}} \in F_i$, where $a_{\boldsymbol{\lambda}} = 0$ for all but finitely many $\boldsymbol{\lambda}$, is said to be *decomposable* if there exist finitely many $v_1, \dots, v_r \in V$ such that

$$\begin{aligned}
\sum_{\boldsymbol{\lambda}\in\mathcal{P}} a_{\boldsymbol{\lambda}}\Phi_{i+\boldsymbol{\lambda}} &= v_1 \wedge v_2 \wedge \cdots \wedge v_r \wedge b_{i-r} \wedge b_{i-r-1} \wedge b_{i-r-2} \wedge \cdots \\
&= v_1 \wedge v_2 \wedge \cdots \wedge v_r \wedge \Phi_{i-r} = \det(\mathcal{M})\Phi_i,
\end{aligned}$$

where $\mathcal{M}$ is the unique element of $Gl_\infty(\mathbb{Q})$ such that $\mathcal{M}b_{i-j+1+\lambda_j} = v_j$ for $1 \le j \le r$ and $\mathcal{M}b_j = b_j$ otherwise. The locus of polynomials of B that correspond to decomposable tensors in F_i (for all $i \in \mathbb{Z}$) belongs then to the $Gl_\infty(\mathbb{Q})$-orbit Ω of $1 \in B$.

From now on, to fix ideas, we choose $i = 0$ and seek the equations describing Ω inside B. An arbitrary element of $\xi \in F_0$ is a finite linear combination

$$\xi = \sum_{i=1}^{n} a_i \tilde{\eta}_i,$$

where $a_i \in \mathbb{Q}$ and $\tilde{\eta}_i \in F_0$. By definition of F_0, there exists a large enough integer r and $\eta_1, \eta_2, \ldots, \eta_n \in \bigwedge^r V$ such that

$$\tilde{\eta}_i = \eta_i \wedge \Phi_{-r} := \eta_i \wedge b_{-r} \wedge b_{-r-1} \wedge b_{-r-2} \wedge \cdots$$

for any $1 \leq i \leq n$. Thus ξ is decomposable if and only if $\sum_{i=1}^{n} a_i \eta_i$ is a decomposable vector in $\bigwedge^r V$.

Now we invoke Theorem 6.1.7, which ensures $\eta \in \bigwedge^r V$ is decomposable if and only if

$$\sum_{i \in \mathbb{Z}} (b_i \wedge \eta) \otimes (\beta_i \lrcorner \eta) = 0, \tag{1.22}$$

where $\beta_i(b_j) = \delta_{ij}$ and $\beta_i \lrcorner \eta$ is the *contraction* of η against $\beta_i \in V^*$. Note that (1.22) makes sense: it is obviously finite because $\beta_i \lrcorner \eta = 0$ for all but finitely many i. Define formal Laurent series

$$\mathbf{b}(z) = \sum_{i \in \mathbb{Z}} b_i z^i \qquad \text{and} \qquad \boldsymbol{\beta}(z) = \sum_{i \in \mathbb{Z}} \beta_i z^{-i-1}.$$

A simple inspection shows that (1.22) holds precisely when

$$\operatorname{Res}_z (\mathbf{b}(z) \wedge \eta) \otimes (\boldsymbol{\beta}(z) \lrcorner \eta) = 0. \tag{1.23}$$

Using the boson–fermion correspondence $\varphi_0 : F_0 \to B$, in Chapters 6 and 7, we will show that equation (1.23) translates into

$$\operatorname{Res}_z \Gamma(z) \varphi_0(\eta) \otimes \Gamma^*(z) \varphi_0(\eta) = 0. \tag{1.24}$$

It follows that a τ function corresponds to a decomposable tensor in F_0 if and only if (1.24) holds, which is the set of Plücker equations defining the (infinite-dimensional) *Grassmann cone* of F_0 in B. This is due to the Sato's School.

1.4.4. Equation (1.24) lives in the tensor product $B \otimes B \cong \mathbb{Q}[\mathbf{x}', \mathbf{x}'']$, where $\mathbf{x}' = (x_1', x_2', \ldots)$ and $\mathbf{x}'' = (x_1'', x_2'', \ldots)$. This implies that $\tau \in B$ is a tau function if and only if

$$\operatorname{Res}_{z=0} \exp\left(\sum_{i \geq 1} (x_i' - x_i'') z^i\right) \exp\left(-\sum_{i \geq 1} \frac{1}{iz} \left(\frac{\partial}{\partial x_i'} - \frac{\partial}{\partial x_i''}\right)\right) \tau(\mathbf{x}') \tau(\mathbf{x}'') = 0. \tag{1.25}$$

1.4.5. At this juncture the reader might be interested in seeing more details, for which we refer to [78, pp. 72–75], the introduction of [79] or [5, Section 4]. We briefly summarize the arguments of these references for the sake of self-containedness, with no claim of originality with respect to either content or exposition. If we switch to the variables

$$x_i' = x_i - y_i \qquad \text{and} \qquad x_i'' = x_i + y_i,$$

formula (1.25) reads

$$\exp\left(\sum_{i\geq 1}(-2y_i z^i) z^i\right) \exp\left(-\sum_{i\geq 1}\frac{1}{iz^i}\left(\frac{\partial}{\partial y_i}\right)\right) \tau(\mathbf{x}-\mathbf{y})\tau(\mathbf{x}+\mathbf{y}),$$

where $\mathbf{y} := (y_1, y_2, \ldots)$. Put otherwise, using the definition of the polynomial S_j,

$$\left(\sum_{i\geq 1} S_i(-2\mathbf{y})z^i \cdot \sum_{i\geq 1} S_j(\tilde{\partial}_{\mathbf{y}})z^{-j}\right) \tau(\mathbf{x}-\mathbf{y})\tau(\mathbf{x}+\mathbf{y}),$$

where $\tilde{\partial}_{\mathbf{y}} = \left(\frac{\partial}{\partial y_1}, \frac{1}{2}\frac{\partial}{\partial y_2}, \frac{1}{3}\frac{\partial}{\partial y_3}, \ldots\right)$, whose residue at $z = 0$ is

$$\sum_{i\geq 1} S_i(-2\mathbf{y})S_{i+1}(\tilde{\partial}_{\mathbf{y}})\tau(\mathbf{x}-\mathbf{y})\tau(\mathbf{x}+\mathbf{y}),$$

or equivalently,

$$\sum_{i\geq 1} S_i(-2\mathbf{y})S_{i+1}(\tilde{\partial}_{\boldsymbol{\zeta}})\tau(\mathbf{x}-\mathbf{y}-\boldsymbol{\zeta})\tau(\mathbf{x}-\mathbf{y}+\boldsymbol{\zeta})_{|\boldsymbol{\zeta}=0},$$

where $\boldsymbol{\zeta} := (\zeta_1, \zeta_2, \ldots)$ are auxiliary variables. Using the Taylor formula for polynomials, we conclude that τ corresponds to a point in the $Gl_\infty(\mathbb{Q})$-orbit of $1 \in B$ if and only if

$$\sum_{i\geq 1} S_i(-2\mathbf{y})S_{i+1}(\tilde{\partial}_{\boldsymbol{\zeta}}) \exp\left(\sum_{j\geq 1} y_j \frac{\partial}{\partial \zeta_j}\right) \tau(\mathbf{x}-\boldsymbol{\zeta})\tau(\mathbf{x}+\boldsymbol{\zeta})_{|\boldsymbol{\zeta}=0} = 0. \tag{1.26}$$

In particular, all the coefficients of $y_{i_1}^{j_1} y_{i_2}^{j_2} \cdots y_{i_k}^{j_k}$ in the expansion of (1.26) in formal power series of $\mathbf{y}$-monomials have to vanish. It is easy to check that the coefficients of y_1 and y_2 are identically zero (cf. [78, pp. 72–74]), while killing the coefficient of y_3 gives the equation

$$\left(\frac{1}{3}\frac{\partial^2}{\partial\zeta_1\partial\zeta_2} - \frac{1}{4}\frac{\partial^2}{\partial\zeta_2^2} - \frac{1}{12}\frac{\partial^4}{\partial\zeta_1^4} - \frac{1}{2}\left(\frac{\partial}{\partial\zeta_4} + \frac{\partial^3}{\partial\zeta_1^2\partial\zeta_2}\right)\right) \tau(\mathbf{x}+\boldsymbol{\zeta})\tau(\mathbf{x}-\boldsymbol{\zeta})_{|u=0} = 0,$$

which is equivalent to

$$\left(\frac{\partial^4}{\partial\zeta_1^4}+3\frac{\partial^2}{\partial\zeta_2^2}-4\frac{\partial^2}{\partial\zeta_1\partial\zeta_3}\right)\tau(\mathbf{x}+\boldsymbol{\zeta})\tau(\mathbf{x}-\boldsymbol{\zeta})_{|_{u=0}}=0, \tag{1.27}$$

because all odd partial derivatives in ζ_i, $i \geq 1$, of the product $\tau(\mathbf{x}-\boldsymbol{\zeta})\tau(\mathbf{x}+\boldsymbol{\zeta})$ vanish at $\boldsymbol{\zeta} = (0, 0, \ldots)$. We leave to the patient reader the task of expanding (1.27) to show that then $\tau(\mathbf{x})$ must satisfy the Hirota bilinear form (1.18) of the KP equation.

1.5 Notes and References

- The massive interest raised over the years by the KdV and KP equations has spawned an equally enormous inflation of thrilling research output, due to multiple interdisciplinary connections ranging from integrable systems to algebraic geometry. What links the subject of this book to the KP equation is the Hirota bilinear form [62, 63], which can be interpreted as one of the equations defining the Plücker embedding of the infinite Grassmannian. The Hirota bilinear form of the KP equation is satisfied by suitable solutions of linear ODEs of infinite order, as explained in Exercise 7.10.11. In turn, this fact indicates a relationship with the basic algebra governing linear recurrence sequences.
- Readers seeking a pleasant, friendly and not too technical introduction, yet rigorous, to soliton theory are recommended to consult the beautiful book by A. Kasman [82], which is rich in examples, all fully worked out. The survey by Gesztesy and Weikard [53] has a slightly more analytical point of view, approaching inverse scattering theory. A compelling general reference is Segal and Wilson's milestone paper [140]: it offers an analytical construction of Sato's infinite Grassmannian, without neglecting the algebraic and algebro-geometric viewpoints, especially in connection with the deep theory of loop groups and loop algebras. See also [123] on this.
- Arbarello's survey [5] collects a multitude of interesting aspects, including the relationship with the Schottky problem (characterizing Jacobians of curves among abelian varieties) and the presentation of the KP hierarchy via vertex operators. As in Mulase's [117], relationships with the intersection theory on the Deligne–Mumford moduli spaces of stable pointed curves, after Witten and Kontsevich [88, 150], are described.
- The Japanese school led by Sato was tremendously influential in the theory of infinite-dimensional integrable systems. We must cite the foundational papers [18, 19, 133]. Integrable systems via algebraic geometry is heavily based on the study of spectral curves, for which we refer to the beautiful comprehensive survey by Previato [124]. See also Krichever [90]. A prototype of spectral curve is the elliptic curve arising from the simple pendulum of Section 1.1.4.
- For readers wishing to know more on vertex operators and vertex algebras, based on our personal experience as beginners, we heavily recommend the

books by Kac and Kac–Raina–Rozhkovskaya [73, 78]. Strong motivations to study this theory can be found in Ben-Zvi, I. Frenkel, E. Frenkel, Lepowsky, Li and Meurman [30, 32, 103], not to mention the paper [10]. Beside Borcherd's rephrasing of the axioms using the formal analogy with a commutative ring with a group action [11], another important, and somehow more economical, framework to deal with vertex algebras is that provided by *conformal Lie algebras*. For these aspects we have greatly profited from notes by R. Heluani [61] and the lecture notes [74]. The latter have been partly reworked in the second half of [78].

Chapter 2
Generic Linear Recurrence Sequences

Let A be a commutative ring with unit and M any A-module. An M-valued linear recurrence sequence (LRS) generalizes the sequence of powers of the roots of a given monic polynomial with A-coefficients. A generic LRS is a linear recurrence sequence with indeterminate coefficients. A main character of this chapter, as of the entire book, is the ring $B_r := \mathbb{Z}[e_1, \ldots, e_r]$, thought of as a free polynomial algebra generated by the coefficients of a generic LRS of order $r \geq 1$. The name of the indeterminates, $(e_1, \ldots, e_r)$, is reminiscent of the elementary symmetric polynomials in r variables. The ring B_r will later be bestowed the more ambitious task of approximating the bosonic Fock representation of the oscillator Heisenberg algebra.

The present chapter is organized as follows. Section 2.1 is just an opening, to fix notation and to agree to identify M-valued sequences with M-valued formal power series. Section 2.2 gives a closer look at the ring B_r: the notion of partition and Schur polynomials are introduced in 2.2.2 to describe in details its *weight graduation*, defined by assigning degree i to each indeterminate e_i.

Section 2.4 focuses on the universal formula expressing the solution to the Cauchy problem for linear ODEs with constant coefficients and analytic forcing term. The expression is attained by means of a distinguished sequence $(u_i)_{i\in\mathbb{Z}}$ of *universal* B_r-valued generic LRSs. They will then go on to play an important role throughout the text. What ties generic LRSs to generic linear ordinary differential equations (ODEs) is the notion of *formal Laplace*[1] *transform*, introduced in a purely algebraic manner in Section 2.5, following [42]. Here, the Cauchy problem for linear ODEs is discussed, and an example to illustrate the method is fully worked out. Section 2.6 is about *generalized Wronskians* as in [50]. The terminology perhaps is

[1] Pierre-Simon, marquis de Laplace, Beaumont-en-Auge, Normandy, 1749–Paris, 1897. Mathematician and astronomer, he wrote the celebrated *celestial mechanics*. Asked by Napoleon why there was no mention to God in his treatise, as was customary at the time, he answered: je n'avais pas besoin de cette hypothèse-là: I didn't need that hypothesis.

L. Gatto, P. Salehyan, *Hasse-Schmidt Derivations on Grassmann Algebras*,
IMPA Monographs, DOI 10.1007/978-3-319-31842-4_2

not standard, but the same notion has been used by several authors in a variety of different contexts, e.g. [144], and is related with the *higher Wronskians* introduced and used by A. Iarrobino [67, 68] with diverse motivations.

This chapter is elementary in nature. It may be seen as a review of the basic background regarding linear ODEs with constant coefficients. The exposition is however carried out in the perspective of revisiting generalized Wronskians, Schubert calculus for Grassmannians and the boson–fermion correspondence within a unified framework.

2.1 Sequences in Modules over Algebras

2.1.1. Let A be a commutative ring with unit and M be an A-module. For any nonempty subset $\mathcal{S}$ of the integers, we denote by $M^{\mathcal{S}}$ the collection of maps $\mathbf{m} : \mathcal{S} \to M$, i.e. the A-module of M-valued sequences defined over $\mathcal{S}$. The image of $s \in \mathcal{S}$ is denoted by m_s. We shall be interested in the cases $\mathcal{S} = \mathbb{N}$ and $\mathcal{S} = \mathbb{Z}$. If $\mathcal{S} = \mathbb{N}$, elements of $M^{\mathbb{N}}$ will be represented by the array of their images $(m_0, m_1, \ldots)$. Alternatively, let t be any indeterminate over A and call

$$M[[t]] := \left\{ \sum_{j\geq 0} m_j t^j \mid m_j \in M \right\} \tag{2.1}$$

the A-module of *formal power series* with coefficients in M. Because of the natural identification of $M[[t]]$ with $M^{\mathbb{N}}$, we shall also speak of M-valued formal power series. If $M = A$, then $A[[t]]$ is an A-algebra with respect to the usual product

$$\sum_{i\geq 0} a_i t^i \cdot \sum_{j\geq 0} a_j t^j = \sum_{k\geq 0} \left(\sum_{i=0}^{k} a_i a_{k-i} \right) t^k, \qquad a_i \in A. \tag{2.2}$$

An obvious generalization of (2.2) endows $M[[t]]$ with a structure of $A[[t]]$-module:

$$\sum_{i\geq 0} a_i t^i \cdot \sum_{j\geq 0} m_j t^j = \sum_{k\geq 0} \left(\sum_{i=0}^{k} a_i m_{k-i} \right) t^k, \qquad (a_i, m_j) \in A \times M.$$

2.1.2. If X is another indeterminate over A and $p \in A[X]$ has degree r, we write

$$p(X) = e_0(p)X^r - e_1(p)X^{n-1} + \cdots + (-1)^r e_r(p).$$

This means that we denote by $(-1)^i e_i(p)$ the coefficient of X^{r-i}, $0 \leq i \leq r$. The polynomial p is *monic* if $e_0(p) = 1$.

Examples 2.1.3. i) If $p := X^4 - 3X^2 + \sqrt{2}X - 3 \in \mathbb{Z}[\sqrt{2}][X]$, then $e_0(P) = 1$, $e_1(P) = 0$, $e_2(P) = -3$, $e_3(P) = -\sqrt{2}$ and $e_4(P) = -3$;

ii) If $P \in A[X]$ splits as product of r distinct linear factors:

$$p(X) = (X - x_1)\cdots(X - x_r),$$

then $e_1(p), e_2(p), \ldots, e_r(P)$ are precisely the elementary symmetric polynomials in the roots $x_1, \ldots, x_n$ of p. In this way we follow the convention of [109] for denoting symmetric functions.

Definition 2.1.4. A sequence $\mathbf{m} := (m_0, m_1, \ldots) \in M^{\mathbb{N}}$ is called a *linear recurrence sequence* (LRS) of order r if there exists a monic $p \in A[X]$ of degree r such that

$$m_{r+j} - e_1(p)m_{r+j-1} + \cdots + (-1)^r e_r(p)m_j = 0, \qquad \forall j \geq 0. \tag{2.3}$$

The polynomial p is the *characteristic polynomial* of the LRS.

Expression (2.3) is also called a *linear difference equation*. All LRSs associated with one polynomial form a submodule of $M^{\mathbb{N}}$ whose elements solve a homogeneous *linear difference equation*.

Examples 2.1.5. i) Let $\alpha \in A$ be a root of $p := X^2 - e_1(p)X + e_2(p) \in A[X]$. Then $(1, \alpha, \alpha^2, \alpha^3, \ldots)$ is a LRS of degree 2 having p as characteristic polynomial;

ii) Let $\mathcal{M}$ be a square $n \times n$ matrix with entries in some commutative associative $\mathbb{Z}$-algebra. By abuse of notation denote by $(-1)^j e_j(\mathcal{M})$ the coefficient of X^{n-j} in the polynomial

$$p_{\mathcal{M}}(X) := \det(X \cdot \mathbb{1}_{n\times n} - \mathcal{M}). \tag{2.4}$$

where $\mathbb{1}_{n\times n}$ denotes the diagonal identity matrix. The *Cayley–Hamilton theorem* (Corollary 4.2.15 to Theorem 4.2.12) says that $p_{\mathcal{M}}(\mathcal{M}) = 0$, namely, $\mathcal{M}$ is a root of (2.4). Hence $(\mathbb{1}_{n\times n}, \mathcal{M}, \mathcal{M}^2, \ldots)$ is a LRS of order n, with $p_{\mathcal{M}}(X)$ as *characteristic polynomial*, where $\mathbb{1}_{n\times n}$ is the square identity matrix;

iii) Let $M = A = \mathbb{Z}$ and $p(X) = X^2 - X - 1$. The unique LRS $(\mathfrak{f}_0, \mathfrak{f}_1, \ldots)$ having characteristic polynomial p and initial conditions $\mathfrak{f}_0 = \mathfrak{f}_1 = 1$ is the *Fibonacci sequence* $(1, 1, 2, 3, 5, 8, \ldots)$. The equality

$$\sum_{n\geq 0} \mathfrak{f}_n t^n = \frac{1}{1 - t - t^2}$$

explains that the nth coefficient of the formal Taylor expansion of the right-hand side is precisely $\mathfrak{f}_n$.

2.1.6. Let

$$M((t)) := M[t^{-1}, t]] = \left\{ \sum_{j\geq -i} m_j t^j \,\middle|\, m_j \in M, i \in \mathbb{N} \right\}$$

be the A-module of M-valued *formal Laurent series*, i.e. elements of $M((t))$ with only finitely many non-zero negative powers of t. A quick check shows that the kernel of the A-epimorphism $M((t)) \mapsto M[[t]]$ mapping $\sum_{j\geq -i} m_j t^j$ to its 'holomorphic' part $\sum_{j\geq 0} m_j t^j$ is the submodule $t^{-1}M[t^{-1}]$. Let

$$\mathbb{J}_M : \frac{M((t))}{t^{-1}M[t^{-1}]} \to M[[t]] \tag{2.5}$$

be the induced isomorphism. For all $j \geq 0$, define

$$D^j \mathbf{m}(t) := \mathbb{J}_M \left(\frac{\mathbf{m}(t)}{t^j} + t^{-1}M[t^{-1}] \right), \tag{2.6}$$

or more explicitly

$$D^j \sum_{i\geq 0} m_i t^i = \sum_{i\geq 0} m_{i+j} t^i. \tag{2.7}$$

Clearly $D^j \in \mathrm{End}_A(M[[t]])$ is the composite of $D := D^1$ with itself j times. We set $D^0 = \mathbb{1}_M$, the identity endomorphism of M. Via the identification $M[[t]] \cong M^{\mathbb{N}}$, the endomorphism D^j is nothing but the *shift operator* $m_i \mapsto m_{i+j}$.

Example 2.1.7. The unique solution to the equation $D\mathbf{a}(t) = \mathbf{a}(t)$ with initial condition a_0 is

$$\mathbf{a}(t) = \frac{a_0}{1-t} = a_0(1 + t + t^2 + \cdots).$$

In other words $\dfrac{a_0}{1-t}$ is related to D as $\exp(a_0 t)$ is related to the formal derivative ∂_t. See Section 2.5.

2.2 Generic Polynomials, Partitions and Schur Determinants

2.2.1. Generic polynomials formalize the idea of polynomials with indeterminate coefficients. There is just one monic generic polynomial in each given degree. For each positive integer r, let $(e_1, \ldots, e_r)$ denote a finite sequence of indeterminates over $\mathbb{Z}$, and let B_r be the polynomial ring in $e_1, \ldots, e_r$ with integral coefficients

$$B_r := \mathbb{Z}[e_1, \ldots, e_r].$$

Henceforth we shall denote by $\mathfrak{p}_r(X)$ the *generic monic polynomial of degree* r:

$$\mathfrak{p}_r(X) := X^r - e_1 X^{r-1} + \cdots + (-1)^r e_r. \tag{2.8}$$

It specializes to any other monic polynomial of the same degree

$$p(X) := X^r - e_1(p)X^{r-1} + \cdots + (-1)^r e_r(p), \qquad e_i(p) \in A,$$

with coefficients in a B_r-algebra A, via the unique ring homomorphism $e_i \mapsto e_i(p)$.

Partitions 2.2.2. Regard the ring B_r as a graded $\mathbb{Z}$-algebra by declaring i to be the degree of the indeterminate e_i. For $w \geq 0$, let $(B_r)_w$ be the submodule of elements of degree w. For example, $(B_r)_1 = \mathbb{Z}e_1$, $(B_r)_2 = \mathbb{Z}e_1^2 \oplus \mathbb{Z}e_2$ and $(B_r)_3 = \mathbb{Z}e_1^3 \oplus \mathbb{Z}e_1e_2 \oplus \mathbb{Z}e_3$. In general, to describe monomials of degree w of B_r, it is useful to use *partitions* of *length* w. A *partition* is a monotone non-increasing sequence of non-negative integers

$$\boldsymbol{\lambda} : \quad \lambda_1 \geq \lambda_2 \geq \cdots \geq 0,$$

such that all the *parts* λ_i are zero but finitely many ([33, 109]).

Denote by $\mathcal{P}$ the set of all partitions. If $\boldsymbol{\lambda} := (\lambda_1, \lambda_2, \ldots) \in \mathcal{P}$, its *length* $\ell(\boldsymbol{\lambda}) := \sharp\{i \,|\, \lambda_i \neq 0\}$ is the number of its non-zero *parts*, and its *weight* is $|\boldsymbol{\lambda}| = \sum_i \lambda_i$. If $\ell(\boldsymbol{\lambda}) = r$, we simply write $\boldsymbol{\lambda} = (\lambda_1, \ldots, \lambda_r)$, omitting the infinite sequence of the zero parts. The *null partition*, $(0) := (0, 0, 0, \ldots)$, having only null parts, will be simply denoted by 0. The set of partitions of length $k \leq r$ is denoted by $\mathcal{P}_r$: if $\boldsymbol{\lambda} \in \mathcal{P}_r$ has length strictly less than r, we may, if convenient, add a string of $r - k$ zeros to it.

Each $\boldsymbol{\lambda} \in \mathcal{P}$ is a partition of the integer $w := |\boldsymbol{\lambda}|$. The *Young diagram* of $\boldsymbol{\lambda} := (\lambda_1, \ldots, \lambda_r) \in \mathcal{P}$ is an array $Y(\boldsymbol{\lambda})$ of left-justified rows of boxes such that the jth row has λ_j boxes, for $1 \leq j \leq r$. The *conjugate* of a partition $\boldsymbol{\lambda}$ is the partition $\boldsymbol{\lambda}'$ whose Young diagram is obtained from $Y(\boldsymbol{\lambda})$ by exchanging columns with rows: the jth column of $Y(\boldsymbol{\lambda}')$ has as many boxes as the jth row of $Y(\boldsymbol{\lambda})$. For example, the conjugate to $(3, 2, 2, 1)$ is $(4, 3, 1)$, as shown in Figure 2.1.

Fig. 2.1 The Young diagram of the partition $(3, 2, 2, 1)$ and its conjugate $(4, 3, 1)$

2.2.3. A partition $\boldsymbol{\lambda} \in \mathcal{P}$ can be also expressed as $\boldsymbol{\lambda} := (1^{i_1} 2^{i_2} \cdots n^{i_n})$ to mean that it has i_j parts equal to j, $1 \leq j \leq n$, so that it has length $i_1 + \cdots + i_n$ and weight $i_1 + 2i_2 + \cdots + ni_n$. For example, the partition $\boldsymbol{\lambda} = (3, 2, 2, 1)$ can be equivalently written as $(1^1 2^2 3^1)$. Analogously, $\boldsymbol{\mu} := (2^4 3^2 4^1)$ denotes the partition that in standard notation is $(4, 3, 3, 2, 2, 2, 2)$.

2.2.4. For all integers $n \geq r$, we denote by $\mathcal{P}_{r,n}$ the set of partitions whose Young diagram is contained in an $r \times (n - r)$ rectangle, i.e. in the diagram of the partition $(n - r, \ldots, n - r)$. We set $\mathcal{P}_{r,\infty} := \mathcal{P}_r$ (Fig. 2.2).

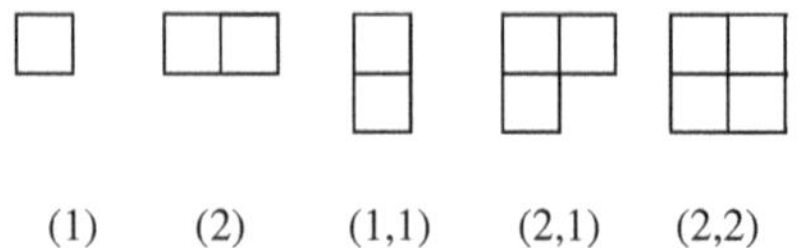

Fig. 2.2 The non-null partitions of $\mathcal{P}_{2,4}$

2.2.5. If $\mathbf{a} := (a_i)_{i\in\mathbb{Z}}$ is a bilateral sequence taking values in an arbitrary commutative associative algebra A and $\boldsymbol{\lambda} \in \mathcal{P}_r$,

$$\Delta_{\boldsymbol{\lambda}}(\mathbf{a}) = \det(a_{\lambda_j+i-j})_{1\le i,j\le r} \in A$$

is the *Schur determinant* associated with the given sequence and partition. In practical terms, one allocates $a_{\lambda_1}, \dots, a_{\lambda_r}$ along the principal diagonal from top-down. Then *above* each a_{λ_i}, in the same column, the index decreases by one unit for each row; *below* a_{λ_i}, in the same column, the index increases by one unit per row. For instance,

$$\Delta_{(211)}(\mathbf{a}) = \begin{vmatrix} a_2 & a_0 & a_{-1} \\ a_3 & a_1 & a_0 \\ a_4 & a_2 & a_1 \end{vmatrix}.$$

2.2.6. As anticipated, partitions provide a useful combinatorial description of the grading of B_r. In fact

$$(B_r)_w = \bigoplus_{\boldsymbol{\lambda}\in\mathcal{P}_r, |\boldsymbol{\lambda}|=w} \mathbb{Z}\cdot \mathbf{e}_r^{\boldsymbol{\lambda}'},$$

where $\boldsymbol{\lambda}'$ denotes the partition conjugate to $\boldsymbol{\lambda}$ and $\mathbf{e}^{\boldsymbol{\lambda}'} := e_1^{i_1}\cdot \ldots \cdot e_r^{i_r}$, once $\boldsymbol{\lambda}'$ is written in the form $(1^{i_1}2^{i_2}\cdots)$. Then

$$B_r = \bigoplus_{w\in\mathbb{N}} (B_r)_w \tag{2.9}$$

defines what we shall refer to as the *weight graduation* of B_r.

Example 2.2.7. Let $r \ge 3$. Then $e^{(4,1,1)'} = e_1^3 e_3$, $e^{(2,2,2)'} = e_3^2$, $e^{(321)'} = e_1e_2e_3$, $e^{(1,1,1)'} = e_3$ and $e^{(3)'} = e_1^3$.

2.2.8. Besides the monic generic polynomial $\mathfrak{p}_r(X)$, it will be also convenient to use

$$E_r(t) := 1 - e_1t + \cdots + (-1)^r e_r t^r \in B_r[t]. \tag{2.10}$$

Considering the rings $B_r((t)) = B_r[t^{-1}, t]]$ and $B_r((X)) = B_r[X^{-1}, X]]$ of *formal Laurent series* in t and X, respectively, the relationship between $E_r(t)$ and $\mathfrak{p}_r(X)$ is the obvious one:

$$E_r(t) = t^r \mathfrak{p}_r\left(\frac{1}{t}\right) \qquad \text{and} \qquad \mathfrak{p}_r(X) = X^r E_r\left(\frac{1}{X}\right).$$

Let $H_r(t) := \sum_{n\in\mathbb{Z}} h_n t^n \in B_r[[t^{-1}, t]]$ be given by

$$H_r(t) := \frac{1}{E_r(t)} = 1 + \sum_{n\geq 1}(1 - E_r(t))^n. \tag{2.11}$$

Because of the definition $h_j = 0$ if $j < 0$, $h_0 = 1$, while for every $i > 0$, the term h_i is an explicit polynomial in $e_1, \ldots, e_r$, which is homogeneous of degree i with respect to the grading (2.9) of B_r. For example,

$$h_1 = e_1, \qquad h_2 = e_1^2 - e_2, \qquad h_3 = e_1^3 - 2e_1e_2 + e_3.$$

In general, for $n \geq 0$, h_n can be computed recursively using the equation $H_r(t)E_r(t) = 1$, equivalent to (2.11). Indeed

$$E_r(t)\sum_{n\geq 0} h_n t^n = 1 \tag{2.12}$$

which holds if and only if $h_0 = 1$ and the coefficient of t^n, in the left-hand side of (2.12), vanishes for all $n \geq 1$, i.e. if and only if

$$h_n - e_1h_{n-1} + \cdots + (-1)^n e_n h_0 = 0, \tag{2.13}$$

with the usual convention that $e_n = 0$ if $n > r$. The Schur polynomial $\Delta_{\boldsymbol{\lambda}}(H_r)$ (cf. 2.2.5) associated with the sequence $H_r := (h_j)_{j\in\mathbb{Z}}$ of the coefficients of $H_r(t)$ and to the partition $\boldsymbol{\lambda} \in \mathcal{P}_r$ is an element of $(B_r)_{|\boldsymbol{\lambda}|}$. For example, for the partition $\boldsymbol{\lambda} = (2, 2, 1)$ of the integer 5,

$$\Delta_{(2,2,1)}(H_r) = \begin{vmatrix} h_2 & h_1 & 0 \\ h_3 & h_2 & 1 \\ h_4 & h_3 & h_1 \end{vmatrix} = h_1h_2^2 - h_1^2h_3 - h_2h_3 + h_1h_4 \in (B_r)_5.$$

Using (2.13) one can easily check (see, e.g. [109]) that

$$e_j = \Delta_{(1^j)}(H_r) \qquad \text{and} \qquad h_n = \Delta_{(1^n)}(E_r) := \det(e_{j-i+1})_{0\leq i,j\leq n}.$$

where we recall from Section 2.2.6 that (1^j) is the partition with j parts equal to 1.

Remark 2.2.9. For the coefficients of the formal power series $u_0 = H_r(t)$, the more careful notation $h_{r,n}$ should be preferred, in place of just h_n, to keep track of their dependence on r. To make the notation less heavy, we decided however to drop the subscript r, hoping for the context being sufficient to avoid confusions.

2.2.10. The sequence $H_r := (h_j)_{j\in\mathbb{Z}}$ of coefficients of the formal power series $H_r(t)$ may be alternatively defined as

$$\sum_{n\geq 0} \frac{h_n}{X^{r+n}} := \frac{1}{\mathfrak{p}_r(X)}. \tag{2.14}$$

More explicitly

$$h_n = \mathrm{Res}_X \frac{X^{r+n-1}}{\mathfrak{p}_r(X)},$$

where Res_X denotes the *residue* of a Laurent series in the indeterminate X, which by definition is the coefficient of X^{-1}. See also [96, 97] and Section 5.8.

Note that the ring B_r can be thought of as a quotient of a polynomial ring with infinitely many indeterminates, by considering the unique $\mathbb{Z}$-module homomorphism

$$\pi : \mathbb{Z}[X_1, X_2, \ldots] \longrightarrow B_r,$$

mapping $X_i \mapsto h_i$. This map is clearly a ring epimorphism, because each element of B_r is a polynomial in $e_1, \ldots, e_r$ and each e_i is a weighted homogeneous polynomial of degree i in $h_1, \ldots, h_i$. The quotient

$$\mathbb{Z}[h_1, h_2, \ldots] := \frac{\mathbb{Z}[X_1, X_2, \ldots]}{\ker(\pi)} = B_r$$

will be denoted by $\mathbb{Z}[H_r]$.

2.3 Generic Linear Recurrence Sequences

The Fundamental Sequence 2.3.1. A special role in the theory is played by the *fundamental sequence* $(u_i)_{i\in\mathbb{Z}}$ of $B_r[[t]]$, which will return in Chapter 7, defined for $i \geq 0$ by

$$u_i := D^i(H_r(t)) \qquad \text{and} \qquad u_{-i} := t^i H_r(t), \tag{2.15}$$

or more explicitly

$$u_i = t^i + \sum_{n\geq 0} h_{n-i} t^n \qquad \text{and} \qquad u_{-i} = h_i + \sum_{n\geq 1} h_{n+i} t^n. \tag{2.16}$$

Equations 2.16 show that the terms of the sequence (u_i) are linearly independent over the integers, because the powers $(t^n)_{n>0}$ and the coefficients $(h_n)_{n\geq 0}$ are. Notice, in particular, that $u_0 = H_r(t)$ and that

$$D^i u_j = u_{i+j}, \qquad \forall (i,j) \in \mathbb{N} \times \mathbb{Z} \tag{2.17}$$

by the very definition of the endomorphism D of $A[[t]]$ (cf. (2.6) for $M = A$).

The same remark as in 2.2.9 applies to the sequence $(u_i)_{i\in Z}$ defined in (2.15), which in turn would deserve to be denoted $(u_{r,i})$

2.3.2. Let now M be a module over a B_r-algebra A, fixed once and for all, and define linear maps $\mathtt{U}_i : M[[t]] \to M$ via

$$\sum_{i\geq 0} \mathtt{U}_i(\mathbf{m}(t))t^i := E_r(t)\mathbf{m}(t), \qquad (\mathbf{m}(t) \in M[[t]]). \tag{2.18}$$

Comparing the coefficients of t^i on both sides of (2.18), we obtain for all $i \geq 0$

$$\mathtt{U}_i(\mathbf{m}(t)) = m_i + \sum_{j=1}^{r} (-1)^j e_j m_{i-j}, \tag{2.19}$$

with the convention that $m_j = 0$ if $j < 0$. So, for example,

$$\mathtt{U}_0(\mathbf{m}(t)) = m_0, \quad \mathtt{U}_1(\mathbf{m}(t)) = m_1 - e_1 m_0,$$

$$\mathtt{U}_2(\mathbf{m}(t)) = m_2 - e_1 m_1 + e_2 m_0, \ \mathtt{U}_3(\mathbf{m}(t)) = m_3 - e_1 m_2 + e_2 m_1 - e_3 m_0.$$

Proposition 2.3.3. Each $\mathbf{m}(t) \in M[[t]]$ admits the unique expansion

$$\mathbf{m}(t) = \sum_{j\geq 0} \mathtt{U}_j(\mathbf{m}(t))u_{-j}. \tag{2.20}$$

In particular, if $M = A$,

$$\mathtt{U}_j(u_{-i}) = \delta_{ji}. \tag{2.21}$$

Proof. Rewrite equality (2.18) by inverting $E_r(t)$ in $B_r[[t]]$:

$$\mathbf{m}(t) = \sum_{j\geq 0} \mathtt{U}_j(\mathbf{m}(t))\frac{t^j}{E_r(t)} = \sum_{j\geq 0} \mathtt{U}_j(\mathbf{m}(t))t^j H_r(t) = \sum_{j\geq 0} \mathtt{U}_j(\mathbf{m}(t))u_{-j}.$$

In particular $t^j = E_r(t)H_r(t)t^j = E_r(t)u_{-j} = \sum_{i\geq 0} \mathtt{U}_i(u_{-j})t^i$ if $M = A$ and (2.21) follows. □

If $M := A[X]$, consider the formal power series

$$\frac{1}{1 - Xt} := \sum_{n\geq 0} X^n t^n \in A[X][[t]]. \tag{2.22}$$

Proposition 2.3.4. For $j \geq 0$,

$$\mathrm{U}_j\left(\frac{1}{1-Xt}\right) = \mathfrak{p}_j(X) := X^j - e_1X^{j-1} + \cdots + (-1)^r e_r X^{j-r}$$

with the agreement that $X^i = 0$ if $i < 0$.

Proof. Since

$$\sum_{j\geq 0} \mathrm{U}_j\left(\frac{1}{1-Xt}\right)t^j = \frac{E_r(t)}{1-Xt} = \left(1+\sum_{j=1}^{r}(-1)e_it^i\right)\sum_{n\geq 0}X^nt^n$$
$$= \sum_{j\geq 0}\left(X^j - e_1X^{j-1} + \cdots + (-1)^r e_r X^{j-r}\right)t^j,$$

the claim follows. □

Definition 2.3.5. The sequence $(m_0, m_1, \ldots) \in M^{\mathbb{N}}$ is a *generic* LRS of order r if

$$\mathrm{U}_{r+j}(\mathbf{m}(t)) = m_{r+j} - e_1m_{r+j-1} + \cdots + (-1)^r e_r m_j = 0$$

for all $j \geq 0$.

The r-tuple $(m_0, m_1, \ldots, m_{r-1})$ is said to be the *initial data* of the generic LRS. The characteristic polynomial of the generic LRS is clearly $\mathfrak{p}_r(X)$, the generic polynomial of degree r. Adopting the language of M-valued formal power series, we can say that $\mathbf{m}(t) = \sum_{j\geq 0} m_jt^j$ is a generic LRS if and only if $\mathrm{U}_{r+j}(\mathbf{m}(t)) = 0$ for all $j \geq 0$.

Proposition 2.3.6. A formal power series $\mathbf{m}(t) \in M[[t]]$ is a generic LRS of order r if and only if $\mathfrak{p}_r(D)\mathbf{m}(t) = 0$, where $\mathfrak{p}_r(D)$ denotes the endomorphism of $M[[t]]$ obtained by evaluating $\mathfrak{p}_r(X)$ at D.

Proof. Note that

$$\mathfrak{p}_r(D)\mathbf{m}(t) = \left(D^r - e_1D^{r-1} + \cdots + (-1)^r e_r\right)\sum_{j\geq 0} m_jt^j$$
$$= \sum_{j\geq 0}\left(m_{r+j} - e_1m_{r+j-1} + \cdots + (-1)^r e_r m_j\right)t^j = \sum_{j\geq 0}\mathrm{U}_{r+j}(\mathbf{m}(t))t^j.$$

Therefore $\mathfrak{p}_r(D)\mathbf{m}(t) = 0$ if and only if $\mathrm{U}_{r+j}(\mathbf{m}(t)) = 0$ for all $j \geq 0$. □

2.3.7. Let K_r be the B_r-submodule $\ker \mathfrak{p}_r(D)$ of $B_r[[t]]$:

$$K_r := \{u \in B_r[[t]] \mid \mathfrak{p}_r(D)u = 0\} \tag{2.23}$$

and define in addition $K_r(A) := K_r \otimes_{\mathbb{Z}} A$ and $K_r(M) := K_r(A) \otimes_A M$.

Proposition 2.3.8. *Each* $\mathbf{m}(t) \in K_r(M)$ *can be uniquely written as*

$$\mathbf{m}(t) = u_0 \mathrm{U}_0(\mathbf{m}(t)) + u_{-1}\mathrm{U}_1(\mathbf{m}(t)) + \cdots + u_{-r+1}\mathrm{U}_{-r+1}(\mathbf{m}(t)). \tag{2.24}$$

Proof. If $\mathbf{m}(t) \in K_r(M)$, then $\mathrm{U}_{r+j}(\mathbf{m}(t)) = 0$ for all $j \geq 0$, and then (2.24) follows from (2.20). □

Corollary 2.3.9. *The* r*-tuple* $(u_0, u_{-1}, \ldots, u_{-r+1})$ *is an* A*-basis of* $K_r(A)$.

Proof. Each $\mathbf{a}(t) \in K_r(A)$ can be uniquely written as $\sum_{j=0}^{r-1} \mathrm{U}_j(\mathbf{a}(t))u_{-j}$ by virtue of Proposition 2.3.8. In addition $\mathrm{U}_{r+j}(u_{-i}) = \delta_{i,r+j}$ is 0 in the range $0 \leq i \leq r-1$, i.e. $u_{-i} \in K_r(A)$. □

Remark 2.3.10. What about u_j if the index j does not satisfy $-r+1 \leq j \leq 0$? First of all, if $j > 0$, then $u_j \in K_r(A)$ for any B_r-algebras A. In fact

$$\mathfrak{p}_r(D)u_j = \mathfrak{p}_r(D)D^j u_0 = D^j \mathfrak{p}_r(D)u_0 = 0,$$

where in the first equality, we used (2.17). In particular

$$u_j = \mathrm{U}_0(u_j)u_0 + \mathrm{U}_1(u_j)u_{-1} + \cdots + \mathrm{U}_{r-1}(u_j)u_{-r+1}, \tag{2.25}$$

where

$$\mathrm{U}_i(u_j) = h_{i+j} - e_1 h_{i+j-1} + \cdots + (-1)^{i-1} e_{i-1} h_j. \tag{2.26}$$

If $j < -r+1$, then $j = -r-i$ for some $i \geq 0$ and

$$\mathfrak{p}_r(D)u_{-r-i} = t^i. \tag{2.27}$$

In fact

$$\mathfrak{p}_r(D)u_{-r} = \mathfrak{p}_r\left(\frac{1}{t}\right)t^r u_0 = E_r(t)u_0 = E_r(t)H_r(t) = 1,$$

and consequently

$$\mathfrak{p}_r(D)u_{-r-i} = \mathfrak{p}_r(D)t^i u_{-r} = t^i \mathfrak{p}_r(D)u_{-r} = t^i,$$

which proves (2.27).

Remark 2.3.11. Observe that $\mathbf{m}(t)$ is a generic LRS if and only if (cf. 2.1.6)

$$\mathbb{J}_M\left(\mathfrak{p}_r\left(\frac{1}{t}\right)\frac{\mathbf{m}(t)}{t} + t^{-1}M[t^{-1}]\right) = 0$$

in $M((t))$. In this case

$$\mathbf{m}(t) = \mathrm{U}_0(\mathbf{m}(t))u_0 + \mathrm{U}_1(\mathbf{m}(t))tu_0 + \cdots + \mathrm{U}_{r-1}(\mathbf{m}(t))t^{r-1}u_0$$

$$= \mathrm{U}_0(\mathbf{m}(t))D^{r-1}u_{-r+1} + \mathrm{U}_1(\mathbf{m}(t))D^{r-2}u_{-r+1} + \cdots + \mathrm{U}_{r-1}(\mathbf{m}(t))u_{-r+1},$$

and
$$\mathrm{U}_j(\mathbf{m}(t)) = \mathrm{Res}_t\left(\mathfrak{p}_j\left(\frac{1}{t}\right)\frac{\mathbf{m}(t)}{t}\right), \qquad (0 \le j \le r-1).$$
The situation is analogous to that of M-valued formal distributions in the module

$$M[[z^{\pm 1}, w^{\pm 1}]] := M[[z^{-1}, z, w^{-1}, w]]$$

that belong to the kernel of the multiplication by a power of $z - w$. In fact, if $\mathbf{m}(z, w) \in M[[z^{\pm 1}, w^{\pm 1}]]$, then the equality$(z - w)^r\mathbf{m}(z, w) = 0$ holds if and only if

$$\mathbf{m}(z, w) = c^0(w)\delta(z - w) + c^{(1)}(w)\partial_w\delta(z - w) + \cdots + c^{(r-1)}(w)\partial_w^{(r-1)}\delta(z - w),$$

where $\delta(z - w) = \sum_{n\in\mathbb{Z}} z^{-n-1}w^n$ is the formal Dirac δ-function,

$$\partial_w^{(j)}\delta(z - w) := \frac{1}{j!}\frac{\partial^j\delta(z - w)}{\partial w^j}$$

and

$$c^{(j)}(w) = \mathrm{Res}_z(z - w)^j\mathbf{m}(z, w)$$

See [73, p. 17] or [74, Theorem 1.6].

2.4 Cauchy Problems for Linear Recurrence Sequences

Definition 2.4.1. Two M-valued formal power series $\mathbf{m}(t)$ and $\tilde{\mathbf{m}}(t)$ share the same *initial conditions* modulo (t^r), if $\mathbf{m}(t) - \tilde{\mathbf{m}}(t) \in t^r M[[t]]$.

The requirement is equivalent to $m_i = \tilde{m}_i$ for $0 \le i \le r-1$. An easy induction argument shows that $\mathbf{m}(t)$ and $\tilde{\mathbf{m}}(t)$ have the same initial conditions modulo (t^r) if and only if $\mathrm{U}_i(\mathbf{m}(t)) = \mathrm{U}_i(\tilde{\mathbf{m}}(t))$ for $0 \le i \le r-1$.

Lemma 2.4.2. If $\mathbf{m}(t) \in K_r(M)$ and $\mathrm{U}_i(\mathbf{m}(t)) = 0$ for $0 \le i \le r-1$, then $\mathbf{m}(t) = 0$.

Proof. Just note that by hypothesis, all coefficients of expansion (2.20) of $\mathbf{m}(t)$ vanish. □

If $\mathbf{f} := (f_0, f_1, \ldots)$ is another sequence of indeterminates over A, set $A[\mathbf{f}] := A[f_0, f_1, \ldots]$, $M[\mathbf{f}] := M \otimes_A A[\mathbf{f}]$, and consider $\mathbf{f}(t) = \sum_{j\geq 0} f_j t^j$. Extend $\mathfrak{p}_r(D)$ to an endomorphism of $M[\mathbf{f}][[t]]$ in the obvious way.

Proposition 2.4.3 (Cauchy theorem for a generic LRS). *Let* $\tilde{\mathbf{m}}(t) \in M[[t]] \subseteq M[\mathbf{f}][[t]]$. *Then*

$$\mathbf{m}(t) := \mathrm{U}_0(\tilde{\mathbf{m}}(t))u_0 + \cdots + \mathrm{U}_{-r+1}(\tilde{\mathbf{m}}(t))u_{-r+1} + \sum_{j\geq 0} u_{-r-j} f_j \tag{2.28}$$

is the unique element of $\mathfrak{p}_r(D)^{-1}(\mathbf{f}(t))$ *sharing with* $\tilde{\mathbf{m}}(t)$ *the same initial conditions modulo* (t^r).

Proof. Apply $\mathfrak{p}_r(D)$ to both sides of (2.28), exploiting its 'M-linearity':

$$\mathfrak{p}_r(D)\mathbf{m}(t) = \sum_{j=0}^{r-1} \mathrm{U}_j(\tilde{\mathbf{m}}(t))\mathfrak{p}_r(D)u_{-j} + \sum_{j\geq 0} f_j \mathfrak{p}_r(D)u_{-r-j}.$$

The equality $\mathfrak{p}_r(D)u_{-j} = 0$ for $0 \leq j \leq r-1$ together with (2.27) implies $\mathfrak{p}_r(D)\mathbf{m}(t) = \sum_{j\geq 0} f_j t^j$ as required. It is clear that (2.28) shares the same initial condition as $\tilde{\mathbf{m}}(t)$. If $\mathbf{m}'(t)$ were another element of $\mathfrak{p}_r(D)^{-1}(\mathbf{f}(t))$ with the same initial conditions as $\tilde{\mathbf{m}}(t)$, one would obtain $\mathbf{m}(t) - \mathbf{m}'(t) \in K_r(M)$, with the same initial conditions as the null series $0 \in M[[t]]$. Hence $\mathbf{m}(t) = \mathbf{m}'(t)$ because of Lemma 2.4.2. □

Proposition 2.4.3, saying that the map $\mathfrak{p}_r(D) : M[[t]] \to M[[t]]$ is surjective, together with the definition of $K_r(M)$, implies:

Corollary 2.4.4. The following Cauchy sequence

$$0 \longrightarrow K_r(M) \hookrightarrow M[[t]] \xrightarrow{\mathfrak{p}_r(D)} M[[t]] \longrightarrow 0 \tag{2.29}$$

is exact.

Universality 2.4.5. Formula (2.28) is universal in the following sense. Let A be any $\mathbb{Z}$-algebra, M any A-module and $P \in A[X]$ any monic polynomial of degree r. Given $\mathbf{m}(t)$ and $\mathbf{g}(t) := \sum_{j\geq 0} g_j t^j$ two M-valued formal power series, the unique element of $P(D)^{-1}(\mathbf{g}(t))$ sharing the same initial conditions as $\mathbf{m}(t)$ is

$$\sum_{j=0}^{r-1} \mathrm{U}_j(\mathbf{m}(t))u_{-j} + \sum_{j\geq 0} g_j u_{-r-j},$$

where A is regarded as a $B_r[\mathbf{f}]$-module under the unique $\mathbb{Z}$-algebra homomorphism defined by $e_i \mapsto e_i(P)$ and $f_j \mapsto g_j$.

2.5 Formal Laplace Transform and Linear ODEs

Definition 2.5.1. *Let M be a module over a $\mathbb{Z}$-algebra A. We call* formal Laplace transform *the A-linear map $\mathfrak{L} : M[[t]] \to M[[t]]$ given by*

$$\mathfrak{L}\Big(\sum_{j\geq 0} m_j t^j\Big) = \sum_{j\geq 0} j! m_j t^j. \tag{2.30}$$

It is easy to verify that $\mathfrak{L}$ is invertible if and only if A contains the rational numbers. In this case

$$\mathfrak{L}^{-1}\left(\sum_{j\geq 0} m_j t^j\right) = \sum_{j\geq 0} m_j \frac{t^j}{j!}.$$

Define the endomorphism $\partial_t : M[[t]] \to M[[t]]$ as

$$\partial_t\left(\sum_{j\geq 0} m_j t^j\right) = \sum_{j\geq 0} (j+1) m_{j+1} t^j.$$

Proposition 2.5.2. *The following diagram commutes*

$$\begin{array}{ccc} M[[t]] & \xrightarrow{\mathfrak{L}} & M[[t]] \\ {\scriptstyle \partial_t}\downarrow & & \downarrow{\scriptstyle D} \\ M[[t]] & \xrightarrow{\mathfrak{L}} & M[[t]] \end{array}$$

In particular, $\mathfrak{L}$ maps $\ker \mathfrak{p}_r(\partial_t)$ *to $K_r(M)$.*

Proof. Just a matter of routine calculation,

$$\mathfrak{L}\partial_t \sum_{j\geq 0} m_j t^j = \sum_{j\geq 0} (j+1)! m_{j+1} t^j = D\sum_{j\geq 0} j! m_j t^j = D\mathfrak{L}\sum_{j\geq 0} m_j t^j.$$

An easy induction shows that, $\mathfrak{L} \circ \partial_t^i = D^i \circ \mathfrak{L}$ and so $\mathfrak{L}$ maps $\ker \mathfrak{p}_r(\partial_t)$ to $K_r(M)$. $\square$

If $\mathfrak{L}$ is invertible, then $\partial_t = \mathfrak{L}^{-1} D \mathfrak{L}$, and, by induction, $\partial_t^i = \mathfrak{L}^{-1} D^i \mathfrak{L}$, which implies that $\ker \mathfrak{p}_r(\partial_t)$ is isomorphic to $K_r(M)$. In other words, to determine $\ker \mathfrak{p}_r(\partial_t)$, it suffices to determine $K_r(M)$. From now on we assume $M = A$. Then ∂_t is a derivation of $A[[t]]$, i.e.

$$\partial_t(\mathbf{a}(t)\mathbf{b}(t)) = \partial_t \mathbf{a}(t) \cdot \mathbf{b}(t) + \mathbf{a}(t) \cdot \partial_t \mathbf{b}(t)$$

a well-known property left as an exercise.

Lemma 2.5.3. If L is invertible, any $\mathbf{f}(t) \in A[[t]]$ admits an expansion of the type

$$\mathbf{f}(t) = \sum_{j\geq 0} \mathrm{U}_j(\mathcal{L}(\mathbf{f}(t))\mathcal{L}^{-1}(u_{-j}). \tag{2.31}$$

Proof. Since $\mathcal{L}(\mathbf{f}(t)) = \sum_{j\geq 0} \mathrm{U}_j(\mathcal{L}(\mathbf{f}(t))u_{-j}$, taking the inverse transform of both sides and using A-linearity of $\mathcal{L}$ give formula (2.31). □

2.5.4. Given $\mathbf{f}(t) \in A[[t]]$, the preimage $\mathfrak{p}_r(\partial_t)^{-1}(\mathbf{f}(t))$ is, by definition, the set of the solutions to the generic linear *ordinary differential equation* (ODE)

$$y^{(r)} - e_1 y^{(r-1)} + \cdots + (-1)^r e_r y = \mathbf{f}(t), \tag{2.32}$$

where $y^{(i)} := \partial_t^i y$. Proposition 2.4.3 has the following important consequence.

Corollary 2.5.5. The unique element of $\mathfrak{p}_r(\partial_t)^{-1}(\mathbf{f}(t))$ sharing the same initial conditions of a given $\varphi \in A[[t]]$ modulo (t^r) is

$$y = \sum_{i=0}^{r-1} \mathrm{U}_i(\mathcal{L}(\varphi))\mathcal{L}^{-1}(u_{-i}) + \sum_{j\geq 0} j! f_j \mathcal{L}^{-1}(u_{-r-j}). \tag{2.33}$$

Proof. Just apply $\mathfrak{p}_r(\partial_t) = \mathcal{L}^{-1}\mathfrak{p}_r(D)\mathcal{L}$ to both sides of (2.33):

$$\mathfrak{p}_r(\partial_t)y = \mathcal{L}^{-1}\mathfrak{p}_r(D)\mathcal{L}\left(\sum_{i=0}^{r-1} \mathrm{U}_i(\mathcal{L}(\varphi))\mathcal{L}^{-1}(u_{-i})\right) \tag{2.34}$$

$$+ \mathcal{L}^{-1}\mathfrak{p}_r(D)\mathcal{L}\left(\sum_{j\geq 0} j! f_j \mathcal{L}^{-1}(u_{-r-j})\right). \tag{2.35}$$

Using the A-linearity of $\mathcal{L}$ and the fact that $\mathfrak{p}_r(D)u_{-i} = 0$ for $0 \leq i \leq r-1$, the first summand of (2.35) vanishes, and the second one is

$$\mathcal{L}^{-1}\mathfrak{p}_r(D)\left(\sum_{j\geq 0} j! f_j u_{-r-j}\right) = \mathcal{L}^{-1}\left(\sum_{j\geq 0} j! f_j t^j\right) = \mathcal{L}^{-1}(\mathcal{L}(\mathbf{f}(t)) = \mathbf{f}(t).$$

The uniqueness is obvious. □

Example 2.5.6. It is worth to discuss a couple of examples. Suppose one wants to solve the linear ODE

$$y'' + \omega^2 y = t^n. \tag{2.36}$$

Write $n+2$ in the form $4p+q$ where $p \in \mathbb{N}$ and $q \in \{0,1,2,3\}$, so that (2.36) reads

$$y'' + \omega^2 y = t^{4p+q-2}, \qquad (4p+q \geq 2).$$

Solving (2.36) then amounts to solving

$$(D^2+\omega^2)\mathcal{L}(y) = (4p+q-2)!t^{4p+q-2}.$$

There is a unique morphism $B_2 \to \mathbb{Q}$ sending $e_1 \mapsto 0$ and $e_2 \mapsto \omega^2$. Take $\phi = a_0 + a_1 t \in \mathbb{Q}[t]$ and look for the unique solution such that $y - \phi = 0 \mod (t^2)$. Then

$$u_0 = H_2(t) = \frac{1}{1+\omega^2 t^2} = 1 + \sum_{j\geq 1}(-1)^j \omega^{2j} t^{2j}$$

and

$$u_{-1} = tu_0 = \frac{1}{\omega}\left(\omega t + \sum_{j\geq 1}(-1)^j \omega^{2j+1} t^{2j+1}\right).$$

The solution to the given Cauchy problem is then

$$\mathcal{L}(y) = a_0 u_0 + a_1 u_{-1} + (4p+q-2)!u_{-4p-q}$$

Now

$$\begin{aligned} u_{-4p-q} &= t^{4p+q}u_0 = \frac{1}{\omega^{4p+q}}(\omega t)^{4p+q}u_0 \\ &= \frac{1}{3}(2q-1)(q-1)(q-3)u_0 + \frac{1}{3\omega}q(q-2)(q-4)\omega u_{-1} \\ &+ \frac{1}{6}q(q-1)(q-2)\omega t - \frac{1}{2}q(q-1)(q-3) - \sum_{a=0}^{p}(-1)^a \omega^{2a+q} t^{2a+q}, \end{aligned}$$

a formula that the reader can check as exercise. Since

$$\mathcal{L}^{-1}(u_0) = \cos(\omega t) \qquad \text{and} \qquad \mathcal{L}^{-1}(u_{-1}) = \frac{1}{\omega}\sin(\omega t)$$

one finally obtains

$$y = a_0 \cos \omega t + \frac{a_1}{\omega}\sin \omega t + (4p+q-2)!\mathcal{L}^{-1}(u_{-4p-q}),$$

where

$$\mathcal{L}^{-1}(u_{-4p-q}) = \mathcal{L}^{-1}(t^{4p+q}u_0) = \frac{1}{\omega^{4p+q}}\mathcal{L}^{-1}((\omega t)^{4p+q}u_0)$$

$$= \frac{1}{\omega^{4p+q}}\left(\frac{(2q-1)(q-1)(q-3)}{3}\cos\omega t + \frac{q(q-2)(q-4)}{3}\sin\omega t\right.$$

$$\left. - \sum_{a=0}^{p}(-1)^{2a+q}\omega^{2a+q}\frac{t^{2a+q}}{(2a+q)!} + \frac{1}{6}q(q-1)(q-2)\omega t - \frac{1}{2}q(q-1)(q-3)\right)$$

If, for instance, we put $p = 1$ and $q = 3$,

$$y = a_0\cos\omega t + \frac{a_1}{\omega}\sin\omega t + \frac{5!}{\omega^7}\left(\omega t - \frac{\omega^3 t^3}{3!} + \frac{\omega^5 t^5}{5!} - \sin\omega t\right)$$

$$= a_0\cos\omega t + \left(\frac{a_1}{\omega} - \frac{5!}{\omega^7}\right)\sin\omega t + \frac{5!}{\omega^6}\left(t - \frac{\omega^2 t^3}{3!} + \frac{\omega^4 t^5}{5!}\right).$$

is the general solution to

$$y'' + \omega^2 y = t^5.$$

2.6 Generalized Wronskians

2.6.1. Recall that if M, N are A-modules, a multilinear map $\varphi : M^r \to N$ is said to be *alternating* if $\varphi(m_1, \ldots, m_r) = 0$ whenever $m_i = m_j$ for some $1 \leq i < j \leq r$. Let $C_1, \ldots, C_r$ be r columns of A^r, in such a way that $(C_1, \ldots, C_r)$ is an $r\times r$ square matrix. As in the undergraduate courses, by the determinant of $(C_1, \ldots, C_r)$, we understand the unique multilinear alternating form taking the value 1 on the canonical basis of A^n.

Denote by

$$\mathbf{u}_r := (u_0, u_{-1}, \ldots, u_{-r+1})$$

the canonical basis of the A-module $\ker(\mathfrak{p}_r(D)\cdot) \subseteq A[[t]]$.

Definition 2.6.2. Given an ordered r-tuple $\mathcal{F} := (f_1, f_2, \ldots, f_r)$ of arbitrary polynomials in $A[X]$, the $\mathcal{F}$-*Wronskian* of $\mathbf{u}_r$ is the determinant:

$$W_{\mathcal{F}}(\mathbf{u}_r) := \begin{vmatrix} f_r(D)u_0 & f_r(D)u_{-1} & \ldots & f_r(D)u_{-r+1} \\ f_{r-1}(D)u_0 & f_{r-1}(D)u_{-1} & \ldots & f_{r-1}(D)u_{-r+1} \\ \vdots & \vdots & \ddots & \vdots \\ f_1(D)u_0 & f_1(D)u_{-1} & \ldots & f_1(D)u_{-r+1} \end{vmatrix} \in A[[t]]$$

where $f_i(D)u_{-j}$ means applying the operator $f_i(D)$ to u_{-j}.

Remark 2.6.3. The Wronskian is A-multilinear and alternating in $f_1,\dots,f_r$. Since $\mathfrak{p}_r(D)u_{-i} = 0$ for $-r+1 \le i \le 0$, we have that if $f_i \equiv g_i \mod \mathfrak{p}_r(X)$, then

$$W_{\mathcal{F}}(\mathbf{u}_r) = W_{\mathcal{G}}(\mathbf{u}_r),$$

where we set $\mathcal{G} = (g_1,\dots,g_r)$ $\big(\in A[X]^r\big)$.

2.6.4. For $\boldsymbol{\lambda} \in \mathcal{P}_r$, define

$$\mathbf{X}^{\boldsymbol{\lambda}} := (X^{\lambda_r}, X^{1+\lambda_{r-1}},\dots,X^{r-1+\lambda_1}) \in A[X]^r$$

and $\mathbf{X} := \mathbf{X}^{(0)}$, where (0) denotes the null partition $(0,0,\dots,0)$. Then

$$W_{\mathbf{X}^{\lambda}}(\mathbf{u}_r) = \det(D^{i-1+\lambda_{r-i+1}}u_{1-j})_{1\le i,j\le r} = \begin{vmatrix} u_{\lambda_r} & u_{-1+\lambda_r} & \dots & u_{-r+1+\lambda_r} \\ u_{1+\lambda_{r-1}} & u_{\lambda_{r-1}} & \dots & u_{-r+2+\lambda_{r-1}} \\ \vdots & \vdots & \ddots & \vdots \\ u_{r-1+\lambda_1} & u_{r-2+\lambda_1} & \dots & u_{\lambda_1} \end{vmatrix}.$$

As in 2.2.10, the *residue* $\mathrm{Res}_X(g)$ of a Laurent series $g = \sum_{i\le n} g_i X^i$ $(n \in \mathbb{Z})$ is the coefficient g_{-1} of X^{-1}. To ease notation, in this section we shall omit the subscript X to indicate the residue.

The following definition is due to Laksov and Thorup [96, 97]:

Definition 2.6.5. The *residue* of an ordered r-tuple

$$g_i := \sum_{j\le n_i} g_{ij}X^j, \qquad 0 \le i \le r-1.$$

of Laurent series with B_r-coefficients is

$$\mathrm{Res}(g_0, g_1,\dots,g_{r-1}) :=$$

$$= \begin{vmatrix} \mathrm{Res}(g_0) & \mathrm{Res}(g_1) & \dots & \mathrm{Res}(g_{r-1}) \\ \mathrm{Res}(Xg_0) & \mathrm{Res}(Xg_1) & \dots & \mathrm{Res}(Xg_{r-1}) \\ \vdots & \vdots & \ddots & \vdots \\ \mathrm{Res}(X^{r-1}g_0) & \mathrm{Res}(X^{r-1}g_1) & \dots & \mathrm{Res}(X^{r-1}g_{r-1}) \end{vmatrix}. \tag{2.37}$$

Clearly $\mathrm{Res}(g_0, g_1,\dots,g_{r-1})$ is B_r-multilinear and alternating:

$$\mathrm{Res}(g_{\mathfrak{s}(0)}, g_{\mathfrak{s}(1)},\dots,g_{\mathfrak{s}(r-1)}) = \mathrm{sgn}(\mathfrak{s})\mathrm{Res}(g_0, g_1,\dots,g_{r-1}),$$

where $\mathfrak{s} \in S_r$ (the permutations on r elements) and $\mathrm{sgn}(\mathfrak{s})$ denotes its parity ± 1. If at least one of the g_j is a polynomial, the jth row of the determinant (2.37) vanishes, which causes

$$\mathrm{Res}(g_0, g_1,\dots,g_{r-1}) = 0.$$

Lemma 2.6.6. *Let* $f(X) = a_0X^\lambda + a_1X^{\lambda-1} + \cdots + a_\lambda$ *be a polynomial of degree* $\leq \lambda$ *with coefficients in a* B_r*-algebra* A*. Then, for all* $1 \leq i \leq r$,

$$\mathrm{Res}\left(\frac{X^{i-1}f(X)}{\mathfrak{p}_r(X)}\right) = \sum_{j=0}^{\lambda} a_j h_{i-r+\lambda-j}. \tag{2.38}$$

Proof. Equality (2.14) gives

$$h_{i-r+\lambda-j} = \mathrm{Res}\left(\frac{X^{i-1+\lambda-j}}{\mathfrak{p}_r(X)}\right) = \mathrm{Res}\left(\frac{X^{i-1}X^{\lambda-j}}{\mathfrak{p}_r(X)}\right)$$

Since taking the residue is A-linear, we eventually get formula (2.38). □

Theorem 2.6.7 (Cf. [98, Theorem 2.7(2)] and [49, p. 286]). *Let* $\mathcal{F} := (f_1, \ldots, f_r) \in A[X]^r$*. Then*

i) $$W_{\mathbf{X}}(\mathbf{u}_r) \neq 0;$$

ii) $$W_{\mathcal{F}}(\mathbf{u}_r) = \mathrm{Res}\left(\frac{f_1}{\mathfrak{p}_r(X)}, \ldots, \frac{f_r}{\mathfrak{p}_r(X)}\right) W_{\mathbf{X}}(\mathbf{u}_r); \tag{2.39}$$

iii) $$W_{\mathbf{X}^\lambda}(\mathbf{u}_r) = \Delta_{\boldsymbol{\lambda}}(H_r) W_{\mathbf{X}}(\mathbf{u}_r). \tag{2.40}$$

In particular, if the specialization morphism $B_r \to A$ *is injective, the generalized Wronskians* $\{W_{\mathbf{X}^\lambda}(\mathbf{u}_r) \mid \boldsymbol{\lambda} \in \mathcal{P}_r\}$ *are linearly independent over the integers.*

Proof. Given a formal power series, $\mathbf{a} \in A[[t]]$, denote by $\mathbf{a}(0) := \mathbf{a} \mod (t)$ its constant term. Notice that if $\mathbf{a}, \mathbf{b} \in A[[t]]$, then $(\mathbf{a} \cdot \mathbf{b})(0) = \mathbf{a}(0) \cdot \mathbf{b}(0)$, i.e. the 'evaluation at $t = 0$' is a ring homomorphism $A[[t]] \to A$. Thus (i) follows from the fact that

$$W_{\mathbf{X}}(\mathbf{u}_r)(0) = \begin{vmatrix} u_0 & u_{-1} & \ldots & u_{-r+1} \\ u_1 & u_0 & \ldots & u_{-r+2} \\ \vdots & \vdots & \ddots & \vdots \\ u_{r-1} & u_{r-2} & \ldots & u_0 \end{vmatrix}(0) = \begin{vmatrix} u_0(0) & u_{-1}(0) & \ldots & u_{-r+1}(0) \\ u_1(0) & u_0(0) & \ldots & u_{-r+2}(0) \\ \vdots & \vdots & \ddots & \vdots \\ u_{r-1}(0) & u_{r-2}(0) & \ldots & u_0(0) \end{vmatrix} =$$

$$= \begin{vmatrix} 1 & 0 & \ldots & 0 \\ h_1 & 1 & \ldots & 0 \\ \vdots & \vdots & \ddots & \vdots \\ h_{r-1} & h_{r-1} & \ldots & 1 \end{vmatrix} = 1,$$

which proves that $W_{\mathbf{X}}(\mathbf{u}_r)$ is a unit in $A[[t]]$ and hence invertible.

(ii) For $i = 1, \ldots, n$, write

$$f_i(X) = q_i(X)\mathfrak{p}_r(X) + \rho_i(X),$$

with $\deg(\rho_i) < r$. According to Remark 2.6.3, we have $W_{\mathcal{F}}(\mathbf{u}_r) = W_{(\rho_1,\ldots,\rho_r)}(\mathbf{u}_r)$ and

$$\mathrm{Res}\left(\frac{f_1}{\mathfrak{p}_r(X)},\ldots,\frac{f_r}{\mathfrak{p}_r(X)}\right) = \mathrm{Res}\left(\frac{\rho_1}{\mathfrak{p}_r(X)},\ldots,\frac{\rho_r}{\mathfrak{p}_r(X)}\right).$$

We may consequently assume that $\deg(f_i) < r$ for $i = 1,\ldots,r$. Since

$$W_{\mathcal{F}}(\mathbf{u}_r) \qquad \text{and} \qquad \mathrm{Res}\left(\frac{f_1}{\mathfrak{p}_r(X)},\ldots,\frac{f_r}{\mathfrak{p}_r(X)}\right),$$

are both linear in $(f_1,\ldots,f_r)$, we may assume that $f_i(X) = X^{m_i}$ with $0 \le m_i < r$ for $i = 1,\ldots,r$. However both sides of (2.39) are alternating in $f_1,\ldots,f_r$, so we may assume that $0 \le m_r < \cdots < m_2 < m_1 < r$. Hence $m_i = r - i$ for $i = 1,\ldots,r$. In this case the theorem follows from

$$\mathrm{Res}\left(\frac{X^{r-1}}{\mathfrak{p}_r(X)},\ldots,\frac{1}{\mathfrak{p}_r(X)}\right) = 1.$$

(iii) To prove formula (2.40), it suffices to recall that

$$\mathrm{Res}\left(\frac{X^{\lambda_i+r-i}}{\mathfrak{p}_r(X)}\right) = h_{\lambda_i-i+1}.$$

We have

$$\mathrm{Res}\left(\frac{X^{\lambda_1+r-1}}{\mathfrak{p}_r(X)},\ldots,\frac{X^{\lambda_r}}{\mathfrak{p}_r(X)}\right) = \Delta_{\boldsymbol{\lambda}}(H_r) := \begin{vmatrix} h_{\lambda_1} & \cdots & h_{\lambda_1+r-1} \\ \vdots & \ddots & \vdots \\ h_{\lambda_r-r+1} & \cdots & h_{\lambda_r} \end{vmatrix}, \tag{2.41}$$

and then (2.40) follows from item (ii) above. Since the Schur polynomials $\Delta_{\boldsymbol{\lambda}}(H_r)$ are well known to form a $\mathbb{Z}$-basis of B_r (e.g. [109, Section I.3]), as we shall also see in Chapter 5, it follows that the generalized Wronskians $\{W_{\mathbf{X}^{\boldsymbol{\lambda}}}(\mathbf{u}_r) \mid \boldsymbol{\lambda} \in \mathcal{P}_r\}$ are linearly independent over the integers as well, by item (i) and (2.40). □

2.6.8. A quick check shows that $DW_{\mathbf{X}}(\mathbf{u}_r) = W_{\mathbf{X}^{(1)}}(\mathbf{u}_r)$. Equation (2.40) reads in this case as

$$W_{\mathbf{X}^{(1)}}(\mathbf{u}_r) = h_1 W_{\mathbf{X}}(\mathbf{u}_r) = e_1 W_{\mathbf{X}}(\mathbf{u}_r), \tag{2.42}$$

Formula (2.42) is the purely algebraic version of the celebrated theorem by Abel and Liouville ([6, p. 195]), stating that the derivative of the Wronskian of a fundamental system of solutions to a linear ODE is proportional to the Wronskian itself (see also [39]). It implies that if the Wronskian does not vanish at a point, it vanishes nowhere. Formula 2.42 is our prototype of the boson–fermion correspondence, in the sense of Chapter 7.

2.6.9. If A contains the rationals, the formal Laplace transform $\mathcal{L}$ yields the isomorphism $\mathcal{L}^{-1} : A[[t]] \to A[[t]]$ described in Proposition 2.5.2. Then, given $\mathcal{F} := (f_1, \ldots, f_r) \in A[X]^r$ and

$$\mathcal{L}^{-1}(\mathbf{u}_r) = (\mathcal{L}^{-1}(u_0), \ldots, \mathcal{L}^{-1}(u_{r-1}))$$

one can define $W_{\mathcal{F}}(\mathcal{L}^{-1}(\mathbf{u}_r))$, as in 2.6.2, just replacing D by ∂_t and u_i by $v_i := \mathcal{L}^{-1}(u_i)$. Notice that

$$\partial_t v_i = \mathcal{L}^{-1} \circ \mathcal{L}\partial_t \circ \mathcal{L}^{-1}(u_i) = \mathcal{L}^{-1}(Du_i) = \mathcal{L}^{-1}(u_{i+1}) = v_{i+1}.$$

In particular $\mathbf{v}_r := (v_0, v_{-1}, \ldots, v_{-r+1})$ is a basis of the solutions to the linear ODE $\mathfrak{p}_r(\partial_t)y = 0$, and formula (2.39) can be rewritten in the form

$$W_{\mathcal{F}}(\mathbf{v}_r) := W_{\mathcal{F}}(\mathcal{L}^{-1}(\mathbf{u}_r)) = \mathrm{Res}\left(\frac{f_1}{\mathfrak{p}_r(X)}, \ldots, \frac{f_r}{\mathfrak{p}_r(X)}\right) W_{\mathbf{X}}(\mathbf{v}_r),$$

i.e. the 'ratio'

$$\frac{W_{\mathcal{F}}(\mathbf{v}_r)}{W_{\mathbf{X}}(\mathbf{v}_r)}$$

is the same one would get from formula (2.40). Notice that

$$W_{\mathbf{X}}(\mathbf{v}_r) = \begin{vmatrix} v_0 & v_{-1} & \cdots & v_{-r+1} \\ v_1 & v_0 & \cdots & v_{-r+2} \\ \vdots & \vdots & \ddots & \vdots \\ v_{r-1} & v_{r-2} & \cdots & v_0 \end{vmatrix} = \begin{vmatrix} v_0 & v_{-1} & \cdots & v_{-r+1} \\ \partial_t v_0 & \partial_t v_{-1} & \cdots & \partial_t v_{-r+1} \\ \vdots & \vdots & \ddots & \vdots \\ \partial_t^{r-1} v_0 & \partial_t^{r-1} v_{-1} & \cdots & \partial_t^{r-1} v_{-r+1} \end{vmatrix}$$

is the usual Wronskian we learn from the undergraduate calculus.

2.6.10. In particular (2.42) can be phrased, in the hypothesis of Section 2.6.9 by saying that the derivative of the Wronskian is proportional to the Wronskian itself, a fact which is equivalent to the fact that the determinant of the exponential of a square matrix is the exponential of its trace. Cf. Exercise 4.5.3.

Generalized Wronskians are related with derivatives of Wronskians. Let us write $W_\lambda := W_{\mathbf{X}^\lambda}$ for short, so that, e.g.

$$\partial_t W_0(\mathbf{v}_r) = W_{(1)}(\mathbf{v}_r). \tag{2.43}$$

In general, the jth derivative of a Wronskian is a linear combination of generalized Wronskians:

$$\partial_t^j W_0(\mathbf{v}_r) = \sum_{|\lambda|=j} g_\lambda W_\lambda(\mathbf{v}_r). \tag{2.44}$$

This is a consequence of (2.43) and an easy induction based on the following *Pieri rule for generalized Wronskians*:

$$\partial_t W_{\boldsymbol{\lambda}}(\mathbf{v}_r) = \sum_{\mathbf{a}} W_{(\mathbf{a}(1)+\lambda_r, \mathbf{a}(2)+\lambda_{r-1}, \dots, \mathbf{a}(r)+\lambda_r)}(\mathbf{v}_r), \tag{2.45}$$

where the summation is taken over all the sequences $\mathbf{a} : \{0, 1, \dots, r-1\} \to \{0, 1\}$ such that $\sum_{j=0}^{r-1} \mathbf{a}(j) = 1$ but cancelling all the terms for which $\mathbf{a} + \boldsymbol{\lambda}$ is not a partition.

Recall now from [33] that each box of a Young diagram $Y(\boldsymbol{\lambda})$ of a partition $\boldsymbol{\lambda}$ determines a *hook*, consisting of that box and of all the boxes in its row to the right of the box and its column below the box. The *hook length* of the box is the number of boxes in its hook. So, for instance, filling the Young diagram of the partition $(3, 2, 1, 1)$ with the hook length of each box gives the following *Young tableau* (Fig. 2.3):

Fig. 2.3 The hook length filling of $(4, 3, 1, 1)$

7	4	3	1
6	2	1	
1			
1			

Then an amazing fact occurs, observed by I. Scherbak:

Theorem 2.6.11. *(see [49, 50]).* The coefficient $g_{\boldsymbol{\lambda}}$ in (2.44) can be computed via the *hook length* formula:

$$g_{\boldsymbol{\lambda}} = \frac{|\boldsymbol{\lambda}|!}{k_1 \cdots k_j} = \frac{j!}{k_1 \cdots k_j}, \tag{2.46}$$

where k_i is the hook length of each box of the Young diagram of the partition $\boldsymbol{\lambda}$.

A consequence of 2.6.11 is the following remarkable fact, already observed in [38]:

Corollary 2.6.12. *The coefficient $g_{(n-r)^r}$ multiplying $W_{(n-r)^r}(\mathbf{v}_r)$ in the expansion of $\partial_t^{r(n-r)} W_0(\mathbf{v}_r)$ is precisely the Plücker degree (see Section 5.3) of the Grassmannian $G_{r,n}$, parametrizing* r*-dimensional linear subspaces of $\mathbb{C}^n$, i.e.*

$$g_{(d-r)^{r+1}} = \frac{1!2! \cdots r! \cdot r(n-r)!}{(n-r)!(d-r+1)! \cdots d!}.$$

Example 2.6.13. Let $\mathbf{v}_2 = (v_0, v_1)$. Then

$$\begin{aligned} \partial_t^4 W_0(\mathbf{v}_2) &= \partial_t^3 \circ \partial_t W_0(\mathbf{v}_2) = \partial_t^3 W_{(1)}(\mathbf{v}_2) = \partial_t^2 (W_{(2)}(\mathbf{v}_2) + W_{(1,1)}(\mathbf{v}_2)) \\ &= \partial_t (2W_{(2,1)}(\mathbf{v}_2) + W_{(3)}(\mathbf{v}_2)) \\ &= 2W_{(2,2)}(\mathbf{v}_2) + 2W_{(3,1)}(\mathbf{v}_2) + W_{(4)}(\mathbf{v}_2) \end{aligned}$$

and the coefficient 2 multiplying $W_{(2,2)}(\mathbf{v}_2)$ is the Plücker degree of the Grassmannian $G_1(\mathbb{P}^3) := G_{2,4}(\mathbb{C})$ in its Plücker embedding, i.e. the *number of lines of* $\mathbb{P}^3_{\mathbb{C}}$ *meeting four others in general position.* All the other coefficients have enumerative interpretation in either the classical or the quantum cohomology of the Grassmannian (see, e.g. [38]).

2.7 Notes and References

- The Fibonacci[2] sequence $1, 1, 2, 3, 5, 8, \ldots$, whose nth term ($n \geq 3$) is the sum of the preceding two, was born to model the growth of a colony of breeding rabbits and is perhaps the most popular example of LRS. For instance, the Italian artist Mario Merz[3] realized many neon-light installations of the Fibonacci sequence, one of which, *Il volo dei numeri*[4], was draped on the spire of the Mole Antonelliana, the symbol of the city of Torino, which nowadays hosts the National Cinema Museum[5] . The Fibonacci numbers obey to the same recursive law enjoyed by the powers $(1, a, a^2, \ldots)$ of one of the two roots $(1 \pm \sqrt{5})/2$ of the polynomial $X^2 - X - 1$. The equality $a^2 = 1(1 + a)$ proves the link with the famous *golden ratio*, which the ancient Greek architects used to design the planimetry of their temples (Fig. 2.4).
- There is an enormous deal of literature concerning LRSs, for example, [57, 92, 113, 115, 147, 153, 154] just to cite a few. Their applications are countless, starting from algebra and algebraic geometry (see [115]) to industrial mathematics. General excellent expository references are [2, 17, 69], while [113] and [153, 154] describe applications to other sciences.
- Linear ODEs with constant coefficients and analytic source term are entirely governed by the algebra of LRSs: see [49, 50], but also [42] where a generalized formal Laplace transform is associated with any sequence of invertible elements in a ring. For more applications look at the paper [151]. The inverse of the formal Laplace transform of the formal series u_{-r+1} in Corollary 2.3.9 is precisely the *impulse response solution* of the generic linear ODE of order r. See Camporesi [13] for a truly enlightening exposition on that.

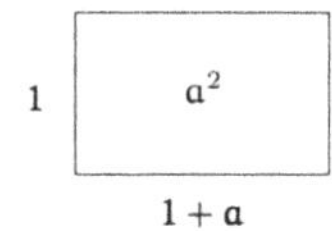

Fig. 2.4 A *golden rectangle* with $a = \frac{1}{2} + \frac{\sqrt{5}}{2}$

[2]Leonardo Pisano, known as Fibonacci (Pisa, 1170–1240), wrote the famous *Liber Abaci* in 1202 where the name 'zero' comes from 'Zephyrus', a wind blowing from the West.

[3]Mario Merz (Milano, 1925–2003) was a painter and sculptor who used poor materials for his creations (http://fondazionemerz.org/en/mario-merz/).

[4]The flight of numbers

[5]http://www.museocinema.it/

- The notions of generic LRS and generic linear ODE were studied in [49] with the purpose to prove a Giambelli (or Jacobi–Trudi) formula for generalized Wronskians, associated with a fundamental system of linear ODEs. That formula generalizes the well-known result whereby the derivative of a Wronskian is proportional to the Wronskian itself; see also [39]. The construction of the universal basis of solutions to the Cauchy problem [49] essentially amounts to the explicit construction of the $\mathcal{D}$-module associated with a generic linear differential operator of order r (see, e.g. [16, Chapter 6] and [65, Example 1.2.4]). This is completely transparent in the algebraic presentation of [42].
- To deal with LRSs, we used the language of formal power series. As we wanted to keep that to a minimum, we did not mention the beautiful calculus of formal distributions as presented in [73, Chapter 2] and [30, 103]. Formal calculus is essential to develop the theory of vertex algebras in a purely algebraic way. One of its most powerful tools, called 'operator product expansion', allows to expand formal Laurent series in two (or more) variables as linear combinations of *formal Dirac deltas* and their derivatives.
- The fact that generalized Wronskians (see the survey [50]) are parametrized by partitions is a consequence of Schubert calculus, not conversely. This is a good point to recall that the generating function of $\mathrm{p}_r(w)$ (the number of partitions of length at most r of the integer w) is given by

$$\sum_{k\geq 0} \mathrm{p}_r(k)q^k = \frac{1}{\prod_{i=1}^r (1-q^i)},$$

a formula essentially due to Euler [78, p. 15]. In turn, this equals the q-rank of B_r:

$$\mathrm{rank}_q B_r := \sum_{w\geq 0} \mathrm{rank}(B_r)_w q^w = \sum_{w\geq 0} \mathrm{p}_r(w)q^w = \frac{1}{\prod_{i=1}^r (1-q^i)}.$$

For instance, $\mathrm{rank}(B_5)_3 = \mathrm{p}_5(3) = 3$ and $\mathrm{rank}(B_5)_8 = \mathrm{p}_5(8) = 18$. Standard references for the combinatorial, representation theoretical and algebro-geometrical aspects of partitions are [33, 109].

2.8 Exercises

2.8.1. Show that a sequence $(a_0, a_i, \ldots)$ of complex numbers is a LRS if and only if its associated formal power series $\sum_{j\geq 0} a_j z^j$ is the formal Taylor expansion of a rational function, i.e. the ratio of two polynomials $f, g \in \mathbb{C}[z]$.

2.8.2. Let A be a $\mathbb{Q}$-algebra, $D : A[[t]] \to A[[t]]$ as in (2.7) and $\partial_t := \mathcal{L} \circ D \circ \mathcal{L}^{-1}$. Prove that ∂_t satisfies the Leibniz rule:

$$\partial_t(\mathbf{a}(t)\mathbf{b}(t)) = \partial_t\mathbf{a}(t) \cdot \mathbf{b}(t) + \mathbf{a}(t) \cdot \partial_t\mathbf{b}(t)$$

although D does not.

2.8.3. Let A be a ring with unit and $A[[t]]$ the algebra of formal power series in the formal variable t. Prove that $\mathbf{a}(t) = \sum_{i\geq 0} a_i t^i$ is invertible if and only if a_0 is. If in addition A is commutative, then

$$\frac{1}{\mathbf{a}(t)} = a_0^{-1}\sum_{i\geq 0}(-1)^i a_0^{-i}\mathbf{a}(t)^i.$$

2.8.4. Let $A := \mathbb{Z}[\mathbf{t}] := \mathbb{Z}[t_1,\ldots,t_r]$ and define

$$1 - e_1(\mathbf{t})z + \cdots + (-1)^r e_r(\mathbf{t})z^r := \prod_{i=1}^{r}(1-t_i z).$$

The coefficient $e_i(\mathbf{t})$ of z^i is called ith elementary symmetric polynomial of degree i in the indeterminates $t_1,\ldots,t_r$. Let $h_j(\mathbf{t})$ denote the sum of all monomials of degree j in $\mathbf{t} := (t_1,\ldots,t_r)$. Show that

$$\sum_{j\geq 0} h_j t^j = \frac{1}{\prod_{i=1}^{r}(1-t_i z)}.$$

The term $h_j(\mathbf{t})$ is called the *complete symmetric polynomial* of degree j.

2.8.5. Write $\mathbf{x} := (x_1,x_2)$, and let $H_2(\mathbf{x})$ be the sequence $(h_1(\mathbf{x}),h_2(\mathbf{x}),\ldots)$ of complete symmetric polynomials in (x_1,x_2). Consider the *generalized Wronskian* associated with $\mathbf{f} := (\exp(x_1 t),\exp(x_2 t))$:

$$W_{(\lambda_1,\lambda_2)}(\mathbf{f}) := \begin{vmatrix} \partial_t^{\lambda_2}\exp(x_1 t) & \partial_t^{\lambda_2}\exp(x_2 t) \\ \partial_t^{1+\lambda_1}\exp(x_1 t) & \partial_t^{1+\lambda_1}\exp(x_2 t) \end{vmatrix}$$

i) Show that $W_{(\lambda_1,\lambda_2)}(\mathbf{f}) := \Delta_{(\lambda_1,\lambda_2)}(H_2(\mathbf{x}))(x_1-x_2)\exp((x_1+x_2)t)$. In particular $W_{(0,0)}(\mathbf{f}) = (x_1 - x_2)\exp((x_1 - x_2)t)$. Thus *Jacobi–Trudy* formula for generalized Wronskians holds:

$$W_{(\lambda_1,\lambda_2)}(\mathbf{f}) = \Delta_{(\lambda_1,\lambda_2)}(H_2)W_{(0,0)}(\mathbf{f}).$$

ii) Show that $\partial_t W_{(0,0)}(\mathbf{f}) = e_1(x)W_{(0,0)}(\mathbf{f})$.

iii) Compute $\partial_t^4 W_{(0,0)}(\mathbf{f})$ as a $\mathbb{Z}$-linear combinations of generalized Wronskians. What is the coefficient of $W_{(2,2)}(\mathbf{f})$?

2.8.6. Prove the Pieri rule (2.45) for generalized Wronskians (hint: write the generalized Wronskian determinant $W_{\boldsymbol{\lambda}}(\mathbf{v}_r)$ in the form

$$\partial_t^{\lambda-r}\mathbf{v}_r \wedge \partial_t^{1+\lambda_{r-1}}\mathbf{v}_r \wedge \cdots \wedge \partial_t^{r-1+\lambda_1}\mathbf{v}_r$$

where $\partial_t^{i+\lambda_{r-i}}\mathbf{v}_r$ represents the ith row of the Wronskian. Taking the derivative of $W_{\boldsymbol{\lambda}}(\mathbf{v}_r)$ then amounts to apply Leibniz rule of ∂_t with respect to the rows).

2.8.7. Show that if $1 \leq r \leq 2$, then $\Delta_{(2,2,1)}(H_r) = 0$.

2.8.8. Let

$$\exp(t) := \sum_{n\geq 0} \frac{t^n}{n!} \in \mathbb{Q}[[t]].$$

It solves the linear ODE $\ddot{y} - 3\dot{y} + 2y = 0$, where $\dot{y}$ denotes the derivative with respect to the indeterminate. Find the expression of $\exp(t)$ as a linear combination of $\mathcal{L}^{-1}(u_0)$ and $\mathcal{L}^{-1}(u_{-1})$, where $\mathcal{L}$ denotes the formal Laplace transform as in 2.5.

2.8.9. Find the set of solutions of $\ddot{y}-3\dot{y}+2y = t^2$ and $\ddot{y}-3\dot{y}+2y = \exp(t)$. Notice that $\exp(t)$ solves the associated homogeneous equation, and so we are in the case of *resonance*.

2.8.10. Consider the linear equation of order two $\partial_t^2 y - (x_1 + x_2)\partial_t y + x_1x_2 \cdot y = 0$, with coefficients in $\widetilde{B}_2 := \mathbb{Q}[x_1, x_2]$. Then $\exp(t_1 z), \exp(t_2 z) \in \widetilde{B}_2[[z]]$ are $\mathbb{Z}$-linearly independent, but they do not form a basis over the integers. Find the expression of $\exp(x_i t)$ as linear combination of a basis of solutions over $\widetilde{B}_2$. Show that $(\exp(x_1 t), \exp(x_2 t))$ form a basis of solutions in the localized ring $\widetilde{B}_2[[z]]_{(x_1 - x_2)}$.

2.8.11. Generalize Exercise 2.8.10. Let $e_i(\mathbf{x}) := e_i(x_1, \ldots, x_r)$ be the elementary symmetric polynomials in $(x_1, \ldots, x_r)$ of degree i. Then $\exp(x_i t)$, $1 \leq i \leq r$, are linearly independent solutions of the linear ODE:

$$\partial_t^r y - e_1(\mathbf{x}_r)\partial_t^{r-1} y + \cdots + (-1)^r e_r(\mathbf{x}_r) y = 0$$

but they are not a basis over $\mathbb{Q}[x_1, \ldots, x_r]$. Show they are a basis up to inverting the *Vandermonde determinant*:

$$V(x_1, \ldots, x_r) = \prod_{1\leq i<j\leq r} (x_i - x_j)$$

2.8.12. Consider the homogeneous linear equation with $B_2 \otimes_{\mathbb{Z}} \mathbb{Q}$-coefficients:

$$\ddot{y} - e_1\dot{y} + e_2 y = 0. \tag{2.47}$$

Let $B_2[\xi] := \dfrac{B_2[X]}{(X^2 - e_1X + e_2)}$, where $\xi := X + (X^2 - e_1X + e_2)$. Then the polynomial $X^2 - e_1X + e_2$ has a root in $B[\xi]$. Show that $\exp(\xi \cdot t)$ is a solution of (2.47), and express it as a $B[\xi]$-linear combination of u_0 and u_{-1}. Deduce the Euler formulas

$$\exp(\pm\sqrt{-1}t) = \cos t \pm \sqrt{-1}\sin t$$

2.8.13. Check that

$$\mathcal{M} := \begin{pmatrix} a & b \\ c & d \end{pmatrix} \in \mathbb{Q}[a,b,c,d]^{2\times 2}$$

is a root of the polynomial $T^2 - (a+d)T + ad - bc = 0$ and, using this fact, that

$$\exp(\mathcal{M}t) = \sum_{n\geq 0} \frac{\mathcal{M}^n t^n}{n!}$$

is a solution of $\ddot{y} - (a+d)\dot{y} + (ad-bc) = 0$. Deduce an expression for the exponential of $\mathcal{M}\cdot t$ by writing it as a linear combination of $v_0 := \mathcal{L}^{-1}(u_0)$ and $v_1 := \mathcal{L}^{-1}(u_{-1})$, with matrix coefficients.

2.8.14. Let $\mathfrak{p}_r(\partial_t) := \partial_t^r - e_1\partial_t^{r-1} + \cdots + (-1)^r e_r$ be the generic linear ODO of order r, so that

$$\{y_1 \in B_r[[t]] \mid \mathfrak{p}_r(\partial_t)y_1 = 0\} \tag{2.48}$$

is a generic linear ODE of order r. For all $2 \leq i \leq r$, define $y_i := \partial_t y_{i-1}$.

i) Show that (2.48) is equivalent to a system of linear ODE of the first order:

$$\partial_t \begin{pmatrix} y_1 \\ \vdots \\ y_r \end{pmatrix} = C(e_1,\ldots,e_r)\cdot \begin{pmatrix} y_1 \\ \vdots \\ y_r \end{pmatrix}$$

where $C(e_1,\ldots,e_r) \in B_r{}^{r\times r}$ is said to be *companion matrix*.

ii) Determine the expression of the companion matrix $C(e_1,\ldots,e_r)$.

iii) For all $n \geq 0$, determine explicit expression for $C(e_1,e_2)^n$ and $C(e_1,e_2,e_3)^n$.

(Companion matrices are very important. They are associated with cyclic endomorphisms of a vector space and cyclic vectors of an endomorphism generate invariant subspaces also known as *Krylov subspaces*. See Tomei [143, Section 2.2].)

2.8.15. Write $H_r(z) = \sum_{n\geq 0} h_n z^n = \exp\left(\sum_{i\geq 1} x_i z^i\right)$, and compute h_n as an explicit polynomial of $(x_1,x_2,\ldots)$ for the first few values of n. Express x_{r+1} as a B_r-linear combination of $(x_1,\ldots,x_r)$.

2.8.16. Let M be a module over a $\mathbb{Z}$-algebra A. Prove that the formal Laplace transform defined in 2.5 is invertible if and only if A contains the rational numbers.

2.8.17. Recall the usual definition of *Laplace transform* of an analytic function f:

$$L(f)(s) = \int_0^\infty e^{-st} f(t)dt,$$

with suitable restrictions on the domain of s depending on the choice of the function. How is L related to the formal Laplace transform defined in Section 2.5?

Chapter 3
Algebras and Derivations

To render the text self-contained and also to set the notation, Section 3.1 recalls the basic notion of tensor and exterior algebra of a module. Readers may refer to [7, p. 24], [99, Ch. XVI, §7] and [99, Ch. XIX] for more complete explanations. Additional combinatorial details are worked out in Section 3.2 to describe the exterior algebra of a free module. Clifford algebras are introduced in Section 3.3 in the customary way, as quotients of the tensor algebra. Derivations on general Grassmann algebras are defined exactly as for other kinds of algebras (like commutative or non-associative): we overview them in Section 3.4 to emphasize that iterating derivations produces higher-order Leibniz rules, the main formal devices characterizing Hasse–Schmidt derivations. The definition of the latter, however, is postponed to the beginning of Chapter 4.1. Throughout this chapter, unless otherwise stated, M will denote a module over a commutative ring A with unit, and all constructions involving M will be understood in the category of A-modules.

3.1 Tensor and Exterior Algebra of a Module

3.1.1. Let $r \geq 1$ be a given integer and

$$M^r := \underbrace{M \times \cdots \times M}_{r \text{ times}}$$

be the r-fold Cartesian product of M. A map $f : M^r \to N$, where N is an A-module, is called *multilinear* if it is linear in each variable separately. The r-fold tensor product of M with itself is a pair $(T^r(M), \phi_r)$ fulfilling the following universal property: for all multilinear maps $f : M^r \to N$, there exists a unique A-module homomorphism $T^r(f) : T^r(M) \to N$ rendering the following diagram commutative:

L. Gatto, P. Salehyan, *Hasse-Schmidt Derivations on Grassmann Algebras*, IMPA Monographs, DOI 10.1007/978-3-319-31842-4_3

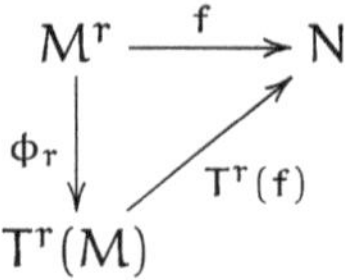

The module $T^r(M)$ is generated by the ϕ_r-images $m_1 \otimes \cdots \otimes m_r$ of r-tuples $(m_1, \ldots, m_r) \in M^r$. By convention $T^0(M) := A$. The *tensor algebra* of M is the pair $(T(M), \otimes)$, where

$$T(M) = \bigoplus_{j \geq 0} T^j(M) \tag{3.1}$$

and '$\otimes$' is the A-bilinear extension of the juxtaposition map

$$(m_{i_1} \otimes \cdots \otimes m_{i_r}) \otimes (m_{j_1} \otimes \cdots \otimes m_{j_s}) = m_{i_1} \otimes \cdots \otimes m_{i_r} \otimes m_{j_1} \otimes \cdots \otimes m_{j_s}.$$

By construction $T(M)$ is an associative graded algebra having 1_A as unit.

3.1.2. If the module M is free of finite or countable rank, with basis $\mathfrak{B} := (b_0, b_1, \ldots)$, say

$$M = \bigoplus_{i \in [0,n] \cap \mathbb{N}} A \cdot b_i, \tag{3.2}$$

for some $n \in \mathbb{N} \cup \{\infty\}$, then $T^r(M)$ itself is a free A-module with basis

$$\otimes^r \mathfrak{B} := \{(b_{i_0} \otimes b_{i_1} \otimes \cdots \otimes b_{i_{r-1}}) \mid (i_0, i_1, \ldots, i_{r-1}) \in \mathbb{N}^r\}.$$

If $\text{rank}_A(M) = n < \infty$, then $\text{rank}_A T^r(M) = n^r$.

Grassmann Algebra of a Module 3.1.3. The *Grassmann* or *exterior*[1] algebra of M is a triple $(\bigwedge M, \wedge, \iota)$, where $\bigwedge M := (\bigwedge M, \wedge)$ is an associative A-algebra with unit $1_{\bigwedge M} = 1_A$ and $\iota : M \to \bigwedge M$ is an A-module monomorphism that satisfies the following universal property: for any associative A-algebra $\mathfrak{A}$ with unit and any homomorphism $f : M \to \mathfrak{A}$ such that $f(m)^2 = 0$ for all $m \in M$, there exists a unique A-algebra homomorphism $\bigwedge f : \bigwedge M \to \mathfrak{A}$ such that $f = \bigwedge f \circ \iota$:

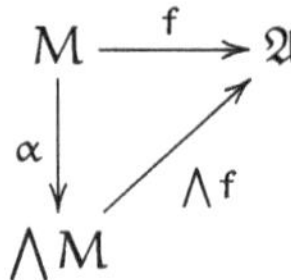

[1]We shall use interchangeably the names 'exterior' and 'Grassmann', when referring to such an algebra.

The exterior algebra of M can be constructed by taking the quotient of the tensor algebra $T(M)$ modulo the two-sided ideal $\mathcal{J}$ generated by elements of $T^2(M)$ of the form $m \otimes m$.

3.1.4. Let $T(M) \to \bigwedge M := T(M)/\mathcal{J}$ be the canonical epimorphism: if $m_1, m_2 \in M$, the image in $\bigwedge M$ of $m_1 \otimes m_2$ is usually denoted by $m_1 \wedge m_2$. It turns out that $\mathcal{J}$ is a homogeneous ideal, i.e. $\mathcal{J} = \bigoplus_{j\geq 0} \mathcal{J}_j$, where $\mathcal{J}_j := \mathcal{J} \cap T^j(M)$. Thus $\bigwedge M := \bigoplus_{j\geq 0} \bigwedge^j M$ is $\mathbb{Z}_{\geq 0}$ graded, where

$$\bigwedge^j M := \frac{T^j(M)}{\mathcal{J}_j}$$

is the jth *exterior power* of M, generated by elements of the form

$$m_1 \wedge \cdots \wedge m_r := m_1 \otimes \cdots \otimes m_r + \mathcal{J}, \qquad (m_i \in M).$$

By construction $\bigwedge^0 M = A$.

3.1.5. If a multilinear map $\varphi : M^r \to N$ is *alternating* (see Section 2.6.1), then for all $\mathfrak{s} \in S_r$, the symmetric group of the permutations on r elements, one has the equality

$$\varphi(m_{\mathfrak{s}(1)}, \ldots, m_{\mathfrak{s}(r)}) = \operatorname{sgn}(\mathfrak{s})\varphi(m_1, \ldots, m_r),$$

where $\operatorname{sgn}(\mathfrak{s})$ denotes the parity ($= \pm 1$) of $\mathfrak{s}$. The natural map $\alpha_r : M^r \to \bigwedge^r M$

$$(m_1, \ldots, m_r) \mapsto m_1 \wedge \cdots \wedge m_r$$

is *universally alternating* in the sense that for all alternating maps $\varphi : M^r \to N$, there exists a unique $\bigwedge^r \varphi$ making commutative the diagram below:

$$\begin{array}{ccc} M^r & \xrightarrow{\varphi} & N \\ \downarrow{\scriptstyle \alpha_r} & \nearrow{\scriptstyle \bigwedge^r \varphi} & \\ \bigwedge^r M & & \end{array}$$

3.1.6. The algebra structure of $\bigwedge M$ is given by the A-bilinear extension of the juxtaposition

$$(m_{i_1} \wedge \cdots \wedge m_{i_r}) \wedge (m_{j_1} \wedge \cdots \wedge m_{j_s}) = m_{i_1} \wedge \cdots \wedge m_{i_r} \wedge m_{j_1} \wedge \cdots \wedge m_{j_s}. \qquad (3.3)$$

The *wedge product* $\wedge$ turns $\bigwedge M$ into a *super-commutative* associative algebra with unit. In fact $\bigwedge M$ comes equipped with a $\mathbb{Z}_2 := \mathbb{Z}/2\mathbb{Z}$ graduation:

$$\bigwedge M = (\bigwedge M)_{\bar{0}} \oplus (\bigwedge M)_{\bar{1}},$$

where, by definition, $(\bigwedge M)_{\bar{0}} := \bigoplus_{i\geq 0} \bigwedge^{2i} M$ and $(\bigwedge M)_{\bar{1}} := \bigoplus_{i\geq 0} \bigwedge^{2i+1} M$. An element $\eta \in \bigwedge M$ is called *even* if $\eta \in (\bigwedge M)_{\bar{0}}$ and *odd* if $\eta \in (\bigwedge M)_{\bar{1}}$. Super-commutativity means that

$$\eta \wedge \theta = (-1)^{ij}\theta \wedge \eta, \qquad \eta, \theta \in \bigwedge M,$$

i.e. even elements are central while odd elements anti-commute.

Example 3.1.7. Let $m_1, m_2, m_3 \in M$. Then $m_1 \wedge (m_2 \wedge m_3) = (m_2 \wedge m_3) \wedge m_1$ (the even element $m_2 \wedge m_3$ commutes with m_1), while $m_i \wedge m_j = -m_j \wedge m_i$ (m_i, m_j are odd, hence anti-commute).

3.2 Exterior Algebra of a Free *A*-Module

3.2.1. Suppose M is a free A-module. Then $\bigwedge^0 M = A$, $\bigwedge^1 M = M$, and for $r \geq 2$, $\bigwedge^r M$ is the A-module generated by the expressions

$$\{b_{i_0} \wedge \cdots \wedge b_{i_{r-1}} \mid (i_0, i_1, \ldots, i_{r-1}) \in \mathbb{N}^r\}$$

modulo the relations

$$b_{i_{\mathfrak{s}(0)}} \wedge \cdots \wedge b_{i_{\mathfrak{s}(r-1)}} = \mathrm{sgn}(\mathfrak{s}) b_{i_0} \wedge \cdots \wedge b_{i_{r-1}}, \quad \mathfrak{s} \in S_r \tag{3.4}$$

and

$$b_{i_0} \wedge \cdots \wedge b_{i_{r-1}} = 0 \tag{3.5}$$

whenever $b_{i_h} = b_{i_k}$ for some $0 \leq h < k \leq r-1$. Notice that equation (3.4) implies (3.5) if $\mathrm{char}(A) \neq 2$, as in all situations we will be interested in. Moreover, (3.4) also tells us that

$$\bigwedge^r \mathfrak{B}_n := (b_{i_0} \wedge \cdots \wedge b_{i_{r-1}})_{0 \leq i_0 < i_1 < \cdots < i_{r-1}}$$

is an A-basis of M. In particular, if $\mathrm{rank}_A M = n$, then

$$\mathrm{rank}_A \bigwedge^r M = \binom{n}{r}.$$

As in the general case, the *exterior algebra* of M is a pair $(\bigwedge M, \wedge)$, where $\bigwedge M := \bigoplus_{j\geq 0} \bigwedge^j M$ and the *wedge product* $\wedge$ is defined as in (3.3). This can be phrased solely in terms of basis elements of M:

$$(b_{i_1} \wedge \cdots \wedge b_{i_h}) \wedge (b_{j_1} \wedge \cdots \wedge b_{j_k}) = b_{i_1} \wedge \cdots \wedge b_{i_h} \wedge b_{j_1} \wedge \cdots \wedge b_{j_k}.$$

If $\mathrm{rank}_A M = n$, then

$$\mathrm{rank}_A(\bigwedge M) = \sum_{r\geq 0} \mathrm{rank}_A \bigwedge^r M = \sum_{r\geq 0} \binom{n}{r} = 2^n.$$

In the sequel it will be convenient to parametrize basis elements of $\bigwedge^r M$ induced by $\mathfrak{B}_n$ using *partitions*, instead of strictly increasing sequences of non-negative integers. Partitions will be used to measure the 'deviation' from the basis element $b_0 \wedge b_1 \wedge \cdots \wedge b_{r-1}$.

Notation 3.2.2. For the sake of brevity, let us introduce the following shortcut notations. For all $i \geq 0$ and $\boldsymbol{\lambda} \in \mathcal{P}_r$, define

$$[\mathbf{b}]^r_{i+\boldsymbol{\lambda}} : = b_{i+\lambda_r} \wedge b_{i+1+\lambda_{r-1}} \wedge \cdots \wedge b_{i+r-1+\lambda_1}, \tag{3.6}$$

$$[\mathbf{b}]^r_i : = [\mathbf{b}]^r_{i+(0)} := b_i \wedge b_{i+1} \wedge \cdots \wedge b_{i+r-1}. \tag{3.7}$$

Accordingly, we set $[\mathbf{b}]^r_{\boldsymbol{\lambda}} := [\mathbf{b}]^r_{0+\boldsymbol{\lambda}}$. If $n := \mathrm{rank}_A M$, then clearly

$$\bigwedge^r M = \bigoplus_{\boldsymbol{\lambda}\in\mathcal{P}_{r,n}} A \cdot [\mathbf{b}]^r_{\boldsymbol{\lambda}}.$$

3.2.3. Define the *weight* of $[\mathbf{b}]^r_{i+\boldsymbol{\lambda}}$ to be $i+|\boldsymbol{\lambda}|$. For all $r \geq 1$, the *weight graduation* of $\bigwedge^r M$ is

$$\bigwedge^r M = \bigoplus_{w\geq 0} (\bigwedge^r M)_w,$$

where $(\bigwedge^r M)_w := \bigoplus_{|\boldsymbol{\lambda}|=w} A \cdot [\mathbf{b}]^r_{\boldsymbol{\lambda}}$ is the submodule of $\bigwedge^r M$ spanned by the basis elements of weight w.

3.3 Exterior Algebras Versus Clifford Algebras

3.3.1. A *Clifford algebra* can be thought of as a deformation, in a suitable sense, of the exterior algebra. Let A be a commutative ring with unit and $g : M \times M \to A$ a symmetric bilinear form. The Clifford algebra associated with the pair (M, g) is an A-algebra $C_g(M)$ together with an A-module monomorphism $\iota_g : M \to C_g(M)$ satisfying the following universal property. For any A-algebra $\mathfrak{A}$ and any map $j : M \to \mathfrak{A}$ with $j(m)^2 = g(m,m)1$, there exists a unique map $C(g) \to \mathfrak{A}$ rendering commutative the following diagram:

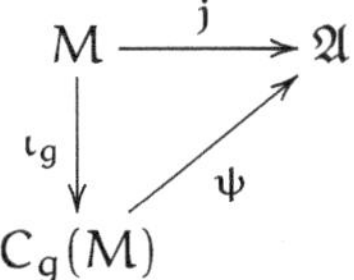

If the Clifford algebra exists, it is unique up to a canonical isomorphism. To prove its existence, one considers the quotient of the tensor algebra $T(M)$ by the two-sided ideal $\mathcal{I}_g$ generated by elements of the form $m \otimes m - g(m, m)1_A$. It can be shown that $\mathrm{rank}_A(C_g(M)) = 2^n$ if $\mathrm{rank}_A(M) = n$, by counting the elements of a basis. For a general argument, see [99, Theorem 4.1]. If $m_1, m_2 \in M$, we have, as usual,

$$g(m_1 + m_2, m_1 + m_2) = g(m_1, m_1) + 2g(m_1, m_2) + g(m_2, m_2).$$

On the other hand, in $C_g(M)$, the equality:

$$\begin{aligned} g(m_1 + m_2, m_1 + m_2)1_A &= (m_1 + m_2)^2 = m_1^2 + m_1 m_2 + m_2 m_1 + m_2^2 \\ &= g(m_1, m_1)1_A + g(m_2, m_2)1_A + m_1 m_2 + m_2 m_1 \end{aligned}$$

implies the relation

$$m_1 m_2 + m_2 m_1 = 2g(m_1, m_2)1_A. \tag{3.8}$$

In particular, if $(b_0, b_1, \ldots, b_{n-1})$ is a g-orthonormal basis for M, i.e. $g(b_i, b_j) = \delta_{ij}$, then (3.8) reads

$$b_i b_j + b_j b_i = 0.$$

Thus $C_g(M)$ is generated by

$$1_A, b_0, \ldots, b_{n-1}, \ (b_i b_j)_{i<j}, \ (b_i b_j b_k)_{0 \le i<j<k \le n}, \ldots, \ b_0 b_1 \cdots b_{n-1},$$

and so its dimension equals

$$1 + n + \binom{n}{2} + \cdots + \binom{n}{n} = (1+1)^n = 2^n$$

provided one has proven that the elements above are linearly independent (this is left as an exercise). Notice that if $g = 0$, the null form, one recovers the exterior algebra.

3.3.2. Consider a free A-module M with a given basis $(b_0, \ldots, b_{n-1})$ and a dual basis $(\beta_0, \ldots, \beta_{n-1})$ (so $\beta_j(b_i) = \delta_{ji}$). On the space $M \oplus M^\vee$, we consider the canonical bilinear pairing

$$g(m_1 \oplus \mu_1, m_2 \oplus \mu_2) = \mu_2(m_1) + \mu_1(m_2).$$

Then, identifying b_i with $b_i \oplus \{0\}$ and $\beta_j = \{0\} \oplus \beta_j$, the $2n$-tuple

$$(b_0, \ldots, b_{n-1}, \beta_0, \ldots, \beta_{n-1})$$

forms a basis of $M \oplus M^\vee$. Hence $g(b_i, b_j) = g(\beta_i, \beta_j) = 0$ and $g(b_i, \beta_j) = g(\beta_j, b_i) = \delta_{ij}$. Let $\mathcal{C}(M \oplus M^\vee)$ be the associated Clifford algebra: hence $\mathcal{C}$ will stand, for us, both for *Clifford* and *canonical* (with respect to the canonical bilinear form). Consider the $\mathcal{C}(M \oplus M^\vee)$-module

$$F := A \cdot |0\rangle \oplus \bigoplus_{I=0}^{n-1} A \cdot b_i|0\rangle \oplus \bigoplus_{0\leq i_1<i_2\leq n} A \cdot b_{i_1}b_{i_2}|0\rangle \oplus \cdots \oplus A \cdot b_0 b_1 \cdots b_{n-1}|0\rangle$$

generated by a *vacuum element* $|0\rangle \in F$ such that $\beta_j|0\rangle = 0$ for all $j \geq 0$. The map $F \to \bigwedge M$ given by $b_{i_1} \cdots b_{i_k}|0\rangle \mapsto b_{i_1} \wedge \cdots \wedge b_{i_k}$ is then an isomorphism of algebras. In particular $\bigwedge M$ is an irreducible representation of $\mathcal{C}(M \oplus M^\vee)$, meaning that any submodule of $\bigwedge M$ that is invariant under the action of $\mathcal{C}(M \oplus M^\vee)$ must be trivial.

3.4 Derivations, in General

3.4.1. Derivations of exterior algebras, which we shall be focusing on, starting from the next section, are defined exactly as on arbitrary, even not associative, algebras (e.g. like Lie algebras). Suppose then $(A, \star)$ is an R-algebra, where R is a commutative ring with unit and $\star$ denotes a binary operation. An R-derivation of A is an R-endomorphism $\mathfrak{D}$ of A satisfying the Leibniz rule:

$$\mathfrak{D}(a \star b) = Da \star b + a \star Db.$$

For example, if $(A, [\,,\,])$ is a Lie algebra, then a derivation of it is an R-endomorphism satisfying

$$\mathfrak{D}[a, b] = [\mathfrak{D}a, b] + [a, \mathfrak{D}b].$$

The following easy result will be employed in Chapter 7.

Proposition 3.4.2. Assume A is an associative commutative algebra over a $\mathbb{Q}$-algebra R and z an indeterminate. Let $\mathfrak{D}_1, \mathfrak{D}_2, \ldots$ be a sequence of R-derivations of A, and consider the generating function $\mathfrak{D}(z) = \sum_{j\geq 1} \mathfrak{D}_j z^j$. Then

$$\exp(\mathfrak{D}(z)) := \sum_{n\geq 0} \frac{\mathfrak{D}(z)^n}{n!} : A \to A[[z]]$$

is an R-algebra homomorphism.

Proof. If $a, b \in A$, it is immediate to see that

$$\mathfrak{D}(z)(ab) = \mathfrak{D}(z)a \cdot b + a \cdot \mathfrak{D}(z)b,$$

just by applying the Leibniz rule to $\mathfrak{D}_j$, for every $j \geq 1$. Since $\mathfrak{D}(z)$ is trivial on R, the same holds for $\mathfrak{D}(z)^n$, $n \geq 1$. Hence

$$\mathfrak{D}(z)^n(ab) = \sum_{j=0}^{n} \binom{n}{i} \mathfrak{D}(z)^i a \cdot \mathfrak{D}(z)^{n-i} b$$

follows from a straightforward induction argument. Consequently

$$\begin{aligned}\exp(\mathfrak{D}(z))(ab) &= \sum_{n\geq 0} \frac{1}{n!}\mathfrak{D}(z)^n(ab) = \sum_{n\geq 0} \frac{1}{n!} \sum_{i=0}^{n} \binom{n}{i} \mathfrak{D}(z)^i a \cdot \mathfrak{D}(z)^{n-i} b \\ &= \sum_{i\geq 0} \frac{\mathfrak{D}(z)^i a}{i!} \sum_{j\geq 0} \frac{\mathfrak{D}(z)^j b}{j!} = \exp(\mathfrak{D}(z)a) \cdot \exp(\mathfrak{D}(z)b),\end{aligned}$$

as desired. □

The following converse statement is true, as well.

Proposition 3.4.3. *If $\psi(z) := 1 + \sum_{j\geq 1} \psi_j z^j : A \to A[[z]]$ is an R-algebra homomorphism, there exists a unique sequence $\mathfrak{D} := (\mathfrak{D}_1, \mathfrak{D}_2, \ldots)$ of R-derivations of A such that $\psi(z) = \exp(\mathfrak{D}(z))$.*

Proof. Define

$$\sum_{j\geq 1} \mathfrak{D}_j z^j = \log(1 + (\psi(z) - 1)) = -\sum_{n\geq 0} \frac{1}{n} \left(\sum_{i\geq 0} \psi_i z^i \right)^n .$$

Then $\mathfrak{D}_1, \mathfrak{D}_2, \ldots$ are R-derivations of A and clearly $\psi(z) = \exp(\mathfrak{D}(z))$. □

Example 3.4.4. Let f be a holomorphic function defined, say, on a disc $U \subset \mathbb{C}$ containing the origin. Consider the derivative

$$\frac{d}{dz} : \mathcal{O}(U) \to \mathcal{O}(U).$$

Then

$$\exp\left(z\frac{d}{dz}\right) : \mathcal{O}(U) \to \mathcal{O}(U)$$

maps f to its Taylor series, and it is an algebra homomorphism because

$$\exp\left(z\frac{d}{dz}\right)(fg) = \exp\left(z\frac{d}{dz}\right) f \cdot \exp\left(z\frac{d}{dz}\right) g.$$

Definition 3.4.5 (cf. [112, p. 207]). Let A be an associative commutative R-algebra. A Hasse–Schmidt derivation is an R-algebra homomorphism $D(z) : A \to A[[z]]$, i.e.

$$D(z)(ab) = D(z)a \cdot D(z)b. \tag{3.9}$$

Write $D(z)a = \sum_{i\geq 0} D_i a \cdot z^i$. A simple exercise shows that (3.9) is equivalent to

$$D_j(ab) = \sum_{i=0}^{j} D_i a \cdot D_{j-i} b, \qquad (j \geq 0).$$

In particular, the operator $\exp(\mathfrak{D}(z))$ of Lemma 3.4.2 is a Hasse–Schmidt derivation on A. The protagonist of this book are Hasse-Schmidt derivations with respect to the ring structure given by the wedge product $\wedge$.

3.5 Derivations on an Exterior Algebra

3.5.1. In the broadest sense, an R-derivation of $\bigwedge M$ is an R-endomorphism $\mathfrak{D}$ of $\bigwedge M$ satisfying

$$\mathfrak{D}(\eta \wedge \theta) = \mathfrak{D}\eta \wedge \theta + \eta \wedge \mathfrak{D}\theta, \quad \forall \eta, \theta \in \bigwedge M. \tag{3.10}$$

Since $\bigwedge M$ possesses both a $\mathbb{Z}_+$ and a $\mathbb{Z}_2$ graduation, it is customary to define *graded derivations*. One says $\mathfrak{D}$ is a (homogeneous) derivation of degree $p \in \mathbb{Z}$ if it maps $\bigwedge^j M$ to $\bigwedge^{j+p} M$ and satisfies the graded Leibniz rule:

$$\mathfrak{D}(\eta \wedge \theta) = \mathfrak{D}\eta \wedge \theta + (-1)^{p \cdot q} \eta \wedge \mathfrak{D}\theta, \quad \forall \eta \in \bigwedge^q M.$$

The definition of (homogeneous) derivation of degree $\bar{i} \in \mathbb{Z}_2$ is completely similar. For instance, a derivation of degree $2p$ is also a derivation of degree $\bar{0} \in \mathbb{Z}_2$, while one of degree $2p + 1$ is a derivation of degree $\bar{1} \in \mathbb{Z}_2$.

Proposition 3.5.2. Let $\mathcal{M}$ be a set of generators of the A-module M. Each (graded) R-derivation of the exterior algebra is uniquely determined by its action on $A := \bigwedge^0 M$ and on the elements of $\mathcal{M} \subseteq M = \bigwedge^1 M$.

Proof. By definition, a derivation is trivial on R, the ring of constants, and satisfies the (graded) Leibniz rule. In particular for each $(a, m) \in A \times \mathcal{M}$, one has $\mathfrak{D}(am) = \mathfrak{D}a \wedge m + a \wedge \mathfrak{D}m$. Moreover, every $\eta \in \bigwedge M$ is a finite A-linear combination of the kind $a_1\eta_{i_1} + \cdots + a_h\eta_{i_h}$, where $\eta_{i_j} := m_1 \wedge \cdots \wedge m_{i_j} \in \bigwedge^{i_j} M$ and $m_{i_j} \in \mathcal{M}$. Therefore we may assume that $\eta = m_1 \wedge \cdots \wedge m_j$. The Leibniz rule and the definition of $\mathfrak{D}$ on M therefore determine the value of $\mathfrak{D}(m_1 \wedge \cdots \wedge m_j)$. The statement on uniqueness is clear. □

Example 3.5.3. Take $\mu \in M^\vee := \mathrm{Hom}_A(M,A)$, and define the contraction $\mu\lrcorner : \bigwedge M \to \bigwedge M$ as follows:

$$\begin{cases} \mu\lrcorner a = 0, \\ \mu\lrcorner m = \mu(m), & \forall m \in M, \\ \mu\lrcorner(m \wedge \eta) = \mu(m)\eta - m \wedge \mu\lrcorner\eta. \end{cases}$$

The operator '*contraction by* μ' is an A-derivation of $\bigwedge M$ of degree -1.

Example 3.5.4. Take an open set U in $\mathbb{R}^n$ with standard coordinates $(x_1,\ldots,x_n)$, and call $A := C^\infty(U)$ the ring of smooth functions on U. Let M be the free A-module generated by the smooth 1-forms $dx_1,\ldots,dx_n$. For each $1 \le k \le n$, a smooth k-form Θ is an element of $\bigwedge^k M$, namely, $\Theta = f_{i_1,\ldots,i_k} dx_{i_1} \wedge \cdots \wedge dx_{i_k}$, where $f_{i_1,\ldots,i_k} \in A$. Define $df = \sum \partial_i f \cdot dx^i$, where $\partial_i := \partial/\partial x_i$ and

$$d\Theta = \sum_{1\le i_1<\cdots<i_k} df_{i_1,\ldots,i_k} \wedge dx_{i_1} \wedge \cdots \wedge dx_{i_k}.$$

Then d defines an $\mathbb{R}$-derivation of degree 1 on the exterior algebra $\bigwedge M$:

$$d(\eta \wedge \theta) = d\eta \wedge \theta + (-1)^{\deg(\eta)}\eta \wedge d\theta.$$

Example 3.5.5. A *vector field* on an open set U of $\mathbb{R}^n$ is a derivation on the ring $A := C^\infty(U)$ and hence an A-linear combination of the coordinate vector fields ∂_i. Let $X := \sum f_i \partial_i$ be a vector field on U. The Lie derivative L_X,

$$L_X\eta = d(X\lrcorner\eta) + X\lrcorner d\eta,$$

is a derivation of degree 0. In particular $L_X(\eta \wedge \theta) = L_X\eta \wedge \theta + \eta \wedge L_X\theta$.

3.5.6. In the following we shall be mainly concerned with A-derivations (e.g. even ones) that enjoy the Leibniz rule as given in (3.10). An easy induction argument, left as an exercise, proves that *Newton's binomial formula* holds:

$$\mathfrak{D}^r(\eta \wedge \theta) = \sum_{k=0}^{r} \binom{r}{k} \mathfrak{D}^k\eta \wedge \mathfrak{D}^{r-k}\theta. \qquad (3.11)$$

Formally, this is the same rule enjoyed by the rth derivative of the product of two smooth maps, so we may as well call it a Leibniz rule of order r. Suppose the base ring A is a $\mathbb{Q}$-algebra and define

$$\mathfrak{D}_i := \frac{\mathfrak{D}^i}{i!}, \qquad i \ge 0. \qquad (3.12)$$

Thus $\mathfrak{D}_1 = \mathfrak{D}$ and

$$\mathfrak{D}_r(\eta \wedge \theta) = \sum_{k=0}^{r} \mathfrak{D}_k\eta \wedge \mathfrak{D}_{r-k}\theta, \qquad (3.13)$$

which essentially rephrases formula (3.11) above, but without binomial coefficients. If $\mathfrak{D}_i$ is defined as in (3.12), formulas (3.11) and (3.13) are of course equivalent.

Consider now the formal power series $\mathfrak{D}(z) = \sum_{j\geq 0}\mathfrak{D}_j z^j$, where z denotes an arbitrary indeterminate on $\bigwedge M$. The Leibniz rules for $\mathfrak{D}_r$, $r \geq 0$, are all simultaneously encoded in the equality

$$\mathfrak{D}(z)(\eta \wedge \theta) = \mathfrak{D}(z)\eta \wedge \mathfrak{D}(z)\theta.$$

Therefore any A-derivation satisfying (3.10) induces an algebra homomorphism $\mathcal{D}(z) : \bigwedge M \to \bigwedge M[[z]]$, which is what in next chapter will be formally defined as a *Hasse–Schmidt derivation* on $\bigwedge M$.

3.6 Notes and References

- We were kept slightly sketchy in this chapter on purpose, assuming the reader already familiar with the material. If that were not the case, one should take the topics we touched as a list of prerequisites for ensuing chapters. Even if we only recalled the definition of a tensor algebra, its importance cannot be stressed enough. Most constructions that we have seen are quotients of tensor algebras of modules by suitable ideals. Two prominent cases are exterior algebras, of course, as a natural environment in which *HS*-derivations thrive, but also Clifford algebras. The irreducible representations of a Clifford algebra, associated with the canonical inner product of the direct sum of an infinite dimensional space with its restricted dual, are precisely the infinite exterior powers we will deal with, e.g. in Chapter 7.
- To keep focused on the aforementioned cases, we neglected to mention other kinds of algebras that would naturally belong to the syllabus of a second lecture course based on the book: symmetric algebras, induced representations of groups and/or Lie algebras, not to mention *universal enveloping algebras*. The latter topic is expertly explained by Wakimoto [148], with the clear intention to reach the widest possible audience: this beautiful book is without doubt also one of the best places for studying the representation theory of the simple Lie algebra $sl_2(\mathbb{C})$, although the main subject is infinite-dimensional Lie algebras.
- A general reference on tensors, exterior and tensor algebras is [99]. There is a vast literature discussing derivations and differentials in the algebraic, algebro-geometrical or differential–geometrical context. The book [112, Chapter 4] discusses in detail the universal module of differentials. The natural continuation of this story would be the theory of $\mathcal{D}$-modules (see, e.g. [16]), prototypes of which are modules over the Weyl algebra.
- Most of the existing literature, especially textbooks, seems not to emphasize adequately one fact, though, which we tried to amend here, namely, that the Taylor series of a local analytic function is an archetypal Hasse–Schmidt derivation. Every single Leibniz rule, of any given order, is encoded in the equality $\exp(z\partial_z)(f \cdot g) = \exp(z\partial_z)f \cdot \exp(z\partial_z)g$, saying that $\exp(z\partial_z)$ is an algebra

homomorphism. In a sense, this property is the incarnation of everybody's Freudian desire whereby the derivative of a product of two maps should be the product of the derivatives.

3.7 Exercises

3.7.1. Let M_0 be a free abelian group of infinite countable rank with basis $(b_0, b_1, \ldots)$, and $\sigma_1 : \bigwedge M_0 \to \bigwedge M_0$ be an even derivation, i.e.

$$\sigma_1(\eta \wedge \theta) = \sigma_1\eta \wedge \theta + \eta \wedge \sigma_1\theta$$

Compute $\sigma_1^4(b_0 \wedge b_1)$ as a linear combination of $[\mathbf{b}]^r_{(\lambda_1,\lambda_2)}$. Compare the outputs with those of Exercise 2.8.5 (iii).

3.7.2. Let $M_{0,n}$ be a free abelian group with basis $(b_0, b_1, \ldots, b_{n-1})$. Write explicitly the direct sum weight decomposition of $M_{0,4}$, $\bigwedge^2 M_{0,4}$ and $\bigwedge^3 M_{0,4}$. For readers aware of singular homology, the rank of $(\bigwedge^2 M_{0,4})_w$ coincides with the $2w$-th *Betti number of the Grassmannian* $G_{2,4}(\mathbb{C})$, i.e. with $b_{2w} := \mathrm{rank}_{\mathbb{Z}} H_{2w}(G_{2,4}(\mathbb{C}), \mathbb{Z})$. In general, $\mathrm{rank}_{\mathbb{Z}} \bigwedge^r (M_{0,n})_w = \mathrm{rank}_{\mathbb{Z}} H_{2w}(G_{r,n}(\mathbb{C}), \mathbb{Z})$. See Section 5.4.

3.7.3. Let $\eta = \sum_{\lambda \in \mathcal{P}_{2,4}} x_{\lambda} [\mathbf{b}]^2_{\lambda} \in \bigwedge^2 M_0 \otimes \mathbb{C}$. Find the equation that the coefficients $x_{\lambda} \in \mathbb{Z}$ must satisfy in order $\eta \wedge \eta = 0$. The *Klein quadric* hypersurface of the five-dimensional complex projective space is the set of all the points

$$(x_0 : x_{(1)} : x_{(1,1)} : x_{(2)} : x_{(2,1)}, x_{(2,2)}) \in \mathbb{P}^5_{\mathbb{C}}$$

satisfying the equation $\eta \wedge \eta = 0$.

3.7.4. Show that $\bigwedge^2 M_{0,n} \otimes \mathbb{C}$ is isomorphic to the $n \times n$ complex-valued skew-symmetric matrices

$$\mathcal{A}^{n\times n}(\mathbb{C}) := \{Q \in \mathbb{C}^{n\times n} \mid Q^T + Q = 0\},$$

and that if $\eta = \sum_{\lambda \in \mathcal{P}_{2,n}} a_{\lambda} [\mathbf{b}]^2_{\lambda} \in \bigwedge^2 M_{0,n}$, then $\eta \wedge \eta$ is a linear combination of elements of $\bigwedge^4 M_{0,n}$ having as coefficients the Pfaffians of the 4×4 principal minors (cf., e.g. [116], where applications to moduli of algebraic curves are shown as well).

Chapter 4
Hasse–Schmidt Derivations on Exterior Algebras

We are going to keep the same notation of Chapter 3, unless otherwise specified. The main goal of this chapter is to introduce the pivotal algebraic notion of *Hasse–Schmidt (HS) derivation on an exterior algebra* and develop its basic formalism, with emphasis on what we call *integration by parts*. The latter is a key tool and will be used heavily throughout. The original motivation comes from Schubert calculus (see, e.g. Chapter 5), a subject concerned with the *intersection theory* on Grassmann varieties, quickly revised in Section 5.4. *HS*-derivations on Grassmann algebras also provide a natural framework to state and prove a generalization (as in [51]) of the classical celebrated Cayley–Hamilton theorem: *any endomorphism is a root of its characteristic polynomial.* This is shown in Section 4.2 and will be used in Chapter 5 to construct a structure of free B_r-module out of any free abelian group of rank $n \geq r$. That construction will enable us to describe the cone of decomposable tensors in an exterior power by means of the same compact and elegant formula we advertised in the introduction. Section 4.3 offers a few simple applications to matrix exponentials and first integrals of linear ODEs with constant coefficients, by means of the same language of Section 2.5.

4.1 Main Definitions and Properties

4.1.1. Let $\bigwedge M[[z]]$ be the algebra of $\bigwedge M$-valued formal power series in z with $\bigwedge M$-coefficients with respect to the product:

$$\sum_{i\geq 0} \eta_i z^i \wedge \sum_{j\geq 0} \theta_j z^j := \sum_{k\geq 0} \sum_{i+j=k} (\eta_i \wedge \theta_j) z^k,$$

whose unit is $1_{\bigwedge M} = 1_A$ sometimes denoted '1' if clear from the context. Let $\mathcal{D} := (D_0, D_1, \ldots)$ be a sequence of endomorphisms of $\bigwedge M$, and let $\mathcal{D}(z)$

L. Gatto, P. Salehyan, *Hasse-Schmidt Derivations on Grassmann Algebras*, IMPA Monographs, DOI 10.1007/978-3-319-31842-4_4

be its generating formal power series:

$$\sum_{i\geq 0} D_i z^i \in \mathrm{End}_A(\bigwedge M)[[z]].$$

4.1.2. Given $\mathcal{D}(z) \in \mathrm{End}_A(\bigwedge M)[[z]]$ and $E \in \mathrm{End}_A(\bigwedge M)$, define

$$\mathcal{D}(z) \circ E := \sum_{i\geq 0}(D_i \circ E)z^i.$$

If $\mathcal{E}(z) = \sum_{i\geq 0} E_i z^i$ is another element of $\mathrm{End}_A(\bigwedge M)[[z]]$, the product

$$\mathcal{D}(z) \cdot \mathcal{E}(z) := \sum_{n\geq 0}(D(z) \circ E_n)z^n := \sum_{n\geq 0}\sum_{i=0}^{n}(D_i \circ E_{n-i})z^n$$

turns $\mathrm{End}_A(\bigwedge M)[[z]]$ into an $\mathrm{End}_A(\bigwedge M)$-algebra whose unit $\mathbb{1}_{\bigwedge M}$ is understood as a constant formal power series. One says that the element $\mathcal{D}(z) \in \mathrm{End}_A(\bigwedge M)$ is *invertible* if there exists $\overline{\mathcal{D}}(z) \in \mathrm{End}_A(\bigwedge M)[[z]]$, called the *inverse* of $\mathcal{D}(z)$, such that $\mathcal{D}(z)\overline{\mathcal{D}}(z) = \overline{\mathcal{D}}(z)\mathcal{D}(z) = \mathbb{1}_{\bigwedge M}$, which is in turn equivalent to D_0 being an automorphism of $\bigwedge M$ (by Exercise 2.8.3). Let us write

$$\overline{\mathcal{D}}(z) = \sum_{i\geq 0}(-1)^i \overline{D}_i z^i,$$

for the inverse of $\mathcal{D}(z)$, denoting by $(-1)^i\overline{D}_i$ the coefficient of z^i.

Remark 4.1.3. The relation $\mathcal{D}(z)\overline{\mathcal{D}}(z) = \mathbb{1}_{\bigwedge M}$ is equivalent to

$$\sum_{i=0}^{j}(-1)^i D_{j-i}\overline{D}_i = 0, \qquad \textit{for all } j \geq 1. \tag{4.1}$$

If $D_0 = \mathbb{1}_{\bigwedge M}$, then $\overline{D}_0 = \mathbb{1}_{\bigwedge M}$ as well and, for example, $D_1 - \overline{D}_1 = 0$, $D_2 - D_1\overline{D}_1 + \overline{D}_2 = 0$.

To continue with, every $\mathcal{D}(z) \in \mathrm{End}_A(\bigwedge M)[[z]]$, which is nothing but that the generating formal power series of a sequence of elements of $\mathrm{End}_A(\bigwedge M)$, induces an A-homomorphism $\bigwedge M \to \bigwedge M[[z]]$ given by

$$\eta \mapsto \mathcal{D}(z)\eta := D_0\eta + D_1\eta \cdot z + \cdots \qquad (\eta \in \bigwedge M).$$

Recall the algebra structure 4.1.1 of $\bigwedge M[[z]]$. Mimicking [112, p. 207], as in [37], we give the following:

Definition 4.1.4. A *Hasse–Schmidt derivation* of $\bigwedge M$ *(HS-derivation)* is a $\bigwedge M$-algebra homomorphism $\mathcal{D}(z) : \bigwedge M \to \bigwedge M[[z]]$, that is

$$\mathcal{D}(z)(\eta \wedge \theta) = \mathcal{D}(z)\eta \wedge \mathcal{D}(z)\theta, \qquad \forall \eta, \theta \in \bigwedge M. \tag{4.2}$$

The set of all Hasse–Schmidt derivations of $\bigwedge M$ will be denoted by $HS(\bigwedge M)$ as in [112, p. 207].

Proposition 4.1.5. *Consider* $\mathcal{D}(z) := \sum_{i\geq 0} D_i z^i \in \mathrm{End}_A(\bigwedge M)[[z]]$. *Then* $\mathcal{D}(z) \in HS(\bigwedge M)$ *if and only if the following Leibniz-like rule holds for all* $i \geq 0$ *:*

$$D_i(\eta \wedge \theta) = \sum_{j=0}^{i} D_j\eta \wedge D_{i-j}\theta. \tag{4.3}$$

Proof. Consider the chain of equalities:

$$\sum_{i\geq 0} D_i(\eta \wedge \theta)z^i = \left(\sum_{i\geq 0} D_i z^i\right)(\eta \wedge \theta) = \sum_{i\geq 0} D_i\eta \cdot z^i \wedge \sum_{j\geq 0} D_j\theta \cdot z^j \tag{4.4}$$

$$= \sum_{i\geq 0}\sum_{j=0}^{i}(D_j\eta \wedge D_{i-j}\theta)z^i. \tag{4.5}$$

Comparing the coefficients of z^i of the first and the last terms of (4.5) gives precisely (4.3). The converse obviously holds. □

4.1.6. We can remark here that putting $i = 1$ in (4.5) yields $D_1(\eta \wedge \theta) = D_1\eta \wedge \theta + \eta \wedge D_1\theta$, which is the ordinary Leibniz rule for the wedge product, i.e. if $\mathcal{D}(z) \in HS(\bigwedge M)$, then D_1 is a derivation of $\bigwedge M$. This partly accounts for the name 'derivation' given to the algebra homomorphism $\mathcal{D}(z)$.

Proposition 4.1.7. *The set* $\mathrm{HS}(\bigwedge M)$ *is a subalgebra of* $\mathrm{End}_A(\bigwedge M)[[z]]$, *i.e.:*

(i) *If* $\mathcal{D}(z), \mathcal{E}(z) \in HS(\bigwedge M)$, *then* $\mathcal{D}(z)\mathcal{E}(z) \in HS(\bigwedge M)$;

(ii) $\mathbb{1}_{\bigwedge M} \in HS(\bigwedge M)$;

(iii) *If* $\mathcal{D}(z) \in HS(\bigwedge M)$ *is invertible, its inverse* $\overline{\mathcal{D}}(z) \in \mathrm{End}_A(\bigwedge M)[[z]]$ *is an HS-derivation.*

Proof. Checking the three properties is a matter of a routine verification. Recall the definition of the product of two $\mathrm{End}_A(\bigwedge M)$-valued formal power series (Section 4.1.2):

$$\begin{aligned}\mathcal{D}(z)\mathcal{E}(z)(\alpha \wedge \beta) &= \mathcal{D}(z)\Big(\sum_{j\geq 0}\sum_{j_1+j_2=j} E_{j_1}\alpha \wedge E_{j_2}\beta\Big)z^j \\ &= \sum_{j\geq 0}\sum_{j_1+j_2=j} \mathcal{D}(z)E_{j_1}\alpha \cdot z^{j_1} \wedge \mathcal{D}(z)E_{j_2}\beta \cdot z^{j_2} = \\ &= \mathcal{D}(z)\mathcal{E}(z)\alpha \wedge \mathcal{D}(z)\mathcal{E}(z)\beta.\end{aligned}$$

proving i). Item ii) is trivial, while iii) follows from i):

$$\begin{aligned}\overline{\mathcal{D}}(z)(\eta \wedge \theta) &= \overline{\mathcal{D}}(z)\big(\mathcal{D}(z)\overline{\mathcal{D}}(z)\eta \wedge \mathcal{D}(z)\overline{\mathcal{D}}(z)\theta\big) = \\ &= \overline{\mathcal{D}}(z)\mathcal{D}(z)(\overline{\mathcal{D}}(z)\eta \wedge \overline{\mathcal{D}}(z)\theta) = \overline{\mathcal{D}}(z)\eta \wedge \overline{\mathcal{D}}(z)\theta. \qquad \square\end{aligned}$$

Definition 4.1.8. If A is a $\mathbb{Q}$-algebra, the *HS*-derivation $\mathfrak{D}(z)$ constructed out of a derivation $\mathfrak{D}$, as in Section 3.5.6, is called *iterative*.

The next result will be crucial in the ensuing sections.

Proposition 4.1.9. *The following formulas, both called* integration by parts, *hold:*

$$\mathcal{D}(z)\eta \wedge \theta = \mathcal{D}(z)(\eta \wedge \overline{\mathcal{D}}(z)\theta); \tag{4.6}$$

$$\overline{\mathcal{D}}(z)\eta \wedge \theta = \overline{\mathcal{D}}(z)(\eta \wedge \mathcal{D}(z)\theta). \tag{4.7}$$

Proof. Clearly $\mathcal{D}(z)\eta \wedge \theta = \mathcal{D}(z)\eta \wedge \mathcal{D}(z)\overline{\mathcal{D}}(z)\theta = \mathcal{D}(z)(\eta \wedge \overline{\mathcal{D}}(z)\theta)$. The proof of (4.7) is analogous. $\square$

Expressing (4.6) and (4.7) in terms of the coefficients $(D_i)_{i\geq 0}$ and $((-1)^j\overline{D}_j)_{j\geq 0}$ of $\mathcal{D}(z)$ and $\overline{\mathcal{D}}(z)$, respectively, one obtains

$$D_i\eta \wedge \theta = D_i(\eta \wedge \overline{D}_0\theta) + \sum_{j=1}^{i}(-1)^j D_{i-j}\eta \wedge \overline{D}_j\theta, \tag{4.8}$$

$$\overline{D}_i\eta \wedge \theta = \overline{D}_i(\eta \wedge D_0\theta) + \sum_{j=1}^{i}(-1)^j \overline{D}_{i-j}\eta \wedge D_j\theta. \tag{4.9}$$

If $D_0 = \overline{D}_0 = \mathbb{1}_{\bigwedge M}$, expressions (4.8) and (4.9) for $i = 1$ reduce to

$$D_1\eta \wedge \theta = D_1(\eta \wedge \theta) - \eta \wedge D_1\theta,$$

and

$$\overline{D}_1\eta \wedge \theta = D_1(\eta \wedge \theta) - \eta \wedge D_1\theta,$$

which motivates the choice of calling (4.6)–(4.7) *integration by parts*, being reminiscent of the renowned formula $\int udv = uv - \int vdu$.

4.1.10. Recall that the multinomial coefficient

$$\binom{j}{i_1, \dots, i_r} = \frac{j!}{i_1!i_2!\cdots i_r!}$$

is the coefficient of $t_1^{i_1}\cdots t_r^{i_r}$ in the expansion of $(t_1+\cdots+t_r)^j$. In particular if $r=2$, one recovers the classical definition of binomial coefficient. The following lemma will be useful in the sequel and is very much related to the paper [14].

Lemma 4.1.11. *For each* $\eta,\theta\in\bigwedge M$, Newton's binomial formula *holds:*

$$D_1^i(\eta\wedge\theta)=\sum_{i_1+i_2=i}\binom{i}{i_1,i_2}D_1^{i_1}\eta\wedge D_1^{i_2}\theta. \tag{4.10}$$

Proof. The verification is a matter of simple induction on the integer i. Since D_1 is an even derivation, (4.10) holds for $i=1$. Suppose now that (4.10) holds for all $1\le j\le i-1$. Then

$$\begin{aligned}D_1^i(\eta\wedge\theta)&=D_1\big(D_1^{i-1}(\eta\wedge\theta)\big)=D_1\sum_{i_1+i_2=i-1}\binom{i-1}{i_1,i_2}D_1^{i_1}(\eta)\wedge D_1^{i_2}\theta\\&=\sum_{i_1+i_2=i-1}\binom{i-1}{i_1,i_2}\big(D_1^{i_1+1}\eta\wedge D_1^{i_2}\theta+D_1^{i_1}(\eta)\wedge D_1^{i_2+1}\theta\big)\\&=\sum_{i_1+i_2=i}\binom{i}{i_1,i_2}D_1^{i_1}\eta\wedge D_1^{i_2}\theta,\end{aligned}$$

having used well-known properties of binomial coefficients. □

Proposition 4.1.12. *For all* $r\ge 2$,

$$D_1^i(m_1\wedge\cdots\wedge m_r)=\sum_{i_1+\cdots+i_r=i}\binom{i}{i_1,\ldots,i_r}D_1^{i_1}(m_1)\wedge\cdots\wedge D_1^{i_r}(m_r). \tag{4.11}$$

Proof. The case $r=2$ follows from (4.10) for $\eta=m_1$, $\theta=m_2$ and the fact that $D_1^j(m)=\overline{D}_1^j(m)$ if $m\in M$. Suppose now that formula (4.11) holds for all $\bigwedge^s M$ with $2\le s\le r-1$. Then

$$\begin{aligned}D_1^i(m_1\wedge\cdots\wedge m_{r-1}\wedge m_r)&=\sum_{j+i_r=i}\binom{i}{j,i_r}D_1^j(m_1\wedge\cdots\wedge m_{r-1})\wedge D_1^{i_r}m_r\\&=\sum_{j+i_r=i}\binom{i}{j,i_r}\sum_{i_1+\cdots+i_{r-1}=j}\binom{j}{i_1,\ldots,i_{r-1}}D_1^{i_1}(m_1)\wedge\cdots\wedge D_1^{i_r}(m_r)\\&=\sum_{i_1+\cdots+i_r=i}\binom{i}{i_1,\ldots,i_r}D_1^{i_1}m_1\wedge\cdots\wedge D_1^{i_r}m_r,\end{aligned}$$

as desired. □

The relevance of the next result lies in that it provides a way to construct a host of *HS*-derivations.

Proposition 4.1.13 (cf. [51]). *Let $f(z) : M \to \bigwedge M[[z]]$ be an A-module homomorphism such that $f(z)m \wedge f(z)m = 0$ for all $m \in M$. Then there exists a unique $\mathcal{D}^f(z) \in HS(\bigwedge M)$ such that $\mathcal{D}^f(z)_{|M} = f(z)$.*

Proof. For $j \geq 1$, consider the linear map $\widehat{\mathcal{D}^f}(z) : T^j(M) \to \bigwedge M[[z]]$ defined as follows: $\widehat{\mathcal{D}^f}(z)m = f(z)m$ for $m \in M$ and, for all $j > 1$,

$$\widehat{\mathcal{D}^f}(z)(m_1 \otimes \cdots \otimes m_j) = f(z)(m_1) \wedge \cdots \wedge f(z)(m_j). \tag{4.12}$$

By construction $\widehat{\mathcal{D}^f}(z)$ is alternating and thus it factorizes through a unique A-module homomorphism

$$\mathcal{D}^f(z) : \bigwedge^j M \to \bigwedge M[[z]].$$

Since each $\eta \in \bigwedge M$ is a finite sum of homogeneous elements, we have thus constructed an A-module homomorphism $\bigwedge M \to \bigwedge M[[z]]$ that extends $f(z)$. It is an *HS*-derivation because, for $\eta = m_{i_1} \wedge \cdots \wedge m_{i_h}$ and $\theta = m_{j_1} \wedge \cdots \wedge m_{j_k}$,

$$\begin{aligned} \mathcal{D}^f(z)(\eta \wedge \theta) &= f(z)(m_{i_1}) \wedge \cdots \wedge f(z)(m_{i_h}) \wedge f(z)(m_{j_1}) \wedge \cdots \wedge f(z)(m_{j_h}) \\ &= (f(z)(m_{i_1}) \wedge \cdots \wedge f(z)(m_{i_h})) \wedge (f(z)(m_{j_1}) \wedge \cdots \wedge f(z)(m_{j_h})) \\ &= \mathcal{D}^f(z)\eta \wedge \mathcal{D}^f(z)\theta, \end{aligned}$$

and the general case follows by expressing arbitrary $\eta, \theta \in \bigwedge M$ as finite sums of homogeneous elements. □

4.2 The Trace Operator Polynomials

In this section, invoking Proposition 4.1.13 we shall attach a distinguished *HS*-derivation to any given endomorphism of a free A-module. The linear dependence of its coefficients will provide a substantial generalization of the Cayley–Hamilton theorem. The main reference for this section is [51].

4.2.1. Assume that $M := \bigoplus_{i=0}^{r-1} Ab_i$ is a free A-module of finite rank r, and take $f \in \mathrm{End}_A(M)$. Consider the formal power series in $End_A(M)[[z]] \subseteq \mathrm{End}_A(\bigwedge M)[[z]]$:

$$f(z) = \mathbb{1}_M + fz + f^2z^2 + \cdots$$

As in Proposition (4.1.13), let $\mathcal{D}^f(z) := \sum_{I\geq 0} D_i^f \cdot z^i$ be the unique *HS*-derivation of $\bigwedge M$ extending $f(z) : M \to \bigwedge M[[z]]$. In particular $D_0^f := \mathbb{1}_{\bigwedge M}$, $D^f_{i\,|M} = f^i$, and we write

$$\overline{\mathcal{D}}^f(z) := \sum_{i\geq 0}(-1)^i\overline{D}_i^f z^i \in HS(\bigwedge M) \tag{4.13}$$

for the inverse of $\mathcal{D}^f(z)$. For reasons that shall be clear in a short while, we call (4.13) the *trace operator polynomial*. The equation $\overline{\mathcal{D}}^f(z) \cdot \mathcal{D}^f(z) = \mathcal{D}^f(z) \cdot \overline{\mathcal{D}}^f(z) = \mathbb{1}_{\bigwedge M}$ implies

$$\overline{\mathcal{D}}^f(z)_{|M} = \frac{1}{f(z)} = \mathbb{1}_M - f \cdot z, \tag{4.14}$$

i.e. for any $m \in M$, $\overline{\mathcal{D}}_1^f(m) = f(m)$ and $\overline{\mathcal{D}}_j^f(m) = 0$ for all $j > 1$.

Proposition 4.2.2. *If* $i > j$, *then* $\bigwedge^j M \subseteq \ker(\overline{D}_i^f)$. *In particular* $\overline{D}_i^f = 0$ *for all* $i > r$.

Proof. Equation (4.14) implies that the property is true for $j = 1$. Suppose now that $\overline{\mathcal{D}}^f_{i\,|\bigwedge^k M} = 0$ for all $1 \leq k \leq j-1$. Each monomial $\eta \in \bigwedge^j M$ can be written as $m \wedge \theta$ for some $m \in M$ and $\theta \in \bigwedge^{j-1} M$. Now for $i \geq j+1$,

$$\overline{D}_i^f(\eta) = \overline{D}_i^f(m \wedge \theta) = \sum_{k=0}^{i} \overline{D}_k^f(m) \wedge \overline{D}_{i-k}^f\theta. \tag{4.15}$$

As $\overline{D}_k^f(m) = 0$ if $k > 1$, equality (4.15) reduces to

$$\overline{D}_i^f(\eta) = \overline{D}_i^f(m \wedge \theta) = m \wedge \overline{D}_i^f(\theta) - \overline{D}_1^f(m) \wedge \overline{D}_{i-1}^f(\theta).$$

By inductive hypothesis $\overline{D}_i^f(\theta) = \overline{D}_{i-1}^f(\theta) = 0$, and then $\eta \in \ker(\overline{D}_i^f)$. □

Definition 4.2.3. The *trace operators* associated with f are the coefficients $\overline{D}_i^f$ of $\overline{\mathcal{D}}^f(z)$.

The rank-one A-module $\bigwedge^r M$ is generated by $[\mathbf{b}]_0^r := b_0 \wedge b_1 \wedge \cdots \wedge b_{r-1}$. Let $e_i(f)$ and $h_i(f)$, $i \geq 0$ be, respectively, the eigenvalues of $\overline{D}_i^f$ and D_i^f restricted to $\bigwedge^r M$, i.e.:

$$\overline{D}_i^f([\mathbf{b}]_0^r) = e_i(f)[\mathbf{b}]_0^r \qquad \text{and} \qquad D_i^f([\mathbf{b}]_0^r) = h_i(f)[\mathbf{b}]_0^r.$$

Clearly $e_0(f) = h_0(f) = 1$ and $e_j(f) = 0$ if $j > r$, because of Proposition 4.2.2.

Definition 4.2.4. *The elements* $e_1(f), \ldots, e_r(f) \in A$ *are called the* traces *of* $f \in \mathrm{End}_A(M)$.

Example 4.2.5. Choose $A = \mathbb{C}$, $M := \mathbb{C}^r$ and $f \in \mathrm{End}_{\mathbb{C}}(M)$. Let $\mathfrak{B}_r := (b_0, \ldots, b_{r-1})$ be any basis of $\mathbb{C}^r$. Then $f(b_i) = \sum_{j=0}^{n-1} f_i^j b_j$, where (f_i^j) is the matrix associated to f in the basis $\mathfrak{B}$. It follows that:

$$
\begin{aligned}
\overline{D}_1^f(b_0 \wedge \cdots \wedge b_{r-1}) &= f(b_0) \wedge b_1 \wedge \cdots \wedge b_{r-1} \\
&+ b_0 \wedge f(b_1) \wedge \cdots \wedge b_{r-1} \\
&\vdots \\
&+ b_0 \wedge b_1 \wedge \cdots \wedge f(b_{r-1}) = \\
&= (f_0^0 + f_1^1 + \cdots + f_{r-1}^{r-1})[\mathbf{b}]_0^r
\end{aligned}
$$

i.e. $e_1(f) := f_0^0 + f_1^1 + \cdots + f_{r-1}^{r-1}$ is the usual *trace* of a matrix. The definition of $e_1(f)$ shows that it does not depend on the choice of a basis. Similarly

$$
\begin{aligned}
\overline{D}_r^f[\mathbf{b}]_0^r = \overline{D}_r^f(b_0 \wedge \cdots \wedge b_{r-1}) &= f(b_0) \wedge f(b_1) \wedge \cdots \wedge f(b_{r-1}) \\
&= \sum_{i_0=0}^{r-1} \cdots \sum_{i_{r-1}=0}^{r-1} f_0^{i_0} f_1^{i_1} \cdots f_{r-1}^{i_{r-1}} b_{i_0} \wedge \cdots \wedge b_{i_{r-1}} \\
&= \sum_{\mathfrak{s} \in S_r} \mathrm{sgn}(\mathfrak{s}) f_0^{\mathfrak{s}(0)} \cdots f_{r-1}^{\mathfrak{s}(r-1)} b_0 \wedge b_1 \wedge \cdots \wedge b_{r-1}
\end{aligned}
$$

i.e.

$$
e_r(f) := \sum_{\mathfrak{s} \in S_r} \mathrm{sgn}(\mathfrak{s}) f_0^{\mathfrak{s}(0)} \cdots f_{r-1}^{\mathfrak{s}(r-1)}
$$

is precisely the *determinant* of the matrix (f_j^i) of f with respect to $\mathfrak{B}_r$, where we have denoted by S_r the symmetric group on r elements and by "sgn" the parity of the permutation $\mathfrak{s}$.

Define

$$
E_r^f(z) := 1 - e_1(f)z + \cdots + (-1)^r e_r(f) z^r \in A[z]
$$

and

$$
H_r^f(z) := 1 + \sum_{i \geq 1} h_i(f) z^i \in A[[z]].
$$

Proposition 4.2.6. *In the algebra* $A[[z]]$, *the equality* $H_r^f(z) E_r^f(z) = 1$ *holds.*

Proof. Let $\zeta := [\mathbf{b}]_0^r$. Then

$$\zeta = \mathbb{1}_{\bigwedge M} \cdot \zeta = \mathcal{D}^f(z)\overline{\mathcal{D}}^f(z)\zeta = \mathcal{D}^f(z)E_r^f(z)\zeta = E_r^f(z)\mathcal{D}^f(z)\zeta$$
$$= E_r^f(z)H_r^f(z)\zeta,$$

which proves that $E_r^f(z)H_r^f(z) = 1$. □

Corollary 4.2.7. *If $f \in \mathrm{End}_A(M)$ and $\mathrm{rank}_A M = r$, then*

$$e_1(f)^i(m_1 \wedge \cdots \wedge m_r) = \sum_{i_1+\cdots+i_r=i} \binom{i}{i_1, \ldots, i_r} f^{i_1}(m_1) \wedge \cdots \wedge f^{i_r}(m_r).$$

Proof. Obvious, since $D_1^f(m_1 \wedge \cdots \wedge m_r) = e_1(f)(m_1 \wedge \cdots \wedge m_r)$, by definition of $e_1(f)$, and applying 4.1.2 □

Proposition 4.2.8. *Let $M[z] := M \otimes_A A[z]$ and regard $\mathbb{1}_M - f \cdot z$ as an $A[z]$-endomorphism of $M[z]$. Then*

$$e_r(\mathbb{1}_M - f \cdot z) = E_r^f(z).$$

Proof. For $m_1 \wedge \cdots \wedge m_r \neq 0$, we have

$$e_r(\mathbb{1}_M - f \cdot z)m_1 \wedge \cdots \wedge m_r = \overline{D}_r^{(\mathbb{1}_M - f\cdot z)} m_1 \wedge \cdots \wedge m_r$$
$$= (\mathbb{1}_M - f \cdot z)m_1 \wedge \cdots \wedge (\mathbb{1}_M - f \cdot z)m_r$$
$$= \sum_{j=0}^{r} (-1)^j \Bigg(\sum_{\substack{i_1+\cdots+i_r=j \\ 0 \le i_k \le 1}} f^{i_1}(m_1) \wedge \cdots \wedge f^{i_r}(m_r)\Bigg) z^j$$
$$= \sum_{j=0}^{r} (-1)^j \overline{D}_j^f(m_1 \wedge \cdots \wedge m_r) z^j$$
$$= \sum_{j=0}^{r} (-1)^j e_j(f)(m_1 \wedge \cdots \wedge m_r) z^j = E_r^f(z) m_1 \wedge \cdots \wedge m_r. \qquad \square$$

Example 4.2.9. Similarly to Proposition 4.2.8, we have

$$p_r^f(z) := e_r(z\mathbb{1}_M - f) = (z\mathbb{1}_M - f)b_0 \wedge (z\mathbb{1}_M - f)b_1 \wedge \cdots \wedge (z\mathbb{1}_M - f)b_{r-1} \tag{4.16}$$
$$= z^r E_r^f\left(\frac{1}{z}\right) \tag{4.17}$$
$$= z^r - e_1 f + \cdots + (-1)^r e_r \mathbb{1}_M, \tag{4.18}$$

traditionally called the *characteristic polynomial* of f.

4.2.10. For all $j \geq 0$, define $\mathrm{U}_j(\mathcal{D}^f(z)) \in End_A(\bigwedge M)$ via

$$\sum_{j\geq 0} \mathrm{U}_j(\mathcal{D}^f(z))z^j := E_r^f(z)\mathcal{D}^f(z) \tag{4.19}$$

Comparing the coefficient of z^j on either side of (4.19), for all $j \geq 0$, one easily obtains

$$\mathrm{U}_0(\mathcal{D}^f(z))) = \mathbb{1}_{\bigwedge M}, \quad \mathrm{U}_1(\mathcal{D}^f(z)) = D_1^f - e_1(f)\cdot \mathbb{1}_{\bigwedge M}$$

$$\mathrm{U}_2(\mathcal{D}^f(z)) = D_2^f - e_1(f)D_1^f + e_2(f)\cdot \mathbb{1}_{\bigwedge M}$$

and, in general,

$$\mathrm{U}_j(\mathcal{D}^f(z)) = D_j^f + \sum_{i\geq 1}(-1)^i e_i(f)D_{j-i}^f, \tag{4.20}$$

with the convention that $D_k^f = 0$ if $k < 0$ and $e_i = 0$ if $i > r$. In particular, for $j = r$,

$$\mathrm{U}_r(\mathcal{D}^f(z)) = D_r^f - e_1(f)D_{r-1}^f + \cdots + (-1)^r e_r(f)\mathbb{1}_{\bigwedge M} \in \mathrm{End}_A(\bigwedge M)$$

so that

$$\begin{aligned}\mathrm{U}_r(\mathcal{D}^f(z))_{|M} &= (D_r^f - e_1(f)D_{r-1}^f + \cdots + (-1)^r e_r(f)\mathbb{1}_M)_{|M}\\ &= f^r - e_1(f)f^{r-1} + \cdots + (-1)e_r(f)\mathbb{1}_M\\ &= p_r^f(f),\end{aligned}$$

i.e. the restriction of $\mathrm{U}_r(\mathcal{D}^f(z))$ to M is the evaluation of the *characteristic polynomial* $p_r^f(z)$ (cf. 4.18) at $z = f$.

Lemma 4.2.11. *Integration by parts (formula (4.6)) implies*

$$\mathrm{U}_i(\mathcal{D}^f(z))\eta \wedge \theta = \sum_{j=0}^{i}(-1)^j \mathrm{U}_{i-j}(\mathcal{D}^f(z))(\eta \wedge \overline{D}_j^f(\theta)). \tag{4.21}$$

Proof. According to (4.19), we write

$$\sum_{i\geq 0}(\mathrm{U}_i(\mathcal{D}^f(z))\eta)z^i \wedge \theta = E_r(z)\mathcal{D}^f(z)\eta \wedge \theta, \tag{4.22}$$

then apply formula (4.6) and use (2.18) again:

$$= E_r(z)\mathcal{D}^f(z)(\eta \wedge \overline{\mathcal{D}}^f(z)\theta)) = \sum_{i\geq 0} \mathrm{U}_i(\mathcal{D}^f(z))(\eta \wedge \overline{\mathcal{D}}^f f(z)\theta)z^j \tag{4.23}$$

$$= \sum_{i\geq 0} \mathrm{U}_i(\mathcal{D}^f(z))(\eta \wedge \sum_{j\geq 0}(-1)^j \overline{D}_j^f(\theta))z^{j+i} \tag{4.24}$$

$$= \sum_{i\geq 0}\left(\sum_{j=0}^{i} \mathrm{U}_{i-j}(\mathcal{D}^f(z))(\eta \wedge (-1)^j \overline{D}_j^f(\theta))\right) z^i. \tag{4.25}$$

Comparing the coefficients of z^i on the left-hand side of (4.22) and the right-hand side of (4.25) proves the claim. □

We are now in the right position to prove a substantial generalization of the classical theorem by Cayley–Hamilton:

Theorem 4.2.12 (Cayley–Hamilton). Let $j \geq 1$ and $\eta \in \bigwedge^j M$. Then for all $k \geq r - j + 1$,

$$\mathrm{U}_k(\mathcal{D}^f(z))\eta = (D_k^f - e_1 D_{k-1}^f + \cdots + (-1)^k e_k)\eta = 0.$$

Proof. If $\eta = 0$ the proposition is trivial, so suppose $0 \neq \eta \in \bigwedge^j M$. We see that for all $\theta \in \bigwedge^{r-j} M$,

$$\mathrm{U}_k(\mathcal{D}^f(z))\eta \wedge \theta = \sum_{i=0}^{k-1} \mathrm{U}_{k-i}(\mathcal{D}^f(z))(\eta \wedge \theta) + (-1)^k(\eta \wedge \overline{D}_k^f\theta).$$

Since $\mathrm{U}_j(\mathcal{D}^f(z))$ vanishes on $\bigwedge^r M$, for all $1 \leq j \leq r$, the first $k-1$ summands cancel out. As $k \geq r - j + 1$ and $\theta \in \bigwedge^{r-j} M$, then $\overline{D}_k^f(\theta) = 0$ (by 4.2.2) and the proof is complete. □

Corollary 4.2.13. *The endomorphisms $\mathrm{U}_{r+i}(\mathcal{D}^f(z))$ vanish identically on the whole exterior algebra $\bigwedge M$, for all $i \geq 0$.* □

Warning 4.2.14. In general, $\mathrm{U}_j(\mathcal{D}(z))$ is not zero on the entire $\bigwedge M$, for $0 \leq j < r$. For example, take $M = \mathbb{R}^2$ and let f be any invertible matrix. If $\mathrm{U}_1(f) = 0$, we have

$$(f - e_1(f)\mathbb{1}_{2\times 2})v = 0$$

for every $v \in \mathbb{R}^2$; in other words the trace $e_1(f)$ is an eigenvalue of f. Then f should have a null eigenvalue, contradicting the invertibility.

Corollary 4.2.15. Every endomorphism $f \in \mathrm{End}_A(M)$ is a root of its own characteristic polynomial.

Proof. Put $j = 1$ in Theorem 4.2.12 . Then

$$0 = \mathrm{U}_k(\mathcal{D}^f(z))m = (f^k - e_1 f^{k-1} + \cdots + (-1)^r e_r f^{k-r})(m)$$

for all $m \in M$, whence $f^k - e_1 f^{k-1} + \cdots + (-1)^r e_r f^{k-r} = 0$ for all $k \geq r$, in particular for $k = r$. □

4.3 The Exponential of an Endomorphism

We retain the notation of previous sections with the exception that we change the name of our indeterminate, from z to t. Pick a map $f \in \mathrm{End}_A(M)$ and let $e_i(f) \in A$ be its ith trace. Equip A with the B_r-module structure induced by the unique $\mathbb{Z}$-algebra homomorphism $B_r \to A$ mapping $e_i \mapsto e_i(f)$. Let

$$(\mathbb{1}_M - tf)^{-1} := \mathbb{1}_M + ft + f^2t^2 + \cdots \in \mathrm{End}_A(M)[[t]].$$

Proposition 4.3.1. *The following equality holds (notation as in 2.3.4):*

$$(\mathbb{1}_M - tf)^{-1} = \mathbb{1}_M u_0 + \mathfrak{p}_1(f)u_{-1} + \cdots + \mathfrak{p}_{r-1}(f)u_{-r-1}. \tag{4.26}$$

Proof. Just compute

$$E_r(t)(\mathbb{1}_M - tf)^{-1} = \sum_{i\geq 0} \mathfrak{U}_i((\mathbb{1}_M - tf)^{-1})t^i = \sum_{i\geq 0} \mathfrak{p}_i(f)t^i. \tag{4.27}$$

According to the Cayley–Hamilton theorem (4.2.12), $\mathfrak{p}_{r+i}(f) = 0$ for all $i \geq 0$, and hence the right-hand side of (4.27) is a finite sum:

$$E_r(t)(\mathbb{1}_M - tf)^{-1} = \sum_{i=0}^{r-1} \mathfrak{p}_i(f)t^i.$$

Dividing both sides by $E_r(t)$ and using the definition of u_{-i}, we finally obtain (4.26). □

We note in passing that

$$\det[(\mathbb{1}_M - tf)^{-1}] = \frac{1}{e_r(\mathbb{1}_M - tf)} = \frac{1}{E_r^f(z)} = H_r^f(t) = u_0 \otimes 1_A.$$

Definition 4.3.2. Assume that A is a $\mathbb{Q}$-algebra. Then $\exp(tf)$ is the formal power series defined by

$$\exp(tf) = \mathcal{L}^{-1}((\mathbb{1}_M - tf)^{-1}) = \mathcal{L}^{-1}(1 + ft + f^2t^2 + \cdots) \tag{4.28}$$

$$= \sum_{j\geq 0} \frac{f^j t^j}{j!} \in \mathrm{End}_A(M)[[z]]. \tag{4.29}$$

Proposition 4.3.3. *One has*

$$\exp(tf) = \mathfrak{p}_0(f)\mathcal{L}^{-1}(u_0) + \cdots + \mathfrak{p}_{r-1}(f)\mathcal{L}^{-1}(u_{-r+1}). \tag{4.30}$$

Proof. In fact:

$$\begin{aligned}\exp(tf) &= \mathcal{L}^{-1}\left((\mathbb{1}_M - tf)^{-1}\right)\\ &= \mathcal{L}^{-1}\left(\sum_{j=0}^{r-1}\mathfrak{p}_j(f)u_{-j}\right) = \sum_{j=0}^{r-1}\mathfrak{p}_j(f)\mathcal{L}^{-1}(u_{-j}),\end{aligned}$$

as claimed. □

The finite sum (4.30) permits to identify $\exp(tf)$ with an endomorphism of $A[[t]] \otimes_A M$. The latter is a free $A[[t]]$-module of rank r, so it is possible to speak of its determinant. The following is a well-known result, here spelt out in a purely algebraic context.

Proposition 4.3.4. *The determinant of the exponential is the exponential of the trace:*

$$e_r(\exp(tf)) = \exp(e_1(f)t).$$

Proof. By definition

$$\begin{aligned} e_r(\exp(tf))m_1 \wedge \cdots \wedge m_r :&= \overline{D}_r(\exp(tf))(m_1 \wedge \cdots \wedge m_r) &(4.31)\\ &= \overline{D}_1(\exp(tf))m_1 \wedge \cdots \wedge \overline{D}_1(\exp(tf))m_r &(4.32)\\ &= \exp(tf)m_1 \wedge \cdots \wedge \exp(tf)m_r. &(4.33)\end{aligned}$$

However, (4.33) is nothing but

$$\sum_{i_1\geq 0} f^{i_1}(m_1)\frac{t^{i_1}}{i_1!} \wedge \cdots \wedge \sum_{i_r\geq 0} f^{i_r}(m_1)\frac{t^{i_r}}{i_r!}$$

which in turn equals

$$\sum_{j\geq 0}\frac{t^j}{j!}\sum_{i_1+\cdots+i_r=j}\binom{j}{i_1,\ldots,i_r} f^{i_1}(m_1) \wedge \cdots \wedge f^{i_r},(m_r).$$

Using Corollary 4.2.7, the above equals

$$\sum_{j\geq 0} e_1^j(f)\frac{t^j}{j!}(m_1 \wedge \cdots \wedge m_r) = \exp(e_1(f)t)m_1 \wedge \cdots \wedge m_r,$$

which proves the desired formula. □

Formula (4.30) and Proposition 4.3.4 give rise to some interesting identities.

Example 4.3.5. Consider

$$f := \begin{pmatrix} a & b \\ c & d \end{pmatrix}$$

thought of as an endomorphism of $M := A^2$ where $A := \mathbb{Q}[a, b, c, d]$. We view A as a B_2-algebra using the mapping $e_1 \mapsto a + d$ and $e_2 \mapsto \det(f) = ad - bc$. Let $u_0 := H_2(t)$ and $u_{-1} = tH_2(t)$. Then

$$(\mathbb{1}_{2\times 2} - tf)^{-1} = \mathbb{1}_{2\times 2}u_0 + (f - e_1\mathbb{1}_{2\times 2})u_{-1} = \begin{pmatrix} u_0 - du_{-1} & bu_{-1} \\ cu_{-1} & u_0 - au_{-1} \end{pmatrix}$$

and

$$\begin{aligned} \exp(tf) &= \mathcal{L}^{-1}((\mathbb{1}_{2\times 2} - tf)^{-1}) = \mathbb{1}_{2\times 2}\mathcal{L}^{-1}(u_0) + (f - e_1\mathbb{1}_{2\times 2})\mathcal{L}^{-1}(u_{-1}) \\ &= \begin{pmatrix} v_0 - dv_{-1} & bv_{-1} \\ cv_{-1} & v_0 - av_{-1} \end{pmatrix}, \end{aligned}$$

where for simplicity we have denoted $\mathcal{L}^{-1}(u_i)$ by v_i. Then

$$e_2(\exp(tf)) = v_0^2 - e_1v_0v_{-1} + e_2v_{-1}^2.$$

On the other hand, $e_2(\exp(tf)) = \exp(e_1(f)t) = \exp(e_1t)$, from which

$$v_0^2 - e_1v_0v_{-1} + e_2v_{-1}^2 = \exp(e_1t).$$

In particular, if $e_1 = 0$, we obtain the relation

$$v_0^2 + e_2v_{-1}^2 = 1,$$

which, for $e_2 = 1$, is the classical formula $\cos^2 t + \sin^2 t = 1$ (cf. Example 2.5.6 and [40, p. 64] for additional details). Taking into account that $\partial_t v_{-1} = v_0$ and putting $y = v_{-1}$, we obtain

$$(y')^2 + e_2y^2 = 1,$$

which is a first integral of the differential equation $y'' + e_2y = 0$.

Example 4.3.6. Let us see how to obtain similar formulas for larger matrices. For instance, let $\mathcal{M}$ be a 3×3 matrix having e_1, e_2 and e_3 as traces. Then

$$\exp(t\mathcal{M}) = v_0\mathbb{1}_{3\times 3} + v_{-1}(\mathcal{M} - e_1\mathbb{1}_{3\times 3}) + v_{-2}(\mathcal{M}^2 - e_1\mathcal{M} + e_2\mathbb{1}_{3\times 3}),$$

where $v_i = \mathcal{L}^{-1}(u_i)$, i.e. (v_0, v_{-1}, v_{-2}) is the canonical basis of solutions to the generic linear ODE $y''' - e_1 y'' + e_2 y' - e_3 = 0$. A straightforward yet tedious computation shows that

$$\det(\exp(t\mathcal{M})) = \exp(e_1 t)$$

can be explicitly phrased as

$$\begin{aligned}\exp(e_1 t) = {} & v_0^3 - 2e_1 v_0^2 v_{-1} + (e_1^2 + e_2) v_0^2 v_{-2} + (e_2 + e_1^2) v_0 v_{-1}^2 + \\ & - (e_1 e_2 + 3e_3) v_0 v_{-1} v_{-2} + e_1 e_3 v_0 v_{-2}^2 + (e_3 - e_1 e_2) v_{-1}^3 + \\ & + (e_1 e_3 + e_2^2) v_{-1}^2 v_{-2} + (e_1 e_2^2 - 2e_2 e_3 - 2e_1^2 e_3) v_{-1} v_{-2}^2 + e_3^2 v_{-2}^3.\end{aligned}$$

For $e_1 = 0$, one obtains

$$\begin{aligned}& v_0^3 + e_2 v_0^2 v_{-2} + e_2 v_0 v_{-1}^2 - 3e_3 v_0 v_{-1} v_{-2} + e_3 v_{-1}^3 + \\ & + e_2^2 v_{-1}^2 v_{-2} - 2e_2 e_3 v_{-1} v_{-2}^2 + e_3^2 v_{-2}^3 = 1.\end{aligned}$$

Putting $e_2 = 0$ and $e_3 = -1$, one gets the relationship

$$v_0^3 + 3v_0 v_{-1} v_{-2} - v_{-1}^3 + v_{-2}^3 = 1 \tag{4.34}$$

holding for the canonical solutions to the ODE $y''' + y = 0$. Formula (4.34) can be seen as a cubic generalization of the renowned $\cos^2 t + \sin^2 t = 1$, enjoyed by the canonical solutions $v_0 = \cos t$ and $v_{-1} = \sin t$ of the second-order equation $y'' + y = 0$. Setting $y := v_{-2}$, equation (4.34) gives

$$(y'')^3 + 3y''y' - (y')^3 + y^3 = 1. \tag{4.35}$$

Notice that differentiating (4.35) gives

$$(y''' + y)((y'')^2 + yy') = 0.$$

In other words the first integral (4.35) is obtained from $y''' + y = 0$ via multiplication by the factor $(y'')^2 + yy'$.

4.4 Notes and References

- Apart from setting up the basic theory, the purpose of this chapter was to show that *HS*-derivations crop up very naturally already in the elementary context of linear algebra. The coefficients of the characteristic polynomial of an endomorphism are concisely defined as eigenvalues of certain trace operators, which are themselves the coefficients of a canonical *HS*-derivation associated with an endomorphism of a free module. In turn, these are linearly dependent on the entire exterior algebra, which is the content of the generalized Cayley–Hamilton theorem [51]. The standard version known to all, that an endomorphism is a zero of its characteristic polynomial, is recovered by looking at the exterior algebra in degree 1.

- It is instructive to supplement Section 4.2, about the Cayley–Hamilton theorem, with more comprehensive texts like Lang [99, p. 561] or Eisenbud [22, p. 120], at least to widen one's horizons on this rather classical subject.
- Our take on *HS*-derivations, keeping into account Section 2, casts a new light on a classical paper by Putzer [125], dating 1966, on how to compute the exponential of a matrix avoiding its reduction to a Jordan normal forms, a method that has been discussed and refined in a nice couple of papers by Leonard and Liz [105, 108]. The more algebraic approach of [51] enables one to even compute exponentials with indeterminate entries. That was used in this chapter to exemplify how to attain nontrivial first integrals of linear ODEs with constant coefficients, generalizing Example 1.1.4. For a nice pedagogical path about these topics, from the point of view of spectral theory, the reader may wish to consult the pleasant exposition of Tomei [143].
- Hasse–Schmidt derivations on exterior algebras appeared first in [37] to develop the formalism of Schubert calculus, to be outlined in Section 5.4. The same formalism applies, with minor modifications, to the cohomology of Grassmann bundles [43]. Quantum and equivariant Schubert calculus [37, Corollary 2.7] can be regarded, formally, as particular cases of the above picture (see also [36] and [48]). This observation was the stepping stone of [47], where the formal algebraic properties of *HS*-derivations were laid out. At that time, the link to vertex operators was yet to be uncovered.
- If $(R, \star)$ is an A-algebra (as in the introduction), denote by the same symbol $\star$ its obvious extension to a product in the module of R-valued formal power series. Homomorphisms of monoids $\mathcal{D}(z) : (R, +) \to (R[[z]], \star)$

$$\mathcal{D}(z)(a + b) = \mathcal{D}(z)a \star \mathcal{D}(z)b$$

are extensively studied in the literature and are called λ-operations (*lambda operation*), as in, e.g. [99, p. 218] or [109, p. 25]. A *lambda* ring is a ring R equipped with such a λ-operation. λ-rings share many formal properties with *HS*-derivations on exterior algebras. The latter were not considered before in the literature (with the obvious exception of [37]), at least to our knowledge. However, since the Schubert derivations to be introduced in the next chapter can be interpreted as Chern polynomials of suitable vector bundles living on Grassmann bundles (which are λ-operations on the Grothendieck group of the vector bundles), the connection with λ-rings becomes more concrete than that suggested by purely formal analogies.

4.5 Exercises

4.5.1. Let $\mathcal{D}(z) \in HS(\bigwedge M)$ and assume that $D_0, D_1, \ldots$ generate a commutative subalgebra of $\mathrm{End}_A(\bigwedge M)$, i.e. $[D_i, D_j] = 0$ for all $i, j \geq 0$. Prove by induction the following generalization of (4.11):

$$D_k^i(\eta \wedge \theta) = \sum_{i_1+\cdots+i_k=i} \binom{i}{i_1, \ldots, i_k} D_k^{i_1} D_{k-1}^{i_2} \cdots D_1^{i_{k-1}} \eta \wedge D_1^{i_2} D_2^{i_3} \cdots D_k^{i_k} \theta$$

More on formulas of this sort can be found in [14].

4.5.2. Let M be a free abelian group of rank n and denote by $M^\vee$ its dual. Let $\mathfrak{B} := (b_0, b_1, \ldots, b_{n-1})$ a basis of M and $\bigwedge^r \mathfrak{B}$ the induced basis of $\bigwedge^r M$. For all $v_1, \ldots, v_r \in M$ and $\alpha_1, \ldots, \alpha_r \in M^\vee$, define

$$\alpha_1 \wedge \cdots \wedge \alpha_r(v_1 \wedge \cdots \wedge v_r) := \det(\alpha_i(v_j)). \tag{4.36}$$

(i) Show that for each $0 \leq r \leq n$ equality (4.36) defines a duality pairing between $\bigwedge^r M$ and $\bigwedge^r M^\vee$.
(ii) If $(\beta_0, \beta_1, \ldots, \beta_{n-1})$ is the basis of $M^\vee$ dual of $(b_0, b_1, \ldots, b_{n-1})$, what is the dual basis of $\bigwedge^r \mathfrak{B}$?
(iii) Let $\mathcal{D}(z)$ be a HS-derivation on $\bigwedge M$. For all $\alpha \in \bigwedge^r M^\vee$, define

$$(\mathcal{D}(z)^T \alpha)(\eta) = \alpha(\mathcal{D}(z)\eta)$$

for all $(\alpha, \eta) \in \bigwedge^r M^\vee \times \bigwedge^r M$. Prove that $\mathcal{D}(z)^T$ is an HS-derivation on $\bigwedge M^\vee$, the *transpose* of $\mathcal{D}(z)$.

4.5.3. Recall the definitions and notation of Exercise 2.8.14. Let $\mathfrak{p}_r(\partial_t)$ be the generic linear ODO of order r, $v_i := \mathcal{L}^{-1}(u_i)$, $-r+1 \leq i \leq 0$, the universal basis of $\ker \mathfrak{p}_r(\partial_t) \subseteq B_r[[t]]$, and $C(e_1, \ldots, e_r)$ be the associated companion matrix.

(i) Using (4.30), find an expression for $\exp(tC(e_1, \ldots, e_r))$ having as entries $B_r \otimes_{\mathbb{Z}} \mathbb{Q}$-linear combinations of $(v_0, v_{-1}, \ldots, v_{-r+1})$ with matrix coefficients (cf. Exercise 2.8.14);
(ii) Show that $\exp(tC(e_1, \ldots, e_r))$ is the Wronskian $W_r(t) := W_{(0)}(\mathbf{v}_r)(t)$, up to a non-zero rational number, of any A-basis of solutions $\mathbf{v}_r$ of the generic linear ODE $\mathfrak{p}_r(\partial_t)y = 0$;
(iii) By general properties of the Wronskian,

$$\partial_t W_r(t) = e_1 W_r(t). \tag{4.37}$$

Conclude from (4.37) that the determinant of the exponential of a companion matrix equals the exponential of its trace;

(iv) Using the fact that all square matrices are conjugated to a block diagonal matrix whose blocks are a companion matrices, conclude that the determinant of the exponential of any matrix is the exponential of its own trace.

For an interesting numerical approach to compute the exponential of a companion matrix, see the 1975 report by Falcidieno and Luvison [28].

Chapter 5
Schubert Derivations

Let A be a commutative ring with unit, $p \in A[X]$ a monic polynomial of degree $n > 0$, M_p the quotient of $A[X]$ modulo (p) and $b_i := X^i + (p)$. The protagonist of this chapter is a distinguished HS-derivation on $\bigwedge M_p$, called *Schubert derivation*. It is the unique HS-derivation $\sigma_+(z) := \sum_{i\geq 0} \sigma_i z^i$ such that $\sigma_i b_j = b_{i+j}$. The subscript '$+$' of σ in the notation is to mark the difference from its relative $\sigma_-(z) := \sum_{i\geq 0} \sigma_{-i} z^{-i}$, to be investigated in Chapter 6. Its connection with Schubert calculus, summarized in Section 5.4, where a few notions of intersection theory are outlined as well, motivates notation and terminology: it is based on Pieri- and Giambelli-type formulas enjoyed by the coefficients of the Schubert derivation. Pieri's rule is treated in Section 5.2, in its quantum version as well, while Giambelli's is proved in Section 5.8, basing on a flexible determinantal formula due to Laksov and Thorup as in [96, 97].

In order to gain practice, Section 5.5 is entirely devoted to examples of manipulations with Schubert derivations, which also carry interpretations in terms of enumerative geometry.

Schubert derivations will be used in Sections 5.6 and 5.7 as a means to equip any free module of infinite countable rank with a structure of free B_r-module of rank r. This will reveal to be an extremely useful tool to compute suitable approximations of the vertex operators, as shown in Chapter 6. The nice interaction between $\sigma_+(z)$ and $\sigma_-(z)$ will pave the way to achieve the elegant characterization, announced in the introduction, of the cone of decomposable tensors in an exterior power.

To end up, Section 5.9 is concerned with the particular, but important, case of Schubert derivations on the exterior algebra of free abelian groups of finite rank.

L. Gatto, P. Salehyan, *Hasse-Schmidt Derivations on Grassmann Algebras*,
IMPA Monographs, DOI 10.1007/978-3-319-31842-4_5

5.1 Generalities on Schubert Derivations

Let $p \in A[X]$ be monic of degree $r \geq 1$, with coefficients in a commutative ring with unit. In the A-module

$$M_p := \frac{A[X]}{(p)}$$

we denote by b_i the residue class $X^i + (p)$. By convention $X^0 = 1_A$. It follows that $(b_0, b_1, \ldots)$ is an M_p-valued LRS with characteristic polynomial p. In particular M_p is a free A-module of rank r with basis $\mathfrak{B}_r := (b_0, b_1, \ldots, b_{r-1})$. The product $X^j(X^i + (p)) := X^{i+j} + (p)$ turns M_p into an $A[X]$-module. By $X^j m$ we shall indicate the product of the ring element X^j with $m \in M_p$.

Definition 5.1.1. The *Schubert derivation* $\sigma_+(z)$ of $\bigwedge M_p$ is the unique *HS*-derivation (Prop. 4.1.13) such that, for all $m \in M_p$,

$$\sigma_+(z)m = \frac{1}{1 - Xz} \cdot m := \sum_{i \geq 0} X^i m \cdot z^i \in M_p[[z]].$$

It will be written in the form

$$\sigma_+(z) = \sum_{i \geq 0} \sigma_i z^i, \tag{5.1}$$

where $\sigma_0 = \mathbb{1}_{\bigwedge M_p}$ and σ_i is the unique endomorphism of $\bigwedge M_p$ satisfying the Leibniz rule (4.15) of order i, with initial conditions $\sigma_i b_j = b_{i+j}$. According to our conventions, the inverse of the *HS*-derivation $\sigma_+(z)$ will be denoted by $\overline{\sigma}_+(z)$, and we set

$$\overline{\sigma}_+(z) = \sum_{i \geq 0} (-1)^i \overline{\sigma}_i z^i.$$

The equality $\sigma_+(z)\overline{\sigma}_+(z) = \overline{\sigma}_+(z)\sigma_+(z) = \mathbb{1}_{\bigwedge M_p}$ implies $\overline{\sigma}_0 = \mathbb{1}_{\bigwedge M_p}$ and the system of quadratic relations

$$\sum_{j \geq 0} (-1)^j \sigma_{i-j} \overline{\sigma}_j = 0, \qquad \forall i \geq 1. \tag{5.2}$$

The latter determine recursively the coefficients of $\overline{\sigma}_+(z)$ in terms of those of $\sigma_+(z)$. For instance, $\overline{\sigma}_1 = \sigma_1$ and $\overline{\sigma}_2 = \sigma_1^2 - \sigma_2$. Once one declares that σ_j has weighted degree j, the general $\overline{\sigma}_i$ will be an explicit homogeneous polynomial in $\sigma_1, \ldots, \sigma_i$ of degree i.

Remarks 5.1.2. a) For every $i,j \geq 0$, $\sigma_i b_j = X^i(X^j + (p)) = X^{i+j} + (p) = b_{i+j}$, so the restriction of σ_i to M_p is the usual *shift operator of step* $i \geq 0$;

b) As $\sigma_i b_j = b_{i+j}$, for all $i,j \geq 0$, it follows that $\sigma_{i|\bigwedge^k M_p}$ is an endomorphism of $\bigwedge^k M_p$, homogeneous of degree i with respect to the weight graduation 3.2.3:

$$\sigma_i(\bigwedge^k M_p)_w \subseteq (\bigwedge^k M_p)_{w+i};$$

c) Let $s_1 := \sigma_{1|M_p}$ be the shift operator of step plus one, i.e. $s_1(b_j) = b_{j+1}$. Then $\overline{\sigma}_+(z)$ is the *trace polynomial operator* (cf. Section 4.2.1) associated with the endomorphism $s_1 \in \mathrm{End}_A(M_p)$. Note that the definition makes sense also when $p = 0$, in which case $M_p = A[X]$;

d) We have $\overline{\sigma}_+(z)m = m - \sigma_1 m \cdot z$ for all $m \in M_p$, because

$$\overline{\sigma}_+(z)m = \frac{1}{\sigma_+(z)} \cdot m = (1 - Xz)m = m - \sigma_1 m \cdot z, \tag{5.3}$$

which in turn implies $\overline{\sigma}_i m = 0$ if $i \geq 2$.

Proposition 5.1.3. *The coefficients of $\sigma_+(z)$ (resp. $\overline{\sigma}_+(z)$) pairwise commute, i.e. for every $\eta \in \bigwedge M_p$ and $i,j \geq 0$,*

$$\sigma_i\sigma_j\eta = \sigma_j\sigma_i\eta, \qquad (\text{resp. } \overline{\sigma}_i\overline{\sigma}_j\eta = \overline{\sigma}_j\overline{\sigma}_i\eta), \tag{5.4}$$

where $\sigma_i\sigma_j$ (resp. $\overline{\sigma}_i\overline{\sigma}_j$) is the composite of σ_i and σ_j, thought of as endomorphisms of $\bigwedge M_p$.

Proof. We have $\sigma_i\sigma_j m = \sigma_1^{i+j} m = \sigma_j\sigma_i m$, for all $i,j \geq 0$ and all $m \in M_p$. Suppose (5.4) holds for all $\eta \in \bigwedge^i M_p$ and all $1 \leq i \leq k-1$. Writing $\eta \in \bigwedge^k M_p$ as $\sum_{\text{finite}} m_h \wedge \theta_h$, it suffices to check that σ_i and σ_j commute on elements of the form $m \wedge \theta$, where $(m, \theta) \in M_p \times \bigwedge^{k-1} M_p$. Using Leibniz's rule (4.3),

$$\begin{aligned}
\sigma_i\sigma_j(m \wedge \theta) &= \sigma_i\Big(\sum_{j_1=0}^{j} \sigma_{j_1} m \wedge \sigma_{j-j_1}\theta\Big) = \sum_{j_1=0}^{j} \sigma_i\big(\sigma_{j_1} m \wedge \sigma_{j-j_1}\theta\big) \\
&= \sum_{j_1=0}^{j}\sum_{i_1=0}^{i} \sigma_{i_1}\sigma_{j_1} m \wedge \sigma_{i-i_1}\sigma_{j-j_1}\theta.
\end{aligned} \tag{5.5}$$

By the inductive hypothesis, the last line of (5.5) equals

$$\sum_{j_1=0}^{j}\sum_{i_1=0}^{i} \sigma_{j_1}\sigma_{i_1} m \wedge \sigma_{j-j_1}\sigma_{i-i_1}\theta = \sigma_j\Big(\sum_{i_1=0}^{i} \sigma_{i_1} m \wedge \sigma_{i-i_1}\theta\Big) = \sigma_j\sigma_i(m \wedge \theta),$$

proving the claim. The proof concerning $\overline{\sigma}_+(z)$ works exactly in the same way. □

5.1.4. Let $A[\mathbf{X}] := A[X_1, X_2, \ldots]$ be the polynomial ring in infinitely many indeterminates over A. Due to Proposition 5.1.3, the map

$$\begin{cases} \mathrm{ev}_{\boldsymbol{\sigma}_+} : A[\mathbf{X}] \longrightarrow \mathrm{End}_A(\bigwedge M_p) \\ \qquad\qquad p(\mathbf{X}) \longmapsto \quad p(\boldsymbol{\sigma}_+), \end{cases}$$

obtained by evaluating p at $X_i = \sigma_i$, is an A-algebra homomorphism. By abuse of notation, we write $A[\boldsymbol{\sigma}_+]$ for the image of $\mathrm{ev}_{\boldsymbol{\sigma}_+}$ in $\mathrm{End}_A(\bigwedge M_p)$. Thus $A[\boldsymbol{\sigma}_+]$ is the minimal commutative A-subalgebra of $\mathrm{End}_A(\bigwedge M_p)$ containing all σ_i $(i \geq 0)$. Similarly, $A[\overline{\boldsymbol{\sigma}}_+]$ arises by evaluating $p \in A[\mathbf{X}]$ on $X_i = \overline{\sigma}_i$. Relations (5.2) imply that each σ_i is a (weighted) homogeneous polynomial expression in $\overline{\sigma}_1, \ldots, \overline{\sigma}_i$ and, conversely, each $\overline{\sigma}_i$ is a (weighted) homogeneous polynomial in $\sigma_1, \ldots, \sigma_i$, so $A[\boldsymbol{\sigma}_+] = A[\overline{\boldsymbol{\sigma}}_+]$. Notice that $A[\boldsymbol{\sigma}_+]$ is a graded A-algebra:

$$A[\boldsymbol{\sigma}_+] = \oplus_{i \geqslant 0} A^i[\boldsymbol{\sigma}_+],$$

where $A^i[\boldsymbol{\sigma}_+]$ is the A-linear span of all monomials of the form $\sigma_1^{i_1} \cdots \sigma_j^{i_j}$ such that $i_1 + 2i_2 + \cdots + ji_j = i$.

Consider the composite map

$$A[\mathbf{X}] \xrightarrow{\mathrm{ev}_{\boldsymbol{\sigma}_+}} \mathrm{End}_A(\bigwedge M_0) \xrightarrow{res} \mathrm{End}_A(\bigwedge^r M_0),$$

where the last map is restriction, and call $A[\boldsymbol{\sigma}_+]_r$ its image. It is just the restriction to $\bigwedge^r M_0$ of the algebra $A[\boldsymbol{\sigma}_+]$ defined above. By abuse of notation, we have denoted by the same symbol (rather than, e.g. $\sigma_{i,r}$ like in Section 5.4) the restriction of σ_i to $\bigwedge^r M_0$.

Proposition 5.1.5. *We have*

$$A[\boldsymbol{\sigma}_+]_r = A[\overline{\sigma}_1, \ldots, \overline{\sigma}_r] := \mathrm{ev}_{\overline{\boldsymbol{\sigma}}_+}(A[X_1, \ldots, X_r]),$$

where we have denoted by the same symbol $\overline{\sigma}_i$ the restriction to $\bigwedge^r M_0$.

Proof. Since $A[\boldsymbol{\sigma}_+]_r$ contains $\overline{\sigma}_1, \overline{\sigma}_2, \ldots, \overline{\sigma}_r$, the inclusion $A[\boldsymbol{\sigma}_+]_r \supseteq A[\overline{\sigma}_1, \ldots, \overline{\sigma}_r]$ is clear. In addition $A[\boldsymbol{\sigma}_+]_r$ is the restriction of a polynomial algebra generated by $(\overline{\sigma}_1, \overline{\sigma}_2, \ldots)$, which implies the reversed inclusion $A[\boldsymbol{\sigma}_+]_r \subseteq A[\overline{\sigma}_1, \ldots, \overline{\sigma}_r]$. In particular $A[\boldsymbol{\sigma}_+]_r$ must be the minimal commutative subalgebra of $\mathrm{End}_A(\bigwedge^r M_0)$ containing $(\sigma_1, \sigma_2, \ldots)$, because $A[\sigma_1, \sigma_2, \ldots]$ is. □

Proposition 5.1.6. *For all integers $i, j \geq 0$ and $\eta \in \bigwedge M_p$, we have*

$$\sigma_i(b_j \wedge \eta) = b_j \wedge \sigma_i \eta + \sigma_{i-1}(b_{j+1} \wedge \eta). \tag{5.6}$$

Proof. By Remark 5.1.2, d), $\overline{\sigma}_i m = 0$ for all $i \geq 2$ and $m \in M_p$. Then

$$\begin{aligned} b_j \wedge \sigma_i \eta &= \sigma_i(b_j \wedge \eta) - \sigma_{j-1}(\overline{\sigma}_1 b_j \wedge \eta) && \text{(integration by parts (4.8))} \\ &= \sigma_i(b_j \wedge \eta) - \sigma_{j-1}(b_{j+1} \wedge \eta). && \text{(definition of } \overline{\sigma}_1 b_j) \end{aligned}$$

Isolating the term $\sigma_i(b_j \wedge \eta)$ yields (5.6). □

Corollary 5.1.7. *The following hold, for all $k \geq 1$ and $i \geq 0$:*

i) $(\overline{\sigma}_+(z)[\mathbf{b}]_0^{k-1}) \wedge b_{k-1} = [\mathbf{b}]_0^k$;
ii) $\sigma_i([\mathbf{b}]_0^k \wedge \eta) = [\mathbf{b}]_0^{k-1} \wedge \sigma_i(b_{k-1} \wedge \eta)$;
iii) $\sigma_i \cdot [\mathbf{b}]_0^k = [\mathbf{b}]_0^{k-1} \wedge b_{k-1+i}$.

Proof. Recall that $\overline{\sigma}_+(z)m = m - \sigma_1 m$ for all $m \in M_p$ again by Remark 5.1.2, d). To prove i) we argue by induction. For $k = 1$, the property is true:

$$(\overline{\sigma}_+(z)b_0) \wedge b_1 = (b_0 - b_1 z) \wedge b_1 = b_0 \wedge b_1.$$

Suppose it holds for all positive integers less than or equal to $k - 1$. Then

$$\begin{aligned} (\overline{\sigma}_+(z)b_0 \wedge \cdots \wedge b_{k-1}) \wedge b_k &= \overline{\sigma}_+(z)(b_0 \wedge \cdots \wedge b_{k-2}) \wedge \overline{\sigma}_+(z)b_{k-1} \wedge b_k \\ &= \overline{\sigma}_+(z)(b_0 \wedge \cdots \wedge b_{k-2}) \wedge (b_{k-1} - b_k z) \wedge b_k \\ &= b_0 \wedge \cdots \wedge b_{k-2} \wedge b_{k-1} \wedge b_k, \end{aligned}$$

as desired. To prove ii), we use integration by parts (4.6) and i):

$$b_0 \wedge \cdots \wedge b_{k-2} \wedge \sigma_+(z)(b_{k-1} \wedge \eta) \tag{5.7}$$

$$\begin{aligned} &= \sigma_+(z)(\overline{\sigma}_+(z)(b_0 \wedge \cdots \wedge b_{k-2}) \wedge b_{k-1} \wedge \eta) \\ &= \sigma_+(z)(b_0 \wedge \cdots \wedge b_{k-1} \wedge \eta). \end{aligned} \tag{5.8}$$

Comparing the coefficients of z^i in (5.7) and (5.8) gives ii). Finally, iii) follows from ii) by putting $\eta = b_{k-1}$. □

Notation 5.1.8. We keep the formal conventions as in 5.1.4 and introduce the following, borrowed from [14]. For each $q \in A[\mathbf{X}]$, let $q(\boldsymbol{\sigma}_+) \in \mathrm{End}_A(\bigwedge M)$ its evaluation at $X_i = \sigma_i$. For all $\boldsymbol{\lambda} \in \mathcal{P}_r$, let

$$\int_n q(\boldsymbol{\sigma}_+)[\mathbf{b}]_{\boldsymbol{\lambda}}^r = \text{coefficient of } [\mathbf{b}]_{((n-r)^r)}^r \text{ in the expansion of } q(\boldsymbol{\sigma}_+)[\mathbf{b}]_{\boldsymbol{\lambda}}^r.$$

So, for instance,

$$\int_4 \sigma_1 [\mathbf{b}]_{(1,1)}^2 = 0 \qquad \text{while} \qquad \int_4 \sigma_1^4 [\mathbf{b}]_0^2 = 2.$$

5.2 Pieri Formula for Schubert Derivations

To introduce the matter of this section, we start with the following:

Example 5.2.1. Let us expand

$$\sigma_1^2[\mathbf{b}]^3_{(1)} \tag{5.9}$$

as linear combination of the basis elements of $\bigwedge^3 M_p$. It must be clearly a linear combination of basis elements of $\bigwedge^3 M_p$ of weight 3, i.e. $[\mathbf{b}]^3_{(1,1,1)}$, $[\mathbf{b}]^3_{(2,1)}$ and $[\mathbf{b}]^3_{(3)}$. The point is to find the coefficients of these elements in the expansion. A straightforward computation using the Leibniz rule tells

$$\begin{aligned}
\sigma_1^2[\mathbf{b}]^3_{(1)} &= \sigma_1^2(b_0 \wedge b_1 \wedge b_3) = \sigma_1(b_0 \wedge \sigma_1(b_1 \wedge b_3)) \\
&= b_1 \wedge \sigma_1(b_1 \wedge b_3) + b_0 \wedge \sigma_1^2(b_1 \wedge b_3) \\
&= b_1 \wedge b_2 \wedge b_3 + b_0 \wedge \sigma_1(b_2 \wedge b_3 + b_1 \wedge b_4) \\
&= b_1 \wedge b_2 \wedge b_3 + b_0 \wedge (2b_2 \wedge b_4 + b_1 \wedge b_5) \\
&= b_1 \wedge b_2 \wedge b_3 + 2b_0 \wedge b_2 \wedge b_4 + b_0 \wedge b_1 \wedge b_5 \\
&= [\mathbf{b}]^3_{(1,1,1)} + 2[\mathbf{b}]^3_{(2,1)} + [\mathbf{b}]^3_{(3)}.
\end{aligned}$$

Similarly $\sigma_2(b_2 \wedge b_3)$ should be a linear combination of $b_3 \wedge b_4$ and $b_2 \wedge b_5$, but the former occurs with null coefficients, because a simple calculation gives

$$\sigma_2(b_2 \wedge b_3) = b_2 \wedge b_5,$$

also as a consequence of 5.1.7 iii).

It is natural to wonder if there is any way to predict which terms survive in an expression $\sigma_i[\mathbf{b}]^r_{\boldsymbol{\lambda}}$, after the cancellations due to the exterior algebra's supercommutativity have taken place. This is answered by the following:

Theorem 5.2.2. *The* Pieri formula *for Schubert derivations holds see [37]:*

$$\sigma_j \cdot [\mathbf{b}]^r_{\boldsymbol{\lambda}} = \sum_{\boldsymbol{\mu} \in \mathcal{P}_r(j,\boldsymbol{\lambda})} [\mathbf{b}]^r_{\boldsymbol{\mu}}, \tag{5.10}$$

where $\mathcal{P}_r(j, \boldsymbol{\lambda}) := \{\boldsymbol{\mu} \in \mathcal{P}_r \mid |\boldsymbol{\mu}| = |\boldsymbol{\lambda}| + j$ and $\mu_1 \geq \lambda_1 \geq \cdots \geq \mu_r \geq \lambda_r\}$.

Proof. Setting $i_{r-j+1} = r - j + \lambda_j$, we write

$$[\mathbf{b}]^r_{\boldsymbol{\lambda}} = b_{i_1} \wedge \cdots \wedge b_{i_r}$$

and define

$$\mathcal{J}_r(j,\boldsymbol{\lambda}) := \left\{ (j_1,\ldots,j_r)\in\mathbb{N}^r \;\middle|\; \begin{array}{c} j_1+\cdots+j_r=j \\ i_1\le i_1+j_1<i_2\le i_2+j_2<\cdots<i_r\le i_r+j_r \end{array} \right\}.$$

Putting $\mu_k = i_{r-k+1}+j_{r-k+1}-r+k = \lambda_k + j_{r-k+1}$, a straightforward computation shows that $(j_1,\ldots,j_r)\in\mathcal{J}_r(j,\boldsymbol{\lambda})$ if and only if $(\mu_1,\ldots,\mu_r)\in\mathcal{P}_r(j,\boldsymbol{\lambda})$. Hence proving (5.10) amounts to showing

$$\sigma_j[\mathbf{b}]^r_{\boldsymbol{\lambda}} = \sum_{(j_1,\ldots,j_r)\in\mathcal{J}_r(j,\boldsymbol{\lambda})} b_{i_1+j_1}\wedge\cdots\wedge b_{i_r+j_r}. \tag{5.11}$$

The formula is obviously true for $r=1$. To show it holds for $r=2$, with $i_1=\lambda_2$ and $i_2=\lambda_1+1$, we argue as follows. Write

$$\sigma_j(b_{i_1}\wedge b_{i_2}) = \sum_{j_1+j_2=j} b_{i_1+j_1}\wedge b_{i_2+j_2} = S+\overline{S},$$

where

$$S := \sum_{(j_1,j_2)\in\mathcal{J}_2(j,\boldsymbol{\lambda})} b_{i_1+j_1}\wedge b_{i_2+j_2}$$

and

$$\overline{S} := \sum_{(j_1,j_2)\notin\mathcal{J}_2(j,\boldsymbol{\lambda})} b_{i_1+j_1}\wedge b_{i_2+j_2}.$$

We claim that $\overline{S}=0$. In fact, on the finite set of integers $i_2-i_1\le a\le i_2-i_1+j$, define the bijection $\rho(a)=i_2-i_1+j-a$. Then

$$2\overline{S} = \sum_{j_1=i_2-i_1}^{h} b_{i_1+j_1}\wedge b_{i_2+j-j_1} + \sum_{j_1=i_2-i_1}^{j} b_{i_1+\rho(j_1)}\wedge b_{i_2+j-\rho(j_1)} =$$

$$= \sum_{j_1=i_2-i_1}^{j} b_{i_2+j-j_1}\wedge b_{i_1+j_1} - \sum_{j_1=i_2-i_1}^{h} b_{i_1+j_1}\wedge b_{i_2+j_2} = 0,$$

hence $\overline{S}=0$ and (5.10) holds for $r=2$. Suppose then (5.10) holds for all $1\le r'\le r-1$. For each $j\ge 0$,

$$\sigma_j(b_{i_1}\wedge\cdots\wedge b_{i_r}) = \sum_{j'_r+j_r=j} \sigma_{j'_r}(b_{i_1}\wedge\cdots\wedge b_{i_{r-1}})\wedge\sigma_{j_r}b_{i_r},$$

and, by inductive hypothesis,

$$\sum_{(j_1,\dots,j_{r-1})} (b_{i_1+j_1} \wedge \cdots \wedge b_{i_{r-2}+j_{r-2}} \wedge b_{i_{r-1}+j_r}) \wedge b_{i_r+j_r}, \tag{5.12}$$

summing over all (j_r) such that $j_1 + \cdots + j_r = j$ and

$$1 \le i_1 + j_1 < i_2 \le \cdots \le i_{r-2} + j_{r-2} < i_{r-1}. \tag{5.13}$$

Formula (5.12) can be written as

$$\sum_{(j_i,j'')} b_{i_1+j_1} \wedge \cdots \wedge b_{i_{r-2}+j_{r-2}} \wedge \sigma_{j''}(b_{i_{r-1}} \wedge b_{i_r}), \tag{5.14}$$

where the sum is over the $(j_1, \dots, j_{r-2}, j'')$ such that $j_1 + \cdots + j_{r-2} + j'' = j$ and satisfying (5.13). Since by assumption

$$\sigma_{j''}(b_{i_{r-1}} \wedge b_{i_r}) = \sum_{\substack{i_{r-1}+j_{r-1}<i_r \\ j_{r-1}+j_r=j''}} b_{i_{r-1}+j_{r-1}} \wedge b_{i_r+j_r},$$

substitution into (5.14) produces exactly (5.11), which is equivalent to (5.10) □

Example 5.2.3. Let $A = \mathbb{Z}[\mathbf{q}]$ and $p := X^4 - \mathbf{q} \in A[X]$, where $\mathbf{q}$ is an indeterminate and M_p as usual. Suppose one wants to compute

$$\sigma_2[\mathbf{b}]^2_{(2,1)} \in \bigwedge^2 M_p.$$

Pieri formula gives

$$\sigma_2[\mathbf{b}]^2_{(2,1)} = [\mathbf{b}]^2_{(4,1)} + [\mathbf{b}]^2_{(3,2)} = b_1 \wedge b_5 + b_2 \wedge b_4. \tag{5.15}$$

Now $b_5 = X^5 + (p) = X\mathbf{q} + (p) = \mathbf{q}b_1$, while $b_4 = X^4 + (p) = \mathbf{q} + (p) = b_0\mathbf{q}$. Substituting into (5.15),

$$\sigma_2[\mathbf{b}]^2_{(2,1)} = b_1 \wedge \mathbf{q}b_1 + b_2 \wedge \mathbf{q}b_0 = -\mathbf{q}b_0 \wedge b_2 = -\mathbf{q}[\mathbf{b}]^2_{(1)}.$$

This $\mathbf{q}$-correction is related to the more general *quantum Pieri formula*:

Theorem 5.2.4. *Let* $A := \mathbb{Z}[\mathbf{q}]$ *and* $M_p := A[X]/(X^n - \mathbf{q})$, $\boldsymbol{\lambda} \in \mathcal{P}_{r,n}$ *and write* $[\mathbf{b}]^r_{\boldsymbol{\lambda}}$ *in the form*

$$b_{i_1} \wedge \cdots \wedge b_{i_r}.$$

Then for every $0 \leq j \leq n-1$,

$$\sigma_j[\mathbf{b}]^r_{\boldsymbol{\lambda}} = \sum_{\boldsymbol{\mu}\in\mathcal{P}_r(j,\boldsymbol{\lambda})\cap\mathcal{P}_{r,n}} [\mathbf{b}]^r_{\boldsymbol{\mu}} + (-1)^{r-1}\mathbf{q}\sum_{\substack{J\in\mathcal{J}_r(j,\boldsymbol{\lambda})\\ i_r+j_r-n<i_1}} b_{i_r+j_r-n}\wedge b_{i_2+j_2}\wedge\cdots\wedge b_{i_{r-1}+j_{r-1}}. \tag{5.16}$$

Proof. Split expression (5.10) as the sum over all partitions $\boldsymbol{\mu}\in\mathcal{P}_r(j,\boldsymbol{\lambda})$ contained in an $r\times(n-r)$ rectangle plus the rest:

$$\sigma_j[\mathbf{b}]^r_{\boldsymbol{\lambda}} = \sum_{\boldsymbol{\mu}\in\mathcal{P}_r(j,\boldsymbol{\lambda})\cap\mathcal{P}_{r,n}} [\mathbf{b}]^r_{\boldsymbol{\mu}} + \sum_{\boldsymbol{\mu}\in\mathcal{P}_r(j,\boldsymbol{\lambda})\setminus\mathcal{P}_{r,n}} [\mathbf{b}]^r_{\boldsymbol{\mu}}.$$

Using the same notation of 5.2.2, it turns out that

$$\sum_{\boldsymbol{\mu}\in\mathcal{P}_r(j,\boldsymbol{\lambda})\setminus\mathcal{P}_{r,n}} [\mathbf{b}]^r_{\boldsymbol{\mu}} = \sum_{J\in\mathcal{J}_r(j,\boldsymbol{\lambda})\,|\,i_r+j_r-n\geq 0} b_{i_1+j_1}\wedge\cdots\wedge b_{i_r+j_r}, \tag{5.17}$$

where J is the multi-index $(j_1,\ldots,j_r)$. Using the $\mathbb{Z}_2$-symmetry of $\wedge$, formula (5.17) can be written as $(-1)^{r-1}\mathbf{q}(C+\overline{C})$, where

$$(-1)^{r-1}\mathbf{q}C := (-1)^{r-1}\mathbf{q}\sum_{\substack{J\in\mathcal{J}_r(j,\boldsymbol{\lambda})\\ i_r+j_r-n<i_1}} b_{i_r+j_r-n}\wedge b_{i_1+j_1}\wedge b_{i_2+j_2}\wedge\cdots\wedge b_{i_{r-1}+j_{r-1}}$$

is exactly the second summand of (5.16), while

$$\begin{aligned}\overline{C} &:= \sum_{\substack{J\in\mathcal{J}(j,\boldsymbol{\lambda})\\ i_r+j_r-n\geq i_1}} b_{i_r+j_r-n}\wedge b_{i_1+j_1}\wedge b_{i_2+j_2}\wedge\cdots\wedge b_{i_{r-1}+j_{r-1}}\\ &= \sum_{j'=0}^{j}\sum_{j_r=i_1+n-i_r}^{h'} b_{i_r+j_r-n}\wedge b_{i_1+j'-j_r}\wedge\sigma_{j-j'}(b_{i_2}\wedge\cdots\wedge b_{i_r-1}).\end{aligned} \tag{5.18}$$

For each $0\leq j'\leq j$, define $\rho_{j'}(a) = i_1+n+j'-i_r-a$, a bijection of the set $\{a\in\mathbb{N} : i_1+n-i_r\leq a\leq j'\}$ on itself. Then expression (5.18) also reads:

$$\begin{aligned}\overline{C} &= \sum_{j'=0}^{j}\sum_{j_r=i_1+n-i_r}^{j'} b_{i_r+\rho_{j'}(j_r)-n}\wedge b_{i_1+j'-\rho_{j'}(j_r)}\wedge\sigma_{j-j'}(b_{i_2}\wedge\cdots\wedge b_{i_r-1}) =\\ &= \sum_{j'=0}^{h}\sum_{j_r=i_1+n-i_r}^{j'} b_{i_1+j_1}\wedge b_{i_r+j_r-n}\wedge\sigma_{j-j'}(b_{i_2}\wedge\cdots\wedge b_{i_r-1}) = -\overline{C},\end{aligned}$$

implying $\overline{C}=0$ and thus completing the proof of (5.16). $\square$

5.3 The Grassmannian and Its Plücker Embedding

5.3.1. In all the rest of the book, we shall denote by

$$G_{r,n} := G_{r,n}(\mathbb{C})$$

the complex Grassmann variety parametrizing r-dimensional subspaces in $\mathbb{C}^n$. It is a smooth complex variety of dimension $r(n\text{-}r)$, and it can be abstractly constructed as the quotient of the (noncompact) *Stiefel manifold*[1] $\mathcal{U}_{r,n}(\mathbb{C})$, of all the $n \times r$ complex matrices of maximal rank, modulo the right action of $Gl_r(\mathbb{C})$. If $\Lambda \in \mathcal{U}_{r,n}$, let $(\mathbf{e}_1, \mathbf{e}_2, \ldots, \mathbf{e}_n)$ be the canonical basis of $\mathbb{C}^n$ and $(\epsilon^1, \epsilon^2, \ldots, \epsilon^n)$ be the basis of $(\mathbb{C}^n)^\vee$, dual of $(\mathbf{e}_i)$. The elements of the latter can be seen as sections of the dual $\mathcal{S}_r^\vee$ of the tautological bundle $\mathcal{S}_r \to G_{r,n}$ (see 5.4.7).

For all $\boldsymbol{\lambda} \in \mathcal{P}_r$, let $[\boldsymbol{\epsilon}]^r_{\boldsymbol{\lambda}} := \epsilon^{1+\lambda_r} \wedge \cdots \wedge \epsilon^{r+\lambda_1} \in \bigwedge^r \mathcal{S}_r^\vee$, and, if $\Lambda \in \mathcal{U}_{r,n}$, let

$$[\boldsymbol{\epsilon}]^r_{\boldsymbol{\lambda}}(\Lambda) = \det\begin{pmatrix} R_{1+\lambda_r}(\Lambda) \\ \vdots \\ R_{r+\lambda_1}(\Lambda) \end{pmatrix} \in \mathbb{C},$$

where $R_j(\Lambda)$ denotes the jth row of the matrix Λ. The map

$$\begin{cases} \widehat{Pl} : \mathcal{U}_{r,n} & \longrightarrow & \mathbb{C}^{\binom{n}{r}} \\ \qquad \Lambda & \longmapsto & ([\boldsymbol{\epsilon}]^r_{\boldsymbol{\lambda}}(\Lambda))_{\boldsymbol{\lambda} \in \mathcal{P}_{r,n}} \end{cases}$$

is $Gl_r(\mathbb{C}) - \mathbb{C}^*$ equivariant, i.e. $\widehat{Pl}(\Lambda \cdot P) = \det(P) \cdot \widehat{Pl}(\Lambda)$, for all $P \in Gl_r(\mathbb{C})$. Thus we are given with a well-defined map

$$\begin{cases} Pl : G_{r,n} \longrightarrow \mathbb{P}^{\binom{n}{r}-1} \\ \qquad [\Lambda] \longmapsto ([\boldsymbol{\epsilon}]^r_{\boldsymbol{\lambda}}(\Lambda))_{\boldsymbol{\lambda} \in \mathcal{P}_{r,n}} \bmod \mathbb{C}^*, \end{cases} \tag{5.19}$$

that can be proved to be an embedding [110, p. 102]. It realizes the Grassmannian as a complex projective subvariety of $\mathbb{P}^{\binom{n}{r}-1}$ whose degree, the number of intersection points with a general linear space of codimension $r(n-r)$, will be expressed in Section 5.4.9 in terms of Schubert calculus.

5.3.2. The map (5.19) is called the *Plücker embedding*. Its most popular example is that of the Grassmannian $G_{2,4}$, realized as a quadric projective hypersurface of $\mathbb{P}^5$ as follows. Let $X_{\boldsymbol{\lambda}}(\Lambda) := [\boldsymbol{\epsilon}]^2_{\boldsymbol{\lambda}}(\Lambda)$, and consider the map

[1] After the Swiss mathematician Eduard Stiefel, 21 April 1909–25 November 1978

$$(X_0 : X_1 : X_{(1,1)} : X_2 : X_{(2,1)} : X_{(2,2)}) : G_{2,4} \longrightarrow \mathbb{P}^5,$$

sending $[\Lambda]$ to $(X_0(\Lambda) : X_1(\Lambda) : X_{(1,1)}(\Lambda) : X_2(\Lambda) : X_{(2,1)}(\Lambda) : X_{(2,2)}(\Lambda))$. It is a standard exercise to see that each $[\Lambda] \in G_{2,4}$ satisfies the equation of *Klein quadric* hypersurface:

$$X_0 X_{(2,2)} - X_1 X_{(2,1)} + X_2 X_{(1,1)} = 0. \tag{5.20}$$

The same equation (5.20) read in the *six*-dimensional affine space is said to be the equation of the *affine Grassmann cone* over $G_{2,4}$.

5.3.3. Another way to construct $G_{r,n}$, which is related to the main content of Section 6.4, is to bring it into being directly inside its Plücker embedding.

Writing an element of $\mathfrak{U}_{r,n}$ as the row of its r columns,

$$\Lambda = (C_1(\Lambda), \ldots, C_r(\Lambda)),$$

one is given with the natural map $\mathfrak{U}_{r,n} \to \bigwedge^r \mathbb{C}^n$:

$$\Lambda \longmapsto C_1(\Lambda) \wedge \cdots \wedge C_r(\Lambda) \in \bigwedge^r \mathbb{C}^n.$$

whose image is the affine cone of $\bigwedge^r \mathbb{C}^n$ formed by all the decomposable tensors. Obviously, this is $Gl_r(\mathbb{C}) - \mathbb{C}^*$ equivariant as well, so that the Grassmannian is the projective variety associated with the cone of decomposable tensors in $\bigwedge^r \mathbb{C}^n$. In Section 6.4 we shall learn how to write equations for the locus of decomposable tensors in the rth exterior power of a free abelian group (and hence those of the Plücker embedding of the Grassmannian) in a very a uniform way. The shape of the equations will make apparent that all the equations are quadratic in the Plücker coordinates.

5.4 Relationship with Schubert Calculus

This section is not a prerequisite to follow the rest of the exposition, so it may be skipped with no harm done. However it anchors to the algebraic results of forthcoming sections. Classical Schubert calculus rules the intersection theory on Grassmann bundles. We give here a quick account to illustrate its relationships with the Schubert derivation on a Grassmann algebra. Detailed references for this section are [34, 145], not to mention the classical renowned paper by Kleiman and Laksov [87]. For relationships with Schubert derivations or with the algebra of exterior powers, see [37, 38, 47, 96, 97].

5.4.1. To begin with, recall that each smooth complex projective variety $\mathcal{X}$ of dimension $n \geq 1$ can be endowed with an *intersection theory* [34, Ch. 8]. Roughly speaking, that amounts to attaching to $\mathcal{X}$ an abelian group $A_*(\mathcal{X})$ and a $\mathbb{Z}$-algebra $A^*(\mathcal{X})$ which are isomorphic as $\mathbb{Z}$-modules (Poincaré duality). The group $A_*(\mathcal{X})$ is generated by *cycles modulo rational equivalence* and is a rank 1 free module over $A^*(\mathcal{X})$, generated by the *fundamental class* $[\mathcal{X}]$. Rational equivalence is a generalization of linear equivalence [34, Sect. 1.3]: so, for instance, two skew lines in $\mathbb{P}^3$ are rationally equivalent but not linearly equivalent (they are not even divisors). The module multiplication $A^*(\mathcal{X}) \otimes_{\mathbb{Z}} A_*(\mathcal{X}) \stackrel{\cap}{\to} A_*(\mathcal{X})$ is called *cap product* and maps $\alpha \otimes c$ to the class $\alpha \cap c \in A^*(\mathcal{X})$, which in many cases (but not all) can be geometrically interpreted as follows: if V_1 and V_2 are representatives of α and c, respectively, i.e. $\alpha = [V_1]$ and $c = [V_2]$, whose intersection is proper[2], then $\alpha \cap c = [V_1] \cap [V_2] = [V_1 \cap V_2]$, the class of the cycle $V_1 \cap V_2$ (an integral linear combination of the irreducible components of the intersection, see [34, 145]) modulo rational equivalence. Denote by $A_i(\mathcal{X}) \subset A_*(\mathcal{X})$ the subgroup of i-dimensional subvarieties of $\mathcal{X}$.

5.4.2. The easiest example of intersection theory is certainly that of the complex projective n-dimensional space. It turns out that all closed subvarieties of the same codimension and degree are rationally equivalent. In fact one can show that each subvariety V of codimension i is rationally equivalent to $d \cdot H^i$, where H^i is the class of any linear subvariety of codimension i. Here H is the class of a hyperplane and d is the *degree* of V, namely, the number of intersections of V with a general linear subspace of complementary codimension (e.g. [34, 145]). We have

$$A_*(\mathbb{P}^n) = \mathbb{Z}[\mathbb{P}^n] \oplus \mathbb{Z} \cdot H \oplus \mathbb{Z} \cdot H^2 \oplus \cdots \oplus \mathbb{Z} \cdot H^n.$$

The class $[\mathbb{P}^n]$ is called the *fundamental class*, and H^n is the *class of a point*. The Chow ring of $\mathbb{P}^n$ is generated as a $\mathbb{Z}$-algebra by the class H:

$$A^*(\mathbb{P}^n) = \frac{\mathbb{Z}[H]}{(H^{n+1})}. \tag{5.21}$$

This ring prescribes the dimension and the degree of the intersection of a hyperplane and a projective subvariety of $\mathbb{P}^n$ in general position. The *cap product* $H \cap [S] = [H_1 \cap S]$ operates by *cutting* S *with a general hyperplane* H_1. The relation $H^{n+1} = 0$ expresses the fact that the intersection of $(n+1)$-hyperplanes in general position is empty.

If $n = 1$, then H is the class of a point, and (5.21) says that two general points on the projective line are distinct! If $n = 2$, it expresses the celebrated *Bézout theorem*, according which two general projective curves in $\mathbb{P}^2$ of degrees d_1 and d_2,

[2] Two subvarieties V_1 and V_2 are said to be properly intersecting if the codimension of each irreducible component of $V_1 \cap V_2$ is the sum of the codimensions.

with no common component, intersect at d_1d_2 points, counted with an appropriate multiplicity. If S_1 and S_2 are closed subvarieties of $\mathbb{P}^n$ in general position, of codimensions i_1 and i_2 and degrees d_1 and d_2, respectively, then $[S_1] = d_1H^{i_1} \cap [\mathbb{P}^n]$, $[S_2] = d_2H^{i_2} \cap [\mathbb{P}^n]$ and

$$[S_1 \cap S_2] = d_1H^{i_1} \cap [S_2] = d_1H^{i_1} \cap (d_2H^{i_2} \cap [\mathbb{P}^n]) = d_1d_2H^{i_1+i_2} \cap [\mathbb{P}^n],$$

that is, two subvarieties of $\mathbb{P}^n$ of degrees d_1 and d_2 in general position do intersect along a subvariety of degree d_1d_2 whose codimension is the sum of the codimensions. This is Bézout's theorem for $\mathbb{P}^n$. The problems it solves are not different in nature from that of determining the *number of lines in complex projective three-space meeting other four in general position*. See below.

5.4.3. Schubert calculus on $G_{r,n}$ generalizes Bézout's theorem for $\mathbb{P}^{n-1} := G_{1,n}$, i.e. it is concerned with the product structure of the intersection ring $A^*(G_{r,n})$. Let $H_*(G_{r,n}, \mathbb{Z})$ and $H^*(G_{r,n}, \mathbb{Z})$ be the singular cohomology and homology of $G_{r,n}$ with integral coefficients. Recall that any complete flag

$$F^\bullet : \ \mathbb{C}^n = F^0 \supset F^1 \supset \cdots \supset F^n = \{\mathbf{0}\}$$

of subspaces of $\mathbb{C}^n$, indexed by the codimension, decomposes $G_{r,n}$ into a disjoint union of affine cells

$$\overset{\circ}{\Omega}_{\boldsymbol{\lambda}}(F^\bullet) := \{\Lambda \in G_{r,n} \mid \dim(F^{r-i+\lambda_i} \cap \Lambda) = i\},$$

parametrized by partitions $\boldsymbol{\lambda} \in \mathcal{P}_{r,n}$ (notation as in Section 2.2.4). This fact guarantees that $A^*(G_{r,n}) \to H^*(G_{r,n}, \mathbb{Z})$, given by $[V]_r \mapsto [V]_c$, where V is a subvariety of $G_{r,n}$, defines a ring isomorphism (see, e.g. [34, Example 19.1.11]) yielding a canonical identification of the intersection theory of the Grassmann variety with its singular cohomology theory. The Chow class $\Omega_{\boldsymbol{\lambda}} \in A_{r(n-r)-|\boldsymbol{\lambda}|}(G_{r,n})$ of the *Schubert variety*

$$\Omega_{\boldsymbol{\lambda}}(F^\bullet) := \overline{\overset{\circ}{\Omega}_{\boldsymbol{\lambda}}(F^\bullet)} := \{\Lambda \in G_{r,n} \mid \dim(F^{r-i+\lambda_i} \cap \Lambda) \geq i\}$$

has codimension $|\boldsymbol{\lambda}|$ and does not depend on the chosen flag. It can be also seen as an element of $H_{2r(n-r)-2|\boldsymbol{\lambda}|}(G_{r,n}, \mathbb{Z})$, where $H_*(G_{r,n}, \mathbb{Z})$ denotes the singular homology. General results in algebraic topology (see, e.g. [55, p. 196]) imply that, in our specific situation,

$$H_*(G_{r,n}, \mathbb{Z}) = A_*(G_{r,n}) = \bigoplus_{\boldsymbol{\lambda} \in \mathcal{P}_{r,n}} \mathbb{Z} \cdot \Omega_{\boldsymbol{\lambda}}.$$

Let $\Omega_0 := [G_{r,n}]$ denotes the rational equivalence class (or homology class) of the whole $G_{r,n}$, called the *fundamental class*. For $r = 1$, one has $\Omega_0 = [\mathbb{P}^{n-1}]$. By *Poincaré duality* the cap product (e.g. [114, p. 276])

$$\begin{cases} A^*(G_{r,n},\mathbb{Z}) \otimes_{\mathbb{Z}} A_*(G_{r,n},\mathbb{Z}) \stackrel{\cap}{\longrightarrow} A^*(G_{r,n},\mathbb{Z}) \\ \\ \sigma \otimes \Omega \qquad\qquad \longmapsto \quad \sigma \cap \Omega \end{cases} \tag{5.22}$$

turns $A_*(G_{r,n})$ into a free $A^*(G_{r,n})$-module of rank 1 generated by Ω_0. If $\sigma_\lambda \in A^*(G_{r,n})$ is such that $\sigma_\lambda \cap \Omega_0 = \Omega_\lambda$, then an important theorem by Kleiman [86] guarantees that given any other partition $\boldsymbol{\mu}$, there exist flags $F^\bullet$ and $G^\bullet$ such that the intersection $\Omega_\lambda(F^\bullet) \cap \Omega_\mu(G^\bullet)$ is proper, so that

$$\sigma_\lambda \cap \Omega_\mu = [\Omega_\lambda \cap \Omega_\mu].$$

5.4.4. With a friendly eye to the reader not particularly familiar with this language, recall that a holomorphic vector bundle $\rho : \mathcal{E} \to \mathcal{X}$ of rank n over a smooth connected projective variety $\mathcal{X}$ of dimension $m \geq 0$ may be roughly thought of as a (in general nontrivial) holomorphically varying family $\{\mathcal{E}_x \,|\, x \in \mathcal{X}\}$ of complex vector spaces parametrized by the points of $\mathcal{X}$. The family is required to be locally trivial, e.g. in the holomorphic category. This simply means that around each point $x \in \mathcal{X}$, there is a neighbourhood U_x of x such that $\rho^{-1}(U_x)$ is bi-holomorphically equivalent to $U_x \times \mathbb{C}^n$. We shall overlook the compatibility conditions relative to overlapping trivializing neighbourhoods. If m=0, the variety $\mathcal{X}$ is a point and $\mathcal{E}$ a honest complex vector space of dimension n with trivial projection to a point. For all $0 \leq r \leq n$, a standard construction permits to associate to $\rho : \mathcal{E} \to \mathcal{X}$ the *Grassmann bundle* $\rho_r : G_r(\mathcal{E}) \to \mathcal{X}$. It parametrizes r-dimensional subspaces in the fibres of ρ:

$$G_r(\mathcal{E})_x = G_r(\mathcal{E}_x).$$

Notice that $\rho_0 = \rho$. The usual Grassmannian is recovered by taking $\mathcal{X}$ to be a point. On $G_r(\mathcal{E})$ sits the *universal exact sequence*,

$$0 \longrightarrow \mathcal{S}_r \longrightarrow \rho_r^*\mathcal{E} \longrightarrow \mathcal{Q}_r \longrightarrow 0,$$

understood in the category of vector bundles over $G_r(\mathcal{E})$. Here $\rho_r^*\mathcal{E}$ is the vector bundle of rank n on $G_r(\mathcal{E})$, defined in such a way that its fibre over each $[\Lambda] \in G_r(\mathcal{E}_x)$ is $\mathcal{E}_x$, while $\mathcal{S}_r$ is its *universal (or tautological) subbundle*: the fibre of $\mathcal{S}_r$ over a point $[\Lambda] \in G_r(\mathcal{E}_x)$ is precisely the r-dimensional subspace $[\Lambda]$ of $\mathcal{E}_x$. Let $\mathbb{P}(\mathcal{E}) := G_1(\mathcal{E})$, and $\xi \in A^*(\mathbb{P}(\mathcal{E}))$ an $A^*(\mathcal{X})$-endomorphism of $A_*(\mathbb{P}(\mathcal{E}))$, operating on classes $c \in A_*(\mathbb{P}(\mathcal{E}))$ returning the class of a general representative of c fibrewise cut with a hyperplane. The fact that the fibres of $\rho_1 : \mathbb{P}(\mathcal{E}) \to \mathcal{X}$ are $(n-1)$-dimensional implies the linear dependence of $1, \xi, \ldots, \xi^n$, satisfying the linear relation below [59, p. 429]:

$$0 = p_{\mathcal{E}}(\xi) := \xi^n - c_1(\mathcal{E})\xi^{n-1} + \cdots + (-1)^n c_n(\mathcal{E})[\mathbb{P}(\mathcal{E})] \in A^*(X)[\xi],$$

which defines the *Chern classes* of the vector bundle $\mathcal{E}$. The module multiplication $c_i(\mathcal{E})\xi^i$ is meant as $(\rho^* c_i(\mathcal{E}))\xi$, where $\rho^* : A^*(\mathcal{X}) \to A^*(\mathbb{P}(\mathcal{E}))$ is a suitable (flat) *pull-back map*. Glossing over important details, it follows that the following generalization of Bézout's theorem holds for $\mathbb{P}(\mathcal{E})$:

$$A^*(\mathbb{P}(\mathcal{E})) = \frac{A^*(\mathcal{X})[\xi]}{(p_{\mathcal{E}}(\xi))}.$$

Let now $p := p_{\mathcal{E}}$ and $A := A^*(\mathcal{X})$ so that $M_p := A^*(\mathcal{X})[\xi]$. Then $b_i := \xi^i \cap [\mathbb{P}(\mathcal{E})]$ are generators and recall that $(b_0, b_1, \ldots, b_n)$ is an $A^*(\mathcal{X})$ basis of M_p. Write formally

$$c_t(\mathcal{Q}_r - \rho_r^* \mathcal{E}) = \sum_{i \geq 0} c_i((\mathcal{Q}_r - \rho^* \mathcal{E}))t^i := \frac{c_t(\mathcal{Q}_r)}{c_t(\rho_r^* \mathcal{E})}$$

$$= (1 + c_1(\mathcal{Q}_r)t + c_2(\mathcal{Q}_r)t^2 + \cdots)(1 - c_1(\rho_r^* \mathcal{E})t + (c_1(\rho_r^* \mathcal{E})^2 - c_2(\rho_r^* \mathcal{E}))t^2 + \cdots)$$

$$= 1 + (c_1(\mathcal{Q}_r)t - c_1(\rho_r^* \mathcal{E}))t + (c_2(\mathcal{Q}_r) - c_1(\mathcal{Q}_r)c_1(\rho_r^* \mathcal{E}) - c_2(\rho_r^* \mathcal{E}))t^2 + \cdots.$$

The difference $\mathcal{Q}_r - \rho_r^* \mathcal{E}$ is to be understood in the *Grothendieck group* of vector bundles over $G_r(\mathcal{E})$, and $c_t(\mathcal{Q}_r - \rho_r^* \mathcal{E})$ is said to be *Chern polynomial*. The reader who feels lost may informally think of $c_i(\mathcal{Q}_r - \rho^* \mathcal{E}) \cap [G(r, \mathcal{E})]$ as the class of the closure of the subvariety fibrewise cutting r-dimensional subspaces that intersect nontrivially a subspace of codimension $r-i+1$. By the basis theorem ([34, Ch. 14]), the Chow group $A_*(G_r(\mathcal{E}))$ is a free $A^*(\mathcal{X})$-module generated by

$$(\Delta_{\boldsymbol{\lambda}}(c_t(\mathcal{Q}_r - \rho^* \mathcal{E})) \cap [G_r(\mathcal{E})] \mid \boldsymbol{\lambda} \in \mathcal{P}_{r,n}),$$

where the Schur determinants above are calculated with respect to the sequence of the coefficients of $c_t(\mathcal{Q}_r - \rho^* \mathcal{E})$. The intersection theory of $G_r(\mathcal{E})$ is ruled by the *Pieri*-type formula enjoyed by Schur determinants (see [34, Appendix A]):

$$c_i(\mathcal{Q}_r - \rho^* \mathcal{E}) \cap \left(\Delta_{\boldsymbol{\lambda}}(c_t(\mathcal{Q}_r - \rho^* \mathcal{E})) \cap [G_r(\mathcal{E})]\right)$$

$$= \sum_{\boldsymbol{\mu} \in \mathcal{P}_{r,n,\boldsymbol{\lambda}}} \Delta_{\boldsymbol{\mu}}(c_t(\mathcal{Q}_r - \rho^* \mathcal{E})) \cap [G_r(\mathcal{E})]. \tag{5.23}$$

The main result of the paper [37], see also [47] and [38], can be phrased as follows:

Theorem 5.4.5. *Let $A[\boldsymbol{\sigma}_+]_r$ be the commutative subring of $\mathrm{End}_A(\bigwedge^r M_p)$ obtained by restricting $A[\boldsymbol{\sigma}_+]$ to $\bigwedge^r M_p$. Denote by $\sigma_{i,r}$ the restriction to $\bigwedge^r M_p$ of the coefficient σ_i of z^i of the Schubert derivation $\sigma_+(z)$. Let $\iota : A^*(G_r(\mathcal{E})) \to \mathbb{Z}[\boldsymbol{\sigma}_+]$ and $j : A_*(G_r(\mathcal{E})) \to \bigwedge^r M_p$ defined as*

$$c_i(\mathcal{Q}_r - \rho_r^* \mathcal{E}) \mapsto \sigma_{i,r} \qquad \text{and} \qquad \Delta_{\boldsymbol{\lambda}}(c_t(\mathcal{Q}_r - \rho^* \mathcal{E})) \mapsto [\mathbf{b}]^r_{\boldsymbol{\lambda}}.$$

Then the diagramme

$$\begin{array}{ccc} A^*(G_r(\mathcal{E}))\otimes_{\mathbb{Z}} A_*(G_r(\mathcal{E})) & \longrightarrow & A_*(G_r(\mathcal{E})) \\ \iota\otimes j\big\downarrow & & j\big\downarrow \\ A[\boldsymbol{\sigma}_+]\otimes_{\mathbb{Z}} \bigwedge^r M_p & \longrightarrow & \bigwedge^r M_p \end{array} \tag{5.24}$$

commutes and the vertical maps are isomorphisms.

Proof. The fact that j is a module isomorphism follows from the basis theorem as in [34, p. 268]. To show that ι is a module isomorphism too, we should invoke the Giambelli formula for Schubert derivations, according to which $[\mathbf{b}]^r_{\boldsymbol{\lambda}} = \Delta_{\boldsymbol{\lambda}}(\sigma_+)[\mathbf{b}]^r_0$, a fact whose proof we postpone to Section 5.8. Showing that $\iota\otimes j$ is a $A^*(G_{r,n}) - \mathbb{Z}[\boldsymbol{\sigma}_+]$-module homomorphism is easy and amounts to proving the commutativity of diagramme (5.27), for which we invoke Pieri's formula for Schubert derivations (5.10) and (5.23), that we assumed known. In fact, through the bottom horizontal homomorphism, that maps $\sigma_h\otimes[\mathbf{b}]^r_{\boldsymbol{\lambda}}$ to $\sigma_h[\mathbf{b}]^r_{\boldsymbol{\lambda}}$:

$$\iota\otimes j\big(c_h(\mathcal{Q}_r-\rho_r^*\mathcal{E})\otimes(\Delta_{\boldsymbol{\lambda}}(c_t(\mathcal{Q}-\rho_r^*\mathcal{E})\cap[G_r(\mathcal{E})])\big)\mapsto\sigma_h[\mathbf{b}]^r_{\boldsymbol{\lambda}}=\sum_{\boldsymbol{\mu}\in\mathcal{P}_r(h,\boldsymbol{\mu})}[\mathbf{b}]^r_{\boldsymbol{\mu}}.$$

On the other hand

$$\begin{aligned} & j(c_h(\mathcal{Q}_r-\rho_r^*\mathcal{E})\cap\Delta_{\boldsymbol{\lambda}}(c_t(\mathcal{Q}_r-\rho_r^*\mathcal{E})\cap[G_r(\mathcal{E})])) \\ =\; & j\Big(\sum_{\boldsymbol{\mu}\in\mathcal{P}_r(h,\boldsymbol{\lambda})}c_{\boldsymbol{\mu}}(\mathcal{Q}_r-\rho_r^*\mathcal{E})\cap[G_r(\mathcal{E})]\Big) \\ =\; & \sum_{\boldsymbol{\mu}\in\mathcal{P}_r(h,\boldsymbol{\lambda})}[\mathbf{b}]^r_{\boldsymbol{\mu}}, \end{aligned}$$

and the commutativity of diagramme (5.24) is proven. □

Remark 5.4.6. The main pedagogical outcome of the above theorem is that at the top level of diagram (5.24) lives a well-established and deep theory, which alas needs many prerequisites and requires one to work out several details. Theorem 5.4.5 says that making computations at the 'difficult' top level is the same as performing them at the bottom level, which is elementary, given that the basic computational recipe is the Leibniz rule enjoyed by Schubert derivations. Notice that the bottom line of (5.24) is the algebraic translation of the *Poincaré duality* for Grassmann bundles, for it phrases the fact that $A^*(G_r(\mathcal{E}))$ is isomorphic as $A^*(\mathcal{X})$-module to $A_*(G_r(\mathcal{E}))$ via the *cap product*. The theorem also shows that $\bigwedge^r M_p\cong\bigwedge^r A^*(\mathbb{P}(\mathcal{E}))$ is a module over the intersection ring of the Grassmann bundle, which is the content of the so-called *Satake identification* in the sense of [36, 54].

5.4.7. If $\mathcal{X} = \{\bullet\}$ is a point, we obtain classical Schubert calculus for the Grassmannians. In this case, in fact, $A^*(\bullet) = \mathbb{Z}$, $\mathcal{E} \to \{\bullet\}$ is the same as an n-dimensional vector space, $G_r(\mathcal{E})$ is the Grassmannian $G_{r,n}$ (up to the identification $\mathcal{E} \cong \mathbb{C}^n$) and $\rho_r^*\mathcal{E}$ is the trivial vector bundle $G_{r,n} \times \mathbb{C}^n$. The tautological exact sequence reads

$$0 \longrightarrow \mathcal{S}_r \longrightarrow \mathcal{O}_{G_{r,n}}^{\oplus n} \longrightarrow \mathcal{Q}_r \longrightarrow 0,$$

where $\mathcal{S}_r := \{([\Lambda], \mathbf{v}) \in G_{r,n} \times \mathbb{C}^{n+1}) \mid \mathbf{v} \in [\Lambda]\}$ is the *universal* rank r subbundle of the trivial bundle $\mathcal{O}_{G_{r,n}}^{\oplus n}$ and $\mathcal{Q}_r := \mathcal{O}_{G_{r,n}}^{\oplus n}/\mathcal{S}_r$ is the *universal quotient bundle* over $G_{r,n}$. The cap product $c_i(\mathcal{Q}_r) \cap \Omega_0$ is, in this instance, precisely the class $\Omega_{(i)}$ of all $\Lambda \in G_{r,n}$ intersecting nontrivially a fixed linear subspace of codimension $r - 1 + i$. The product of two Chern classes in $A^*(G_{r,n})$ can be defined via

$$c_i(\mathcal{Q}_r)c_j(\mathcal{Q}_r) \cap \Omega_{\boldsymbol{\lambda}} = c_i(\mathcal{Q}_r) \cap (c_j(\mathcal{Q}_r) \cap \Omega_{\boldsymbol{\lambda}}).$$

Clearly $c_j(\mathcal{Q}_r) = 0$ if $j > n - r$, as $\Omega_{(j)}$ would correspond to the class of r-dimensional subspaces intersecting nontrivially a fixed subspace of codimension $r - 1 + j \geq r - 1 + n - r + 1 = n$, which is empty (the class of the empty set is 0 by definition).

Let $\Delta_{\boldsymbol{\lambda}}(c(\mathcal{Q}_r)) \in A^*(G_{r,n})$ be the Schur polynomial associated with the sequence $c(\mathcal{Q}_r) = (c_0(\mathcal{Q}_r), c_1(\mathcal{Q}_r), \ldots)$ and to $\boldsymbol{\lambda}$.

Schubert calculus for $G_{r,n}$ handles the constants $L_{\boldsymbol{\lambda}\boldsymbol{\mu}}^{\boldsymbol{\nu}} \in \mathbb{Z}$, named *Littlewood–Richardson coefficients*, occurring in the expansion

$$\Delta_{\boldsymbol{\lambda}}(c(\mathcal{Q}_r)) \cap \Omega_{\boldsymbol{\mu}} = \sum L_{\boldsymbol{\lambda}\boldsymbol{\mu}}^{\boldsymbol{\nu}} \Omega_{\boldsymbol{\nu}},$$

The structure constants $L_{\boldsymbol{\lambda}\boldsymbol{\mu}}^{\boldsymbol{\nu}}$ can be algorithmically determined by the celebrated *Pieri formula*:

$$c_i(\mathcal{Q}_r) \cap \Omega_{\boldsymbol{\lambda}} = \sum_{\boldsymbol{\mu} \in \mathcal{P}_r(i, \boldsymbol{\lambda})} \Omega_{\boldsymbol{\mu}}, \tag{5.25}$$

where the sum is taken over all $\boldsymbol{\mu} \in \mathcal{P}_{r,n}$, such that $\mu_1 \geq \lambda_1 \geq \cdots \geq \mu_r \geq \lambda_r$ and $|\boldsymbol{\mu}| = |\boldsymbol{\lambda}| + i$, together with the *Giambelli formula*:

$$\Delta_{\boldsymbol{\lambda}}(c(\mathcal{Q}_r) \cap \Omega_0 = \Omega_{\boldsymbol{\lambda}}. \tag{5.26}$$

Formula (5.25) is formally the same as (5.10) up to writing $[\mathbf{b}]^r$ instead of Ω. The analogue of (5.26) in the realm of Schubert derivations will be proven in Section 5.8. Giambelli formula (5.26), in particular, implies that

$$A^*(G_{r,n}) = \bigoplus_{\boldsymbol{\lambda} \in \mathcal{P}_{r,n}} \mathbb{Z}\Delta_{\boldsymbol{\lambda}}(c(\mathcal{Q}_r)),$$

and that the Chern classes $c_1(\mathcal{Q}_r), \ldots, c_r(\mathcal{Q}_r)$, which cut on $G_{r,n}$ the *special Schubert cycles* $\Omega_{(j)}$, generate $A^*(G_{r,n})$ as a $\mathbb{Z}$-algebra. Theorem 5.4.5 applied in the case where $\mathcal{X}$ reduces to a point yields:

Corollary 5.4.8. *Let $p = X^n \in \mathbb{Z}[X]$, M_p as in 5.1, with basis $(b_0, b_1, \ldots, b_{n-1})$ and $\mathbb{Z}[\boldsymbol{\sigma}_+]_r$ the commutative subring of* $\mathrm{End}_{\mathbb{Z}}(\bigwedge^r M_p)$ *obtained by restricting $\mathbb{Z}[\boldsymbol{\sigma}_+]$ to $\bigwedge^r M_p$. Then the following diagram*

$$\begin{array}{ccc} A^*(G_{r,n}) \otimes_{\mathbb{Z}} A_*(G_{r,n}) & \longrightarrow & A_*(G_{r,n}) \\ \iota \otimes j \big\downarrow & & j \big\downarrow \\ \mathbb{Z}[\boldsymbol{\sigma}_+]_r \otimes_{\mathbb{Z}} \bigwedge^r M_p & \longrightarrow & \bigwedge^r M_p \end{array} \tag{5.27}$$

commutes, and the vertical maps are isomorphisms. □

By virtue of 5.4.8, one would be tempted to denote $c_i(\mathcal{Q}_r)$ by $\sigma_{i,r}$ or σ_i *tout court.* This is exactly what is done in the literature, and $\sigma_i := c_i(\mathcal{Q}_r)$ are called *special Schubert cocycles.*

This ends the account of the notation σ_i we have chosen for the coefficients of the Schubert derivation, to emphasize that they are indeed special Schubert cycles but seen as operators on an exterior power of the Chow group of the projective space.

5.4.9. Given a partition $\boldsymbol{\lambda} \in \mathcal{P}_{r,n}$, the cap product of $c_1(\mathcal{Q}_r)^{r(n-r)-|\boldsymbol{\lambda}|} \cap \Omega_{\boldsymbol{\lambda}}$ lands in $A_0(G_{r,n})$ which is generated, over the integers, by $\Omega_{((n-r)^r)}$, the class of any point of $G_{r,n}$. Then

$$c_1(\mathcal{Q}_r)^{r(n-r)-|\boldsymbol{\lambda}|} \cap \Omega_{\boldsymbol{\lambda}} = d_{\boldsymbol{\lambda}} \cdot \Omega_{((n-r)^r)}, \quad d_{\boldsymbol{\lambda}} \in \mathbb{N}.$$

The integer $d_{\boldsymbol{\lambda}}$ is said to be the *degree* of the Schubert variety $\Omega_{\boldsymbol{\lambda}}(F^\bullet)$, where $F^\bullet$ is any complete flag of $\mathbb{C}^n$, because it is the number of its intersection points with a general linear subspace of complementary codimension in the Plücker embedding (see Section 5.3). It is customary to write

$$\int_{G_{r,n}} \sigma_1^{r(n-r)-|\boldsymbol{\lambda}|} \cap \Omega_{\boldsymbol{\lambda}} := d_{\boldsymbol{\lambda}}$$

and, by virtue of the dictionary (5.27), this is the same as

$$\int_n \sigma_1^{r(n-r)-|\boldsymbol{\lambda}|} [\mathbf{b}]^r_{\boldsymbol{\lambda}}.$$

For example,

$$\int_{G_{2,4}} c_1(\mathcal{Q}_2)^4 \cap \Omega_0 = \int_4 \sigma_1^4 [\mathbf{b}]^2_0 = 2,$$

is the degree of the Grassmannian $G_{2,4}$ in its Plücker embedding, or the number of 2-planes meeting nontrivially four 2-planes in general position or the number of lines in $\mathbb{P}^3$ intersecting other four in general position.

5.5 Examples

To familiarize readers with the formalism of Schubert derivations and, at the same time, make them appreciate its versatility, we present a few examples.

5.5.1. If $A = \mathbb{Z}$ and $p = X^n$, the module M_p is isomorphic to the Chow group $A_*(\mathbb{P}^{n-1})$ (singular homology) of the complex projective space $\mathbb{P}^{n-1}$. The Chow ring is generated by $\sigma_1 := c_1(O_{\mathbb{P}^n}(1))$, the first Chern class of the hyperplane bundle:

$$A^*(\mathbb{P}^{n-1}) := \frac{\mathbb{Z}[\sigma_1]}{(\sigma_1^n)},$$

where $\sigma_1 \cap [\mathbb{P}^{n-1}] = [\mathbb{P}^{n-2}]$, i.e. σ_1 is the operator cutting out a hyperplane section. The degree of $\mathbb{P}^{n-1}$ embedded in itself via the identity morphism is

$$1 = \int_{\mathbb{P}^{n-1}} \sigma_1^{n-1} \cap [\mathbb{P}^{n-1}] := \text{ coefficient of } [\mathbb{P}^0] \text{ in the expansion of } \sigma_1^{n-1} \cap [\mathbb{P}^{n-1}].$$

5.5.2. Let $A = \mathbb{Z}$, $p := X^4$ and $b_i := X^i + (X^4)$. Then M_p is the free abelian group of rank 4 generated by $\mathfrak{B}_4 := (b_0, b_1, b_2, b_3)$. Recall that $\sigma_i b_j = X^{i+j} + (X^4)$, i.e. $\sigma_i b_j$ vanishes if $i + j \geq 4$. The module $\bigwedge^2 M_p$ has rank 6:

$$\bigwedge^2 M_p = (\bigwedge^2 M_p)_0 \oplus (\bigwedge^2 M_p)_1 \oplus (\bigwedge^2 M_p)_2 \oplus (\bigwedge^2 M_p)_3 \oplus (\bigwedge^2 M_p)_4,$$

where $(\bigwedge^2 M_p)_i$ is the free $\mathbb{Z}$-submodule of $\bigwedge^2 M_p$ spanned by the basis elements $\bigwedge^2 \mathfrak{B}_4$, parametrized by partitions of $\mathcal{P}_{2,2}$ of weight i. More explicitly

$$(\bigwedge^2 M_p)_0 = \mathbb{Z} \cdot b_0 \wedge b_1, \qquad (\bigwedge^2 M_p)_1 = \mathbb{Z} \cdot b_0 \wedge b_2,$$
$$(\bigwedge^2 M_p)_2 = \mathbb{Z} \cdot b_0 \wedge b_3 \oplus \mathbb{Z} b_1 \wedge b_2,$$
$$(\bigwedge^2 M_p)_3 = \mathbb{Z} \cdot b_1 \wedge b_3, \qquad (\bigwedge^2 M_p)_4 = \mathbb{Z} \cdot b_2 \wedge b_3.$$

The evaluation $q(\boldsymbol{\sigma}_+)$ of a degree 4 polynomial $q \in \mathbb{Z}[T_1, T_2, \ldots]$ at $T_i = \sigma_i$ maps the submodule $(\bigwedge^2 M_p)_0$ to the rank 1 submodule $(\bigwedge^2 M_p)_4$, and hence the evaluation of the integral

$$\int_4 q(\sigma_+) b_0 \wedge b_1$$

is not trivially zero. For instance, to check that

$$\int_4 \sigma_1^4 [\mathbf{b}]_0^2 = 2,$$

one computes $\sigma_1^4(b_0 \wedge b_1)$ by iterating the Leibniz rule for σ_1. So

$$\begin{aligned}\sigma_1^4 \cdot [\mathbf{b}]_0^2 &= \sigma_1^4(b_0 \wedge b_1) = \sigma_1^3(b_0 \wedge b_2) = \sigma_1^2(b_1 \wedge b_2 + b_0 \wedge b_3)\\ &= \sigma_1(2b_1 \wedge b_3) = 2\sigma_1(b_1 \wedge b_3) = 2b_2 \wedge b_3\\ &= 2 \cdot [\mathbf{b}]_{(2,2)}^2.\end{aligned}$$

It turns out that computing σ_1^4 on $b_0 \wedge b_1$ amounts to determining the number of lines $(= 2)$ in projective 3-space that intersect four others in general position. Similarly,

$$\int_4 \sigma_1^2\sigma_2[\mathbf{b}]_0^2 = 1 \qquad \text{and} \qquad \int_4 \sigma_2^2[\mathbf{b}]_0^2 = 1.$$

5.5.3. We retain the notation of Example 5.5.2. The map $A[\boldsymbol{\sigma}_+] \to \bigwedge^2 M_p$ given by

$$p(\boldsymbol{\sigma}_+) \mapsto p(\boldsymbol{\sigma}_+)[\mathbf{b}]_0^2,$$

is a $\mathbb{Z}$-module epimorphism. Put otherwise, for each $\boldsymbol{\lambda} \in \mathcal{P}_{2,2}$, there exists at least one $p_{\boldsymbol{\lambda}} \in \mathbb{Z}[\mathbf{X}]$ such that $[\mathbf{b}]_{\boldsymbol{\lambda}}^2 = \mathfrak{p}_{\boldsymbol{\lambda}}(\boldsymbol{\sigma}_+)[\mathbf{b}]_0^2$. We can check this fact by hand in this simple case. We have $b_0 \wedge b_2 = \sigma_1(b_0 \wedge b_1)$ and $b_0 \wedge b_3 = \sigma_2(b_0 \wedge b_1)$ by a simple application of Corollary 5.1.7. Moreover, integrating by parts gives

$$b_1 \wedge b_2 = (\sigma_1 b_0) \wedge b_2 = \sigma_1(b_0 \wedge b_2) - b_0 \wedge \overline{\sigma}_1 b_2.$$

But $\overline{\sigma}_1 = \sigma_1$ and so

$$b_1 \wedge b_2 = \sigma_1^2(b_0 \wedge b_1) - b_0 \wedge b_3 = (\sigma_1^2 - \sigma_2)b_0 \wedge b_1 = \begin{vmatrix}\sigma_1 & 1\\ \sigma_2 & \sigma_1\end{vmatrix} b_0 \wedge b_1.$$

Similarly

$$b_0 \wedge b_3 = \sigma_2[\mathbf{b}]_0^2, \quad b_1 \wedge b_3 = \sigma_1\sigma_2[\mathbf{b}]_0^2, \quad b_2 \wedge b_3 = \sigma_2^2[\mathbf{b}]_0^2.$$

It follows that σ_i vanishes on $\bigwedge^2 M_p$ if $i > 2$. In fact

$$\sigma_i[\mathbf{b}]_0^2 = \sigma_i(b_0 \wedge b_1) = b_0 \wedge b_{1+i} = [\mathbf{b}]_{(i)}^2 = 0, \quad \forall i \geq 3,$$

because $b_{2+i} = 0$, for all $i \geq 3$. In general

$$\sigma_i[\mathbf{b}]_{\boldsymbol{\lambda}}^2 = \sigma_i \mathfrak{p}_{\boldsymbol{\lambda}}(\boldsymbol{\sigma}_+)[\mathbf{b}]_0^2 = \mathfrak{p}_{\boldsymbol{\lambda}}(\boldsymbol{\sigma}_+)\sigma_i[\mathbf{b}]_0^2 = 0,$$

where in the second equality, we used the commutativity of $A[\sigma_+]$.

5.5.4. Let us explain how Example 5.5.2 can be generalized. Let $p = X^n$, so that M_p is free of rank n and $\bigwedge^2 M_p$ is free of rank $\binom{n}{2}$ generated by $[\mathbf{b}]^2_{\boldsymbol{\lambda}}$, where $\boldsymbol{\lambda} \in \mathcal{P}_{2,n-2}$. Notice that $\mathrm{rank}_{\mathbb{Z}}(\bigwedge^2 M_p)_0 = \mathrm{rank}_{\mathbb{Z}}(\bigwedge^2 M_p)_{2(n-2)} = 1$. Then

$$\sigma_1^{2(n-2)} b_0 \wedge b_1 = \left(\int_n \sigma_1^{2(n-2)} [\mathbf{b}]^2_0\right) \cdot b_{n-2} \wedge b_{n-1}.$$

The binomial formula (4.10) tells us

$$\begin{aligned}
\sigma_1^{2(n-2)} b_0 \wedge b_1 &= \sum_{j=0}^{2(n-2)} \binom{2(n-2)}{j} \sigma_1^j b_0 \wedge \sigma_1^{2n-4-j} b_1 \\
&= \sum_{j=0}^{2(n-2)} \binom{2(n-2)}{j} b_j \wedge b_{2n-3-j}.
\end{aligned} \tag{5.28}$$

As $(\bigwedge^2 M_p)_{2(n-2)}$ is generated by $b_{n-2} \wedge b_{n-1}$, all summands in (5.28), but those corresponding to $i = n-2$ and $i = n-1$, vanish. Hence:

$$\begin{aligned}
\sigma_1^{2(r-2)} b_0 \wedge b_1 &= \binom{2(n-2)}{r-2} b_{n-2} \wedge b_{n-1} + \binom{2(n-2)}{n-1} b_{n-1} \wedge b_{n-2} \\
&= \left[\binom{2(n-2)}{n-2} - \binom{2(n-2)}{r-1}\right] b_{n-2} \wedge b_{n-1} \\
&= \frac{(2n-4)!}{(n-1)!(n-2)!} b_{n-2} \wedge b_{n-1}.
\end{aligned} \tag{5.29}$$

The numbers

$$C_n := \frac{(2n-4)!}{(n-1)!(n-2)!}$$

are classically known as *Catalan numbers* and count the number of distinct ways a convex polygon with $n+2$ sides can be decomposed into triangles by drawing diagonals.

5.5.5 (cf. [34, 64]). Let $\boldsymbol{\lambda} \in \mathcal{P}_{r,n-r}$ and $F^\bullet$ any complete flag of $\mathbb{C}^n$. The purpose of this example is to compute a formula for the degree of the Schubert variety $\Omega_{\boldsymbol{\lambda}}(F^\bullet)$ in the Plücker embedding of $G_{r,n}$. It amounts to compute the 'integral':

$$\int_{G_{r,n}} c_1(\mathcal{Q}_r)^{r(n-r)-|\boldsymbol{\lambda}|} \cap \Omega_{\boldsymbol{\lambda}}. \tag{5.30}$$

This is done, e.g. in [64, Ch. XIV] and in [34, p. 274]. The proof exhibited in both references relies on an expression of the class of a hyperplane section of a Schubert variety, namely, formula (i) in [34, Example 14.7.10] and formula (6) in [64, p. 364]. Those coincident formulas provide the geometrical interpretation of, and actually they are, the Leibniz rule enjoyed by the coefficient σ_1 of the Schubert derivation $\sigma_+(z)$. The latter will be heavily used below to compute (5.30), by invoking the dictionary (5.27).

To get into our language, let us take $A = \mathbb{Z}, p = X^n \in \mathbb{Z}[X]$, so that $\mathrm{rank}_{\mathbb{Z}} M_P = n$ with basis $\mathfrak{B} := (b_0, b_1, \ldots, b_{n-1})$, and let $\boldsymbol{\lambda} \in \mathcal{P}_{r,n-r}$. We set out to compute

$$d_{\boldsymbol{\lambda}} = \int_n \sigma_1^{r(n-r)-w} [\mathbf{b}]^r_{\boldsymbol{\lambda}},$$

where $0 \leq w := |\boldsymbol{\lambda}| \leq r(n-r)$. It is convenient, solely for the purposes of this example, to set

$$i_{k-1} = \lambda_{r+1-k} + k - 1, \qquad 1 \leq k \leq r,$$

so that

$$[\mathbf{b}]^r_{\boldsymbol{\lambda}} = b_{i_0} \wedge b_{i_1} \wedge \cdots \wedge b_{i_{r-1}},$$

with $i_0 + \cdots + i_{r-1} = w + \frac{1}{2}(r-1)r$. By applying Newton's formula (4.11), we obtain

$$\sigma_1^{r(n-r)-w} [\mathbf{b}]^r_{\boldsymbol{\lambda}} = \sum \binom{r(n-r)-w}{j_0, j_1, \ldots, j_{r-1}} b_{i_0+j_0} \wedge b_{i_1+j_1} \wedge \cdots \wedge b_{i_{r-1}+j_{r-1}}, \tag{5.31}$$

where the summation is over all non-negative integers $(j_0, \ldots, j_{r-1})$ such that $j_0 + \cdots + j_{r-1} = r(n-r) - w$. The only terms surviving in (5.31) are those for which

$$(i_0 + j_0, i_1 + j_1, \ldots, i_{r-1} + j_{r-1})$$

is a permutation of $(n-r, n-r+1, \ldots, n-1)$, i.e.

$$i_{k-1} + j_{k-1} = n - r + \mathfrak{s}(k-1) \qquad k = 1, \ldots, r$$

and where $\mathfrak{s}$ belongs to the group S_r of bijections of $\{0, 1, \ldots, r-1\}$. Therefore

$$d_{\boldsymbol{\lambda}} = \sum_{\mathfrak{s} \in S_r} \mathrm{sgn}(\mathfrak{s}) \binom{r(n-r)-w}{n-r+\mathfrak{s}(0)-i_0, n-r+\mathfrak{s}(1)-i_1, \ldots, n-r+\mathfrak{s}(r-1)-i_{r-1}},$$

and a simple inspection shows that

$$d_\lambda = \sum_{\mathfrak{s}\in S_r} \operatorname{sgn}(\mathfrak{s}) \binom{r(n-r)-w}{n-r-i_{\mathfrak{s}(0)}, n-r+1-i_{\mathfrak{s}(1)}, \dots, n-r+(r-1)-i_{\mathfrak{s}(r-1)}}$$
$$= \sum_{\mathfrak{s}\in S_r} \operatorname{sgn}(\mathfrak{s}) \binom{r(n-r)-w}{n-r-i_{\mathfrak{s}(0)}, n-r+1-i_{\mathfrak{s}(1)}, \dots, n-1-i_{\mathfrak{s}(r-1)}}. \quad (5.32)$$

Since $0 \le i_0 < i_1 < \cdots < i_{r-1}$, the common denominator of the alternating sum (5.32) is precisely

$$(n-1-i_0)! \cdot (n-1-i_1)! \cdots (n-1-i_{r-1})!,$$

or

$$(n-1-i_{\mathfrak{s}(0)})! \cdot (n-1-i_{\mathfrak{s}(1)})! \cdots (n-1-i_{\mathfrak{s}(r-1)})! = \prod_{j=0}^{r-1} (n-1-i_{\mathfrak{s}(j)})!$$

for each $\mathfrak{s} \in S_r$. Therefore (5.32) can be written as[3]

$$\sum_{\mathfrak{s}\in S_r} \operatorname{sgn}(\mathfrak{s}) \frac{(r(n-r)-w)!}{(n-1-i_0)!(n-1-i_1)!\cdots(n-1-i_{r-1})!} \cdot \frac{\prod_{j=0}^{r-1}(n-1-i_{\mathfrak{s}(j)})!}{\prod_{j=0}^{r-1}(n-r-j-i_{\mathfrak{s}(j)})!}$$

$$= \frac{(r(n-r)-w)!}{(n-i_0)!(n-i_1)!\cdots(n-i_{r-1})!} \sum_{\mathfrak{s}\in S_r} \operatorname{sgn}(\mathfrak{s}) \prod_{k=0}^{r-1} \prod_{j_k=0}^{r-1-k} (n-i_{\mathfrak{s}(k)}-j_k). \quad (5.33)$$

The point is, now, that the sum over the permutations of the products in formula (5.33) is a homogeneous polynomial of degree $r(r-1)/2$ in $(n-1, n-2, \dots, n-r+1)$ and $(i_0, i_1, \dots i_{r-1})$; furthermore, it changes sign if one permutes $i_a \leftrightarrow i_b$, for $a \neq b$. Therefore

$$\sum_{\mathfrak{s}\in S_r} \operatorname{sgn}(\mathfrak{s}) \prod_{k=0}^{r-1} \prod_{j_k=0}^{r-1-k} (n-i_{\mathfrak{s}(k)}-j_k) = m \prod_{a>b} (i_a - i_b) \quad (5.34)$$

for some integer m. To determine the proportionality factor, it suffices to compare the coefficients of one common monomial. In the expansion of the left-hand side, the monomial $(-1)^{\frac{r(r-1)}{2}} i_0^{r-1} i_1^{r-2} i_2^{r-3} \cdots i_{r-1}$ occurs, corresponding to the identical

[3] The lack of space allowed by the book's template prevented us from being more explicit, so we recommend the reader to follow the proofs of the cases $r = 2, 3$ to check that indeed the computations are correct and to sense what is going on.

substitution $\mathfrak{s} = (12\ldots r)$; the same term, with the same sign, occurs on the right-hand side, proving that (5.34) holds with $m = 1$. To sum up, the final formula is

$$d_{\boldsymbol{\lambda}} = \frac{(r(n-r)-w)!\prod_{b<a}(i_a - i_b)}{(n-1-i_0)!(n-1-i_1)!\cdots(n-1-i_{r-1})!}$$

$$= \frac{(r(n-r)-w)!\prod_{b<a}(\lambda_b - \lambda_a + a - b)}{(n-1-\lambda_r)!(n-2-\lambda_{r-1})!\cdots(n-r-\lambda_1)!} \qquad (5.35)$$

In particular, for $\boldsymbol{\lambda} = 0$,

$$d_{(0)} = \frac{(r(n-r))!\prod_{h<k}(k-h)}{(n-1)!(n-2)!\cdots(n-r)!} = \frac{1!2!\cdots k!(r(n-r))!}{(n-1)!(n-2)!\cdots(n-r)!},$$

which by our vocabulary (5.27) is the degree of the Grassmannian $G_{r,n}$ in its Plücker embedding (Section 5.3).

5.5.6. Let $p := X^n$, $2 \le r \le n$ and $\boldsymbol{\lambda} \in \mathcal{P}_{r-2}$. Then

$$\sigma_i([\mathbf{b}]^{r-2}_{\boldsymbol{\lambda}} \wedge b_{n-1}) = \sigma_i([\mathbf{b}]^{r-2}_{\boldsymbol{\lambda}}) \wedge b_{n-1} \in \bigwedge^r M_p.$$

In fact the Leibniz rule for the operator σ_i will give

$$\sigma_i([\mathbf{b}]^{r-1}_{\boldsymbol{\lambda}} \wedge b_{n-1}) = \sigma_i[\mathbf{b}]^{r-1}_{\boldsymbol{\lambda}} \wedge b_{n-1} + \sum_{j=1}^{i} \sigma_j[\mathbf{b}]^{r-1}_{\boldsymbol{\lambda}} \wedge b_{n-1+j}.$$

But $b_{n-1+j} = 0$ for all $j > 0$.

5.5.7 (See [55, p. 206] and [20, Lemma 1.1.]). Let $p_n = X^{2n+2} \in \mathbb{Z}[X]$ so that $\mathrm{rank}_{\mathbb{Z}} \bigwedge^2 M_{p_n} = \binom{2n+2}{2}$. We wish to compute

$$\mathfrak{L}_n := \int_{4n} \sigma_n^4 [\mathbf{b}]_0^2$$

$$= \text{ coefficient of } b_{2n} \wedge b_{1+2n} \in \bigwedge^2 M_{p_n} \text{ in the expansion of } \sigma_n^4(b_0 \wedge b_1).$$

We shall prove that:

$$\sigma_n^4(b_0 \wedge b_1) = b_{2n} \wedge b_{1+2n} + \sigma_{n-1}^4(b_2 \wedge b_3). \qquad (5.36)$$

Since computing $\sigma_{n-1}^4(b_2 \wedge b_3)$ in $\bigwedge^2 M_{p_n}$ is formally the same as evaluating $\sigma_{n-1}^4(b_0 \wedge b_1)$ in $\bigwedge^2 M_{(p_{n-1})}$ (prove it!), (5.36) would imply:

$$\mathfrak{L}_n = 1 + \mathfrak{L}_{n-1}.$$

By induction $\mathfrak{L}_n = n + \mathfrak{L}_0$, and then $\mathfrak{L}_n = n+1$, because $\mathfrak{L}_0 = \int_0 \sigma_0^4[\mathbf{b}]_0^2 = 1$! One is then left to prove formula (5.36). First we have by Corollary 5.1.7 iii)

$$\sigma_n(b_0 \wedge b_1) = b_0 \wedge b_{1+n}.$$

Then

$$\begin{aligned}\sigma_n^2(b_0 \wedge b_1) &= \sigma_n(b_0 \wedge b_{1+n}) = b_n \wedge b_{1+n} + \sigma_{n-1}(b_1 \wedge b_{1+n}) \\ &= b_n \wedge b_{1+n} + \sigma_{n-1}\overline{\sigma}_2(b_0 \wedge b_1).\end{aligned} \tag{5.37}$$

Therefore

$$\sigma_n^4(b_0 \wedge b_1) = \sigma_n^2(b_n \wedge b_{1+n}) + \sigma_{n-1}\overline{\sigma}_2\sigma_n^2(b_0 \wedge b_1) \tag{5.38}$$

having composed both summands of (5.37) with σ_n^2 and the commutativity in $A[\boldsymbol{\sigma}_+]$. Substituting (5.37) into (5.38), using 5.1.7 iii), Example 5.5.6 and the definition of $\overline{\sigma}_2$, we eventually obtain

$$\sigma_n^4(b_0 \wedge b_1) = \sigma_n(b_n \wedge b_{2n+1}) + \sigma_{n_1}^2\overline{\sigma}_2^2(b_0 \wedge b_1) = b_{2n} \wedge b_{1+2n} + \sigma_{n-1}^2(b_2 \wedge b_3)$$

which is precisely formula (5.36), as desired.

5.6 Module Structures Induced by Schubert Derivations, I

From this section onward, the main actor will be the free abelian group of countably infinite rank

$$M_0 := \mathbb{Z}[X] = \bigoplus_{i \in \mathbb{N}} \mathbb{Z}b_i,$$

corresponding to the $\mathbb{Z}$-module M_p defined in Section 5.1 for $p = 0 \in \mathbb{Z}[X]$. In this case $b_i = X^i$, but we shall still keep using 'b_i' for uniformity and denote by $\mathfrak{B}_0$ the basis $(b_0, b_1, \ldots)$ of M_0. We have a direct sum decomposition

$$M_0 = M_{0,n} \oplus \sigma_n M_0, \qquad (n \geq 1)$$

where we put

$$M_{0,n} := \bigoplus_{i=0}^{n-1} \mathbb{Z}b_i \qquad \text{and} \qquad \sigma_n M_0 = \sigma_1^n M_0 := \bigoplus_{i \geq n} \mathbb{Z} \cdot b_i. \tag{5.39}$$

The basis $(b_n, b_{n+1}, \ldots)$ of $\sigma_n M_0$ will be denoted by $\sigma_n \mathfrak{B}_0$ (in particular $\sigma_0 \mathfrak{B}_0 = \mathfrak{B}_0$). The goal of this section is to learn how to equip M_0, $M_{0,n}$ and $\sigma_n M_0$ with natural structures of free B_r-modules of rank r, for any chosen integer $r \geq 1$, for the purposes of Chapter 6.

5.6.1. The first easy step is to define a B_r-module structure on $\bigwedge^r M_0$. This is achieved by imposing the relations

$$e_i \cdot \alpha = \overline{\sigma}_i \eta, \qquad \forall \eta \in \bigwedge^r M_0. \tag{5.40}$$

By construction, $\bigwedge^r M_0$ is an eigenmodule of every $\overline{\sigma}_j$, with e_j as eigenvalue. Consider the natural $\mathbb{Z}$-module evaluation homomorphism:

$$\begin{cases} \mathrm{ev}_{[\mathbf{b}]_0^r} : \qquad B_r & \longrightarrow \bigwedge^r M_0 \\ \qquad p(e_1, \ldots, e_r) & \longmapsto p(\overline{\boldsymbol{\sigma}}_+)[\mathbf{b}]_0^r \end{cases}, \tag{5.41}$$

where $p(\overline{\boldsymbol{\sigma}}_+)$ means evaluation of p at $e_i = \overline{\sigma}_i$.

Proposition 5.6.2. *The map (5.41) is a $\mathbb{Z}$-module epimorphism.*

Proof. Given that σ_i is a weighted homogeneous polynomial of degree i in $\overline{\sigma}_1, \ldots, \overline{\sigma}_r$, the claim amounts to showing that for all $\boldsymbol{\lambda} \in \mathcal{P}_r$, there exists $G_{\boldsymbol{\lambda}} \in \mathbb{Z}[\mathbf{X}] := \mathbb{Z}[X_1, X_2, \ldots]$ such that

$$[\mathbf{b}]_{\boldsymbol{\lambda}}^r = G_{\boldsymbol{\lambda}}(\boldsymbol{\sigma}_+)[\mathbf{b}]_0^r. \tag{5.42}$$

We argue by induction on the length of the partition. If $\boldsymbol{\lambda} = (\lambda)$, a partition of length at most 1, we have

$$[\mathbf{b}]_{(\lambda)}^r = b_0 \wedge b_1 \wedge \cdots \wedge b_{r-1+\lambda} = \sigma_\lambda (b_0 \wedge b_1 \wedge \cdots \wedge b_{r-1}),$$

because of 5.1.7 ii). In this case $G_{(\lambda)}(\boldsymbol{\sigma}_+) = \sigma_\lambda$. Assume that the property holds for all partitions of length at most $k - 1 \leq r - 1$. Then

$$\begin{aligned} & b_0 \wedge \cdots \wedge b_{r-k+\lambda_k} \wedge \cdots \wedge b_{r-1+\lambda_1} \\ = {} & \sigma_{\lambda_k}(b_0 \wedge \cdots \wedge b_{r-k}) \wedge b_{r-k+1+\lambda_{k-1}} \wedge \cdots \wedge b_{r-1+\lambda_1} \end{aligned}$$

i.e. by applying integration by parts (4.6)

$$= \sum_{i=0}^{\lambda_k} (-1)^i \sigma_{\lambda_k - i}([\mathbf{b}]_0^{r-k+1} \wedge \overline{\sigma}_i (b_{r-k+1+\lambda_{k-1}} \wedge \cdots \wedge b_{r-1+\lambda_1})).$$

Each term $[\mathbf{b}]_0^{r-k+1} \wedge \overline{\sigma}_i(b_{r-k+\lambda_k} \wedge \cdots \wedge b_{r-1+\lambda_1})$ is a linear combination of elements of the form $[\mathbf{b}]^r_{\boldsymbol{\lambda}_{ij}}$ where $\boldsymbol{\lambda}_{ij}$ is a partition of length at most $k-1$. By induction, there exists $G_i(\boldsymbol{\sigma}_+)$ such that

$$[\mathbf{b}]_0^{r-k} \wedge \overline{\sigma}_i(b_{r-k+\lambda_k} \wedge \cdots \wedge b_{r-1+\lambda_1}) = G_i(\boldsymbol{\sigma}_+)[\mathbf{b}]_0^r,$$

and

$$G_{\boldsymbol{\lambda}}(\boldsymbol{\sigma}_+) = \sum_{i=0}(-1)^i \sigma_{\lambda_k - i} G_i(\boldsymbol{\sigma}_+) \in \mathbb{Z}[\boldsymbol{\sigma}_+]_r$$

is a polynomial with the required property. □

Denote by $G_{\boldsymbol{\lambda}}(H_r) \in B_r$ the eigenvalue of the endomorphism $G_{\boldsymbol{\lambda}}(\boldsymbol{\sigma}_+)$ of $\bigwedge^r M_p$, from the proof of Proposition 5.6.2. Its eigenmodule is the entire $\bigwedge^r M_p$. It is a well-known fact (see, e.g. [109]) that any $\mathbb{Z}$-polynomial ring in r indeterminates possesses a $\mathbb{Z}$-basis parametrized by partitions of length at most r. We are going to prove this fact anew using our formalism. Recall the weight graduation of B_r (Section 2.2.6).

Theorem 5.6.3. *The data $(G_{\boldsymbol{\lambda}}(H_r) \mid |\boldsymbol{\lambda}| = w)$ form a $\mathbb{Z}$-basis for $(B_r)_w$. In particular the epimorphism (5.41) is a $\mathbb{Z}$-module isomorphism.*

Proof. It is clear that $G_{\boldsymbol{\lambda}}(H_r) \in B_r$ for all $\boldsymbol{\lambda} \in \mathcal{P}_r$. We claim that

$$(G_{\boldsymbol{\lambda}}(H_r) \mid |\boldsymbol{\lambda}| = w)$$

are linearly independent over the integers. In fact $\sum_{|\boldsymbol{\lambda}|=i} a_{\boldsymbol{\lambda}} G_{\boldsymbol{\lambda}}(H_r) = 0$ implies

$$0 = \sum_{|\boldsymbol{\lambda}|=w} a_{\boldsymbol{\lambda}} G_{\boldsymbol{\lambda}}(H_r)[\mathbf{b}]_0^r = \sum_{|\boldsymbol{\lambda}|=w} a_{\boldsymbol{\lambda}}[\mathbf{b}]^r_{\boldsymbol{\lambda}},$$

which in turn forces all $a_{\boldsymbol{\lambda}}$ to be zero because $([\mathbf{b}]^r_{\boldsymbol{\lambda}} \mid |\boldsymbol{\lambda}| = w)$ is a $\mathbb{Z}$-basis of $(\bigwedge^r M_0)_w$. There remains to show that $B_r = \bigoplus_{|\boldsymbol{\lambda}|=w} \mathbb{Z} G_{\boldsymbol{\lambda}}(H_r)$. To this purpose we observe that both $(e^{\boldsymbol{\lambda}'})_{|\boldsymbol{\lambda}|=w}$ (notation as in 2.2.6) and $(G_{\boldsymbol{\lambda}}(H_r))_{|\boldsymbol{\lambda}|=w}$ are bases of the $\mathbb{Q}$-vector space $(B_r \otimes_{\mathbb{Z}} \mathbb{Q})_w$, so

$$e^{\boldsymbol{\lambda}'} = \sum_{|\boldsymbol{\mu}|=w} a_{\boldsymbol{\lambda}\boldsymbol{\mu}} G_{\boldsymbol{\mu}}(H_r)$$

with $a_{\boldsymbol{\lambda}\boldsymbol{\mu}} \in \mathbb{Q}$. Applying $[\mathbf{b}]_0^r$ to either side gives

$$e^{\boldsymbol{\lambda}'}[\mathbf{b}]_0^r = \overline{\sigma}_1^{i_1} \cdots \overline{\sigma}_r^{i_r}[\mathbf{b}]_0^r = \sum_{|\boldsymbol{\mu}|=w} a_{\boldsymbol{\lambda}\boldsymbol{\mu}} G_{\boldsymbol{\mu}}(H_r)[\mathbf{b}]_0^r$$

Given that the left-hand side is an integral combination of the basis elements of $(\bigwedge^r M_0)_w$, so is the right-hand side. In other words $a_{\lambda\mu} \in \mathbb{Z}$. □

Corollary 5.6.4 (Cf. 5.1.5). *The map* $\mathbb{Z}[\overline{\sigma}_1, \dots, \overline{\sigma}_r] \rightarrow \bigwedge^r M_0$, *given by* $G_\lambda(\sigma_+) \mapsto [\mathbf{b}]^r_\lambda$, *is a* $\mathbb{Z}$*-module isomorphism.*

Proof. By its very construction, the evaluation morphism $\text{ev}_{[\mathbf{b}]^r_0} : B_r \rightarrow \bigwedge^r M_0$ factorizes through $\text{ev}_{\sigma_+} : B_r \rightarrow A[\boldsymbol{\sigma}_+]_r$ (with $A = \mathbb{Z}$), and the claim follows. □

Notation 5.6.5. Via the $\mathbb{Z}$-module isomorphism $B_r \rightarrow \bigwedge^r M_0$ of Corollary 5.6.4, it follows that $\bigwedge^r M_0$ is a free B_r-module of rank 1 generated by $[\mathbf{b}]^r_0$. We provisionally write such a module structure as $B_r \otimes_{\text{ev}_{\sigma_+}} \bigwedge^r M_0$.

5.7 Module Structures Induced by Schubert Derivations, II

The next goal is to construct a free B_r-module M_r of rank r out of M_0, such that $\bigwedge^r M_r = B_r \otimes_{\text{ev}_{\sigma_+}} \bigwedge^r M_0$. This will enable us to prove the relationship $G_\lambda(H_r) = \Delta_\lambda(H_r)$, which will then imply the *Giambelli formula*:

$$[\mathbf{b}]^r_\lambda = \Delta_\lambda(H_r)[\mathbf{b}]^r_0. \tag{5.43}$$

Lemma 5.7.1. *1. Suppose* $m \in M_0$ *satisfies* $m \wedge \eta = 0$ *for all* $\eta \in \bigwedge^{r-1} M_0$. *Then* $m = 0$.
2. If $m_1, m_2 \in M_0$ *are such that* $m_1 \wedge \eta = m_2 \wedge \eta$ *for all* $\eta \in \bigwedge^{r-1} M_0$, *then* $m_1 = m_2$.

Proof. Let

$$m = a_1 b_{i_1} + a_2 b_{i_2} + \cdots + a_k b_{i_k} \in M_0, \qquad (0 \leq i_1 < i_2 < \ldots < i_k),$$

be an arbitrary element of M_0, expressed as a linear combination of the basis $\mathfrak{B}_0$. For each $1 \leq j \leq k$, let

$$\eta_j = b_{i_1} \wedge \cdots \wedge \hat{b}_{i_j} \wedge \cdots \wedge b_{i_k} \in \bigwedge^{r-1} \mathfrak{B}_0,$$

where $\hat{b}_{i_j}$ means that b_{i_j} is missing. Then we have

$$0 = m \wedge \eta_j = a_j b_{i_j} \wedge \eta_j,$$

which implies $a_j = 0$, for all $0 \leq j \leq k-1$, i.e. $m = 0$.

Part ii) follows immediately from i): the assumption is equivalent to $(m_1 - m_2) \wedge \eta = 0$ for all $\eta \in \bigwedge^{r-1} M_0$, whence $m_1 - m_2 = 0$. □

5.7.2. Based on Lemma 5.7.1 define, for all $m \in M$ and $1 \leq i \leq r$,

$$e_i m$$

as the unique element of $M_r := B_r \otimes_{\mathbb{Z}} M_0$ such that

$$e_i m \wedge \eta = e_i(m \wedge \eta) = \overline{\sigma}_i(m \wedge \eta). \tag{5.44}$$

Proposition 5.7.3. *The shift operator $\sigma_1 : M_0 \to M_0$ is B_r-linear and so extends to an endomorphism of M_r.*

Proof. It suffices to show that

$$\sigma_1(e_i m) = e_i(\sigma_1 m), \qquad \forall i \in \{1, \dots, r\}.$$

In fact, using integration by parts (4.6) for $j = 1$,

$$\begin{aligned}\sigma_1(e_i m) \wedge \eta &= \sigma_1(e_i m \wedge \eta) - e_i m \wedge \sigma_1 \eta = e_i \sigma_1(m \wedge \eta) - e_i m \wedge \sigma_1 \eta \\ &= e_i(\sigma_1 m \wedge \eta) + e_i m \wedge \sigma_1 \eta - e_i m \wedge \sigma_1 \eta \\ &= e_i(\sigma_1 m) \wedge \eta,\end{aligned}$$

and the claim is proved. □

Proposition 5.7.4. *The module M_r defined by equation (5.44) is a free B_r-module of rank* r.

Proof. Since σ_1 is B_r-linear and $\overline{\sigma}_+(z)$ is the trace polynomial operator (i.e. $\overline{\sigma}_1, \dots, \overline{\sigma}_r$ are the trace endomorphisms of the B_r-module $\bigwedge^r M_r$ induced by $(\sigma_1)_{|M_r}$), the theorem of Cayley and Hamilton implies

$$\mathfrak{p}_r(\sigma_1) b_j = 0,$$

(recall that $\mathfrak{p}_r$ is the generic monic polynomial of degree r: see formula (2.8)) so $(b_0, b_1, \dots, b_{r-1}, b_r)$ are surely linearly dependent over B_r.

Consider the evaluation map $\mathrm{ev}_{b_0} : B_r[X] \to M_r$ defined by $p(X) \to p(\sigma_1) b_0$. It is clearly surjective. We claim that if p is a non-zero polynomial of degree $\leq r - 1$, then $p(\sigma_1) b_0 \neq 0$. To see this, let

$$q(X) = a_0 X^i + a_1 X^{i-1} + \cdots + a_i, \qquad a_i \in B_r,$$

with $a_0 \neq 0$ and $0 \leq i \leq r - 1$, so that

$$q(\sigma_1) b_0 = a_0 b_i + a_1 b_{i-1} + \cdots + a_i b_0.$$

Then

$$\bigwedge^r M_0 \ni (q(\sigma_1) b_0) \wedge b_0 \wedge \cdots \wedge b_{i-1} \wedge \hat{b}_i \wedge b_{i+1} \wedge \cdots \wedge b_{r-1} = \pm a_0 [\mathbf{b}]_0^r \neq 0,$$

i.e. $q(X)$ does not belong to the kernel of $\mathrm{ev}_{b_0} := \mathrm{ev}_{[\mathbf{b}]_0^1}$. If on the contrary $q \in \ker(\mathrm{ev}_{b_0})$, then $\deg(q) \geq r$, and we could write it as $\mathfrak{p}_r(X)Q(X) + R(X)$, where $\deg(R(X)) < r - 1$. Then

$$0 = q(\sigma_1)b_0 = Q(\sigma_1)\mathfrak{p}_r(\sigma_1)b_0 + r(\sigma_1)b_0 = r(\sigma_1)b_0$$

from which $R(X) = 0$, i.e. $\ker(\mathrm{ev}_{b_0}) = \mathfrak{p}_r(X)$. In other words $M_r := B_r \otimes_{\mathbb{Z}} M_0$, with the module structure defined by (5.44), is a free B_r-module of rank r isomorphic to $B_r[X]/(\mathfrak{p}_r(X))$ and, moreover,

$$\bigwedge^r M_r = B_r \otimes_{\mathrm{ev}_{\sigma_+}} \bigwedge^r M_0. \qquad \square$$

5.8 Giambelli Formula

Recall definitions and notation of Section 2.6.

Theorem 5.8.1. *Let* $f_0, f_1, \ldots, f_{r-1} \in B_r[X]$. *Then*

$$\begin{aligned} &f_0(\sigma_1)b_0 \wedge f_1(\sigma_1)b_0 \wedge \cdots \wedge f_{r-1}(\sigma_1)b_0 \\ &= \mathrm{Res}\left(\frac{f_{r-1}(X)}{\mathfrak{p}_r(X)}, \frac{f_{r-2}(X)}{\mathfrak{p}_r(X)}, \ldots, \frac{f_0(X)}{\mathfrak{p}_r(X)}\right) b_0 \wedge b_1 \wedge \cdots \wedge b_{r-1}. \end{aligned} \tag{5.45}$$

Proof. Either side of (5.45) is B_r-multilinear and alternating in $f_0, f_1, \ldots, f_{r-1}$. If one of the f_j is divisible by $\mathfrak{p}_r(X)$, it follows that $f_j := q_j(X)\mathfrak{p}_r(X)$, but then $f_j(\sigma_1)b_0 = q_j(\sigma_1)\mathfrak{p}_r(\sigma_1)b_0 = 0$. This proves that either side of (5.45) vanish if one f_j is divisible by $\mathfrak{p}_r(X)$. Moreover, they coincide when $f_i(X) = X^i$. If so, in fact, the left-hand side would be just $b_0 \wedge b_1 \wedge \cdots \wedge b_{r-1}$, while

$$\mathrm{Res}\left(\frac{X^{r-1}}{\mathfrak{p}_r(X)}, \frac{X^{r-2}}{\mathfrak{p}_r(X)}, \ldots, \frac{1}{\mathfrak{p}_r(X)}\right) = \begin{vmatrix} 1 & 0 & \ldots & 0 \\ h_1 & 1 & \ldots & 0 \\ \vdots & \vdots & \ddots & \vdots \\ h_{r-1} & h_{r-2} & \ldots & 1 \end{vmatrix} = 1$$

Thus equality (5.45) is verified. $\square$

Corollary 5.8.2. *Giambelli formula holds:*

$$[\mathbf{b}]_\lambda^r = \Delta_\lambda(H_r)[\mathbf{b}]_0^r = \Delta_\lambda(\sigma_+)[\mathbf{b}]_0^r.$$

Proof. Just compute

$$\begin{aligned}[\mathbf{b}]^r_\lambda &= \sigma_1^{\lambda_r} b_0 \wedge \sigma_1^{1+\lambda_{r-1}} b_0 \wedge \cdots \wedge \sigma_1^{r-1+\lambda_r} b_0 \\ &= \operatorname{Res}\left(\frac{X^{\lambda_r}}{\mathfrak{p}_r(X)}, \frac{X^{1+\lambda_{r-1}}}{\mathfrak{p}_r(X)}, \ldots, \frac{X^{r-1+\lambda_1}}{\mathfrak{p}_r(X)}\right) b_0 \wedge b_1 \wedge \cdots \wedge b_{r-1} \\ &= \Delta_\lambda(H_r)[\mathbf{b}]^r_0 = \Delta_\lambda(\boldsymbol{\sigma}_+)[\mathbf{b}]^r_0.\end{aligned}$$

The last equality descends from $\Delta_\lambda(H_r)$ being an eigenvalue of $\Delta_\lambda(\boldsymbol{\sigma}_+)$. □

Corollary 5.8.3. *The Schur polynomials $\Delta_\lambda(H_r)$ form a $\mathbb{Z}$-basis of B_r.*

Proof. It follows from 5.6.3 and the fact that $G_\lambda(H_r) = \Delta_\lambda(H_r)$. □

Notation 5.8.4. Corollary 5.8.3 says that the map $\Delta_\lambda(H_r) \mapsto \Delta_\lambda(H_r)[\mathbf{b}]^r_0 = [\mathbf{b}]^r_\lambda$ extends to a $\mathbb{Z}$-module isomorphism $B_r \to \bigwedge^r M_0$ which we shall call, abusing and anticipating the terminology of Chapter 7, *boson–fermion correspondence*: bosons are polynomials; fermions are elements of exterior powers. The terminology is definitely appropriate in the case $r = \infty$, as we shall see. The inverse isomorphism $\bigwedge^r M \to B_r$ will be denoted

$$\eta \mapsto \frac{\eta}{[\mathbf{b}]^r_0}, \tag{5.46}$$

where by the ratio we simply mean the unique element of B_r which multiplied by $[\mathbf{b}]^r_0$ gives back η.

5.9 Application to Modules of Finite Rank

5.9.1. The notation remains that of (5.39). The truncation map

$$\begin{cases} \gamma_n^0 : \quad M_0 \longrightarrow M_{0,n} \\ \displaystyle\sum_{i\ge 0} n_i b_i \longmapsto \sum_{i=0}^{n-1} n_i b_i \quad (n_i \in \mathbb{Z}) \end{cases}$$

is a homomorphism of abelian groups with kernel $\sigma_n M_0$. We have a canonical graded epimorphism, homogeneous of degree 0, $\bigwedge \gamma_n^0 : \bigwedge M_0 \longrightarrow \bigwedge M_{0,n}$ defined by

$$[\mathbf{b}]^r_\lambda \longmapsto \bigwedge^r \gamma_n^0([\mathbf{b}]^r_\lambda) := \gamma_n^0(b_{\lambda_r}) \wedge \cdots \wedge \gamma_n^0(b_{r-1+\lambda_1})$$

From now on, for brevity we set

$$\gamma_n^r := \bigwedge^r \gamma_n^0. \tag{5.47}$$

Proposition 5.9.2. *The kernel of γ_n^r is precisely the two-sided ideal*

$$\bigwedge M_0 \wedge \sigma_n M_0 := \{[\mathbf{b}]_\lambda^r \mid \lambda_1 \geq n - r + 1\}$$

generated by $\sigma_n M_0$ in $\bigwedge M_0$.

Proof. It suffices to prove that $\ker \gamma_n^r = \bigwedge^{r-1} M_0 \wedge \sigma_n M_0$. It is clear that if $\lambda_1 \geq n - r + 1$, then

$$b_{\lambda_r} \wedge b_{1+\lambda_{r-1}} \wedge \cdots \wedge b_{r-1+\lambda_1} \in \ker(\gamma_n^r).$$

Conversely, suppose that $\gamma_n^r([\mathbf{b}]_\lambda^r) = 0$. As γ_n^0 is an isomorphism when restricted to $M_{0,n}$, the vanishing implies that $r - j + \lambda_j \geq n$ for at least one $j \in \{0, 1, \ldots, r-1\}$, forcing the inequality $\lambda_1 \geq n - r + 1$. Therefore

$$\sum a_\lambda [\mathbf{b}]_\lambda^r \in \ker(\gamma_n^r) \qquad (0 \neq a_\lambda \in \mathbb{Z}) \tag{5.48}$$

if and only if all summands belong to $\bigwedge^{r-1} M_0 \wedge \sigma_n M_0$. □

5.9.3. We have just proved that γ_n^r induces an isomorphism

$$\frac{\bigwedge^r M_0}{\bigwedge^{r-1} M_0 \wedge \sigma_n M_0} \longrightarrow \bigwedge^r M_{0,n}.$$

For brevity denote by $\overline{\mathrm{ev}}_{[\mathbf{b}]_0^r}$ the composite $B_r \to \bigwedge^r M_{0,n}$ of the evaluation map (5.41) with γ_n^r. Let

$$B_{r,n} := \frac{B_r}{\ker \overline{\mathrm{ev}}_{[\mathbf{b}]_0^r}}.$$

Then the epimorphism $B_r \to \bigwedge^r M_{0,n}$ factorizes as

$$B_r \xrightarrow{\pi_{r,n}} B_{r,n} \xrightarrow{\mathrm{ev}_{[\mathbf{b}]_0^r}} \bigwedge^r M_{0,n},$$

where $\pi_{r,n}$ is the canonical projection and the second arrow is an isomorphism denoted, by a reasonable abuse of notation, in the same way as in (5.41). The diagram

$$\begin{array}{ccc} B_r & \xrightarrow{\pi_{r,n}} & B_{r,n} \\ {\scriptstyle \mathrm{ev}_{[\mathbf{b}]_0^r}} \downarrow & \searrow^{\overline{\mathrm{ev}}_{[\mathbf{b}]_0^r}} & \downarrow {\scriptstyle \mathrm{ev}_{[\mathbf{b}]_0^r}} \\ \bigwedge^r M_0 & \xrightarrow[\gamma_n^r]{} & \bigwedge^r M_{0,n} \end{array} \tag{5.49}$$

is obviously commutative, for all $r \in \mathbb{N}^*$.

Proposition 5.9.4. We have $\ker(\overline{\mathrm{ev}}_{[\mathbf{b}]_0^r}) = (h_{n-r+1}, \ldots, h_n)$.

Proof. If $j \geq 0$, then $h_{n-r+1+j} \in \ker \overline{\mathrm{ev}}_{[\mathbf{b}]_0^r}$. In fact

$$\gamma_n^r(h_{n-r+1+j} b_0 \wedge b_1 \wedge \cdots \wedge b_{r-1})$$

$$= \gamma_n^r(\sigma_{n-r+1+j} b_0 \wedge b_1 \wedge \cdots \wedge b_{r-1})$$

$$= \bigwedge^r \gamma_n^0([\mathbf{b}]_0^{r-1} \wedge b_{n+j}) = b_0 \wedge \cdots \wedge b_{r-2} \wedge \gamma_n^0(b_{n+j}) = 0.$$

Conversely, suppose that $\sum_\lambda a_\lambda \Delta_\lambda(H_r) \in B_r$ belongs to $\ker \overline{\mathrm{ev}}_{[\mathbf{b}]_0^r}$. Then

$$0 = \gamma_n^r(\sum_\lambda a_\lambda \Delta_\lambda(H_r)[\mathbf{b}]_0^r) = \sum_\lambda a_\lambda \cdot \gamma_n^r([\mathbf{b}]_\lambda^r),$$

which implies $[\mathbf{b}]_\lambda^r \in \bigwedge^{r-1} M_0 \wedge \sigma_n M_0$ because of Proposition 5.9.2. So all partitions occurring in the sum have $\lambda_1 \geq n - r + 1$. In case $\lambda_1 = n - r + 1$, we have

$$\Delta_\lambda(H_r) \in (h_{n-r+1}, \ldots, h_n).$$

We contend that h_{n+1+j} belongs to $(h_{n-r+1}, \ldots, h_n)$ for all $j \geq 0$. Using the relation

$$h_{n+1+j} + \sum_{i=1}^r (-1)^i e_i h_{n+1+j-i} = 0,$$

it is apparent that $h_{n+1+j} \in (h_{n+j}, \ldots, h_{n+j-r})$. An induction argument then shows $\ker(\overline{\mathrm{ev}}_{[\mathbf{b}]_0^r}) = (h_{n-r+1}, \ldots, h_n)$ as claimed. □

Corollary 5.9.5. *The $\mathbb{Z}$-algebra homomorphism*

$$B_{r,n} \longrightarrow \frac{\mathbb{Z}[\overline{\sigma}_1, \ldots, \overline{\sigma}_r]}{(\sigma_{n-r+1}, \ldots, \sigma_n)}$$

defined by $e_i \mapsto \overline{\sigma}_i$ is an isomorphism.

Proof. This is due to the fact that the evaluation map $\mathrm{ev}_{[\mathbf{b}]_0^r} : B_r \to \bigwedge^r M_0$ factorizes through $\mathbb{Z}[\overline{\sigma}_1, \ldots, \overline{\sigma}_r]$ and that $p(H_r)[\mathbf{b}]_0^r = 0$ if and only if $p(\boldsymbol{\sigma}_+)[\mathbf{b}]_0^r = 0$. □

5.9.6. Use the notation of 5.9.3. It turns out that $\pi_{r,n}(h_{n-r+j}) = 0$ for all $j \geq 1$. Let $H_{r,n}$ be the sequence

$$\pi_{r,n}(H_r) = (1, h_1, \ldots, h_{n-r}, 0, 0, \ldots), \tag{5.50}$$

and

$$H_{r,n}(t) := \sum_{i=0}^{n-r} h_i t^i. \tag{5.51}$$

Therefore $B_{r,n} := \bigoplus_{\boldsymbol{\lambda}\in\mathcal{P}_{r,n}} \mathbb{Z}\Delta_{\boldsymbol{\lambda}}(H_{r,n})$ is obviously isomorphic to the submodule $\bigoplus_{\boldsymbol{\lambda}\in\mathcal{P}_{r,n}} \mathbb{Z}\Delta_{\boldsymbol{\lambda}}(H_r)$ of B_r. The equality $\pi_{r,n}\Delta_{\boldsymbol{\lambda}}(H_r) = \Delta_{\boldsymbol{\lambda}}(\pi_{r,n}H_r)$ holds because $\pi_{r,n}$ is a ring epimorphism. Notice in addition that $H_{r,n} = \pi_{r,n}(H_r)$ is defined by

$$E_r(t)\sum_{i=0}^{n-r} h_i t^i = 1,$$

valid in $B_{r,n} = \pi_{r,n}(B_r)$. In fact $\pi_{r,n}\Delta_{\boldsymbol{\lambda}}(H_r) = 0$ implies $[\mathbf{b}]^r_{\boldsymbol{\lambda}} \in \bigwedge^{r-1} M_0 \wedge \sigma_n M_0$, i.e. $\Delta_{\boldsymbol{\lambda}}(H_r) \in (h_{n-r+1}, \ldots, h_n)$, that is $\lambda_1 \geq n-r+1$ and then the Young diagram of $\boldsymbol{\lambda}$ cannot be contained in an $r \times (n-r)$ rectangle.

Proposition 5.9.7. *The map* $\gamma^r_n : \bigwedge^r M_0 \to \bigwedge^r M_{0,n}$ *is a* B_r*–*$B_{r,n}$*-module homomorphism:*

$$\gamma^r_n(\Delta_{\boldsymbol{\lambda}}(H_r)[\mathbf{b}]^r_0) = (\pi_{r,n}\Delta_{\boldsymbol{\lambda}}(H_r))[\mathbf{b}]^r_0 = \Delta_{\boldsymbol{\lambda}}(H_{r,n})[\mathbf{b}]^r_0. \tag{5.52}$$

Proof. If $\boldsymbol{\lambda} \in \mathcal{P}_r \setminus \mathcal{P}_{r,n-r}$, then $\lambda_1 \geq n-r+1$, i.e. both sides of (5.52) vanish. If $\boldsymbol{\lambda} \in \mathcal{P}_{r,n}$, instead,

$$\Delta_{\boldsymbol{\lambda}}(H_r)[\mathbf{b}]^r_0 = [\mathbf{b}]^r_{\boldsymbol{\lambda}} \in \bigwedge^r M_{0,n}$$

i.e. $[\mathbf{b}]^r_{\boldsymbol{\lambda}} = \Delta_{\boldsymbol{\lambda}}(H_{r,n})[\mathbf{b}]^r_0 = \pi_{r,n}\Delta_{\boldsymbol{\lambda}}(H_r)[\mathbf{b}]^r_0$. □

5.9.8. The $\mathbb{Z}$-algebra homomorphism $\pi_{r,n} : B_r \to B_{r,n}$ has a natural extension to an epimorphism $B_r((z)) \to B_{r,n}((z))$ denoted by the same symbol. The algebra $B_{r,n}$ can also be viewed as

$$B_{r,n} = \frac{\mathbb{Z}[e_1, \ldots, e_r, h_1, \ldots, h_{n-r}]}{(E_r(t)H_{r,n}(t) - 1)}, \tag{5.53}$$

where the quotient is taken with respect to the ideal generated by the coefficients of positive degree in the expansion $E_r(t)H_{r,n}(t)$, where $H_{r,n}(t)$ is the polynomial (5.51).

Corollary 5.9.9. *The isomorphism* $B_{r,n} \mapsto \bigwedge^r M_{0,n}$ *turns* $\bigwedge^r M_{0,n}$ *into a free* $B_{r,n}$*-module* $\bigwedge^r M_{r,n}$ *of rank* 1 *generated by* $[\mathbf{b}]^r_0$*, such that for each* $\boldsymbol{\lambda} \in \mathcal{P}_{r,n}$*,*

$$[\mathbf{b}]^r_{\boldsymbol{\lambda}} = \Delta_{\boldsymbol{\lambda}}(H_{r,n})[\mathbf{b}]^r_0$$

Proof. For all $\boldsymbol{\lambda} \in \mathcal{P}_{r,n}$, define the $B_{r,n}$-module structure through

$$\Delta_{\boldsymbol{\lambda}}(H_{r,n})[\mathbf{b}]^r_{\boldsymbol{\mu}} = \Delta_{\boldsymbol{\lambda}}(H_r)[\mathbf{b}]^r_{\boldsymbol{\mu}},$$

which is well defined for $\boldsymbol{\lambda} \in \mathcal{P}_{r,n}$ because the map $\Delta_{\boldsymbol{\lambda}}(H_{r,n}) \mapsto \Delta_{\boldsymbol{\lambda}}(H_r)$ is a section $B_{r,n} \to B_r$. Moreover, for all $\boldsymbol{\lambda} \in \mathcal{P}_{r,n}$,

$$[\mathbf{b}]^r_{\boldsymbol{\lambda}} = \Delta_{\boldsymbol{\lambda}}(H_r)[\mathbf{b}]^r_0 = \Delta_{\boldsymbol{\lambda}}(H_{r,n})[\mathbf{b}]^r_0.$$ □

Corollary 5.9.10. *Let $B_{r,n} \otimes_{\mathbb{Z}} \bigwedge^r M_0$ be the $B_{r,n}$-structure of $\bigwedge^r M_{0,n}$ described in Corollary 5.9.9. Then $M_{0,n}$ can be equipped with a structure of free $B_{r,n}$-module of rank* r, *say $M_{r,n}$, such that $\bigwedge^r M_{r,n} = B_{r,n} \otimes \bigwedge^r M_{0,n}$.*

Proof. The epimorphism $M_0 \to M_{0,n}$ induces a B_r-module structure on $M_{0,n}$ which factorizes through $B_{r,n}$. Hence $M_{0,n}$ can be given a structure of free $B_{r,n}$-module of rank r that we denote by $M_{r,n}$. Clearly $\bigwedge^r M_{r,n} = B_{r,n} \otimes \bigwedge^r M_{0,n}$, where the tensor product is that induced by ev_{σ_+}. □

Example 5.9.11. The factorization $B_r \to B_{r,n} \to \bigwedge^r M_{0,n}$ allows to simplify some computations. For instance, suppose one wants to evaluate $\Delta_{(3,1)}(H_2)b_0 \wedge b_2 \in \bigwedge^2 M_0$. One has, on one hand,

$$\begin{aligned}(h_3h_1 - h_4)(b_0 \wedge b_2) =& (\sigma_3\sigma_1 - \sigma_4)(b_0 \wedge b_2) \\ =& \sigma_3(b_1 \wedge b_2 + b_0 \wedge b_3) - b_0 \wedge b_6 \\ =& b_1 \wedge b_5 + b_2 \wedge b_4 + b_1 \wedge b_5 + b_0 \wedge b_6 - b_0 \wedge b_6 \\ =& 2b_1 \wedge b_5 + b_2 \wedge b_4.\end{aligned}$$

On the other hand, since the partitions (3, 1) and (1) are both contained in a 2×3 rectangle, computations can be done in the $B_{2,5}$-module $\bigwedge^2 M_{0,5}$, where $h_4 = 0$. In this case

$$\begin{aligned}(h_3h_1 - h_4)(b_0 \wedge b_2) = h_3h_1(b_0 \wedge b_2) &= \sigma_3(b_1 \wedge b_2 + b_0 \wedge b_3) \\ &= 2b_1 \wedge b_5 + b_2 \wedge b_4,\end{aligned}$$

having used the fact that $\gamma_5^0(b_6) = 0$.

5.10 Notes and References

- For a classical take on Schubert calculus, the reader is invited to consult [55, 85], as well as the widely cited [87].
- The paper [14] offers enumerative applications of the Schubert derivation, originally suggested by a question of K. Ranestad [126]. It was about listing the numbers of projectively nonequivalent rational space curves of degree $d + 3$ in $\mathbb{P}^3$, with inflection points at $2d$ given distinct points. The solution simply amounts to compute a suitable power of a special Schubert cycle. A related problem is to determine the number $\widetilde{H}_d$ of projectively nonequivalent curves of degree $d + 3$ possessing *hyperstalls* at $2d$ given distinct points. A *hyperstall* is a smooth point on a space curve C for which there exists a plane H such that $(H \cdot C)_P \geq 4$, that is, H intersects C at P with multiplicity at least 4. From the point of view of Schubert calculus, it boils down to finding the degree

$$\int_{G_{4,d+4}} \sigma_2^{2d} \cap [G_{4,d+4}].$$

Using the dictionary 5.27, one obtains

$$\widetilde{H}_d := \int_{d+4} \sigma_2^{2d}[\mathbf{b}]_0^4 = \sum_{\substack{j_1+\cdots+j_5=2d \\ 0\le l\le j_2+j_3}} \binom{j_2+j_3}{l} \omega_{I(j_1,\ldots,j_5;l)} \tag{5.54}$$

where $\omega_{I(j_1,\ldots,j_5;l)}$ is, up to a sign, the degree of a certain Schubert variety explicitly described in [14, p. 134]. For instance,

$$\widetilde{H}_{42} = 2015171822559430028139548731191434761573293931374576969881230909739979 00.$$

The novelty is that expression (5.54) can be computed using the following formula proven in [14, Theorem 2.5]:

$$\sigma_h^m(b_i \wedge \eta) = \sum_{j=0}^{m} \binom{m}{j} \sigma_{h-1}^j (b_{i+j} \wedge \sigma_h^{m-j}\eta).$$

which cannot be phrased (or at least we do not know how) within classical Schubert calculus. It involves in fact the interaction of Schubert calculus on more than one Grassmannian at once.

- That the Chow ring of $G_{r,n}$ is entirely determined once one knows the intersection product of any Schubert cycle with the class of a divisor only (Chevalley formula) is shown in [14], which also provides a purely algebro-geometric proof of a formula, holding in the cohomology of the Grassmannian of lines, that Inna Scherbak obtained in [136] using techniques from representation theory.
- As a further application of [14], we should also cite the paper by Costa, Marchesi and Miró-Roig [15], who proved the existence of certain indecomposable bundles over $G_{r,n}$, called *Tango bundles*.
- Within a quite different perspective with respect to [49], relationships between linear ODEs and Schubert calculus are faced in the paper [139] by Shapiro and Shapiro, where the authors review basic results and relations between the qualitative theory of linear ordinary differential equations with real time and the reality problems in Schubert calculus. For more connections between linear ODEs and differential equations, we should cite the wide area of research regarding the theory of Heine–Stieltjes, for which we refer to [138] and references therein.
- From the discussion of Section 5.4 and the isomorphism between $\mathbb{Z}[\boldsymbol{\sigma}_+]$ and B_r, we know that there is a $\mathbb{Z}$-algebra isomorphism between $B_{r,n}$ and the singular cohomology of the Grassmann variety $H^*(G_{r,n}, \mathbb{Z})$. The isomorphism is given explicitly by $h_i \mapsto \sigma_i := c_i(\mathcal{Q}_r)$. One of the arguments used to explain it, is based on the known fact that in the cohomology ring of the Grassmannian, the polynomial X^n factorizes universally as product of monic polynomials of degrees

r and $n-r$ (cf. Exercise 5.11.9). This point of view was extensively developed by Laksov–Thorup [96–98] and Laksov [93, 94] in the case of equivariant Schubert calculus.

- The $\mathbb{Z}$-module isomorphisms $B_{1,n} \cong M_{0,n}$ and $B_{r,n} \longrightarrow \bigwedge^r M_{0,n}$ show that there is a natural identification between $\bigwedge^r H_*(\mathbb{P}^{n-1},\mathbb{Z})$ and $H^*(G_{r,n},\mathbb{Z})$. Composing the isomorphism $\bigwedge^r H_*(\mathbb{P}^{n-1},\mathbb{Z}) \to H^*(G_{r,n},\mathbb{Z})$ with the Poincaré isomorphism (5.22) gives the *Satake identification* [36, 54] :

$$\begin{cases} \textbf{Sat}: & \bigwedge^r H^*(\mathbb{P}^{n-1}) \longrightarrow H^*(G_{r,n}) \\ & \sigma_{\lambda_1+r-1}\wedge\sigma_{\lambda_2+r-2}\wedge\cdots\wedge\sigma_{\lambda_r} \longrightarrow \sigma_{\boldsymbol{\lambda}}, \end{cases}$$

where $\sigma_{\lambda_i+r-i} = (\sigma_1^{\lambda_i+r-i})$ is the special Schubert class of $\mathbb{P}^{n-1} = G_{1,n}$, namely, the class of a linear space of dimension $n-r+i-\lambda_i$. The arguments used there are based on the map that 'takes the span', $(\mathbb{P}^{n-1})^r \dashrightarrow G_{r,n}$, and the fact that the cohomology groups of $\mathbb{P}^{n-1}$ and $G_{r,n}$ are representations of $Gl_n(\mathbb{C})$. In the language of [36, 54], $H^*(\mathbb{P}^{n-1})$ is the standard representation of $Gl(H^*(\mathbb{P}^{n-1})) = Gl_n(\mathbb{C})$, while $H^*(G_{r,n})$ is the rth wedge power of it. A quantum version of the Satake identification, essentially as in [36, p. 47–48], was already used in [37, Corollary 2.7] to obtain the quantum Pieri formula. The identification of the $H^*(G_{r,n},\mathbb{Z})$-module $H_*(G_{r,n},\mathbb{Z})$ with the $A[\boldsymbol{\sigma}_+]_r$-module $\bigwedge^r M_{0,n}$ was observed in [37, 47], where small quantum Schubert calculus is addressed, too. It is shown that if $p(X) := X^n - q$, then the Schubert derivation satisfies the *quantum Pieri rule* as in [8, 12].
- The series of papers [96–98] tackle Schubert calculus by considering exterior powers of polynomial rings but avoiding to phrase the results in terms of the entire exterior algebra. The formalism of Schubert calculus is recovered by looking at $\bigwedge^r A[X]$ as a free module over the ring of symmetric polynomials in r indeterminates, generated by $X^{r-1}\wedge\cdots\wedge X\wedge 1$. This is yet another algebraic way to spell out the approximation to finite exterior powers of the boson–fermion correspondence [78, Theorem 6.1]. Laksov and Thorup write in [96, p. 827] that

  > '*In the work of E. Date, M. Jimbo, M. Kashiwara, and T. Miwa* [18] *Schur functions appear in connection with exterior products in another context. See also the work of V. G. Kac and A. K. Raina* [77]",

  while the present book shows that their picture fits in exactly the same framework. Indeed, had they looked their impressive works in terms of exterior algebras would have inevitably led those authors to an earlier emergence of vertex operators in their elegant framework.
- For equivariant Schubert calculus, see [35, 93, 94] and references therein, while for a general introduction to equivariant cohomology in algebraic geometry, the beautiful lecture notes by Anderson [4] are advised. A uniform treatment of equivariant quantum Schubert calculus, also for Grassmannians of other Lie types, is treated in the recent papers [66, 106] by Yongdong Huang, Changzheng Li and V. Ravikumar. See also [149], [93, 94] and references therein.

5.11 Exercises

5.11.1. Cf. Exercise 3.7.3. Prove that a tensor $\eta \in \bigwedge^2 M_0$ is decomposable if and only if $\eta \wedge \eta = 0$. Write such equation for a generic $\eta \in \bigwedge^2 \mathbb{C}^4$, and find the equation of the Plücker embedding of the Grassmannian $G_{2,4}$ in $\mathbb{P}^5$.

5.11.2. Compute $\sigma_3(b_2 \wedge b_4 \wedge b_6)$ using Pieri formula (5.10) or by a direct application of Leibniz rule. Do the same for $\overline{\sigma}_3(b_2 \wedge b_4 \wedge b_6)$ and $\overline{\sigma}_3(b_i \wedge b_j)$.

5.11.3. Consider the Schubert derivation $\sigma_+(z)$ on $\bigwedge M_0$. Using integration by parts, prove the following identity:

$$b_{\lambda_2} \wedge b_{1+\lambda_1} = \sigma_{\lambda_2}(b_0 \wedge b_{1+\lambda_1}) + \sigma_{1+\lambda_1}(b_{\lambda_2} \wedge b_0).$$

Deduce from it the *Giambelli formula*:

$$[\mathbf{b}]^2_{\boldsymbol{\lambda}} = \begin{vmatrix} \sigma_{\lambda_1} & \sigma_{\lambda_2 - 1} \\ \\ \sigma_{\lambda_1+1} & \sigma_{\lambda_2} \end{vmatrix} [\mathbf{b}]^2_0.$$

5.11.4. Show that the Giambelli formula $[\mathbf{b}]^r_{\boldsymbol{\lambda}} = \Delta_{\boldsymbol{\lambda}}(\boldsymbol{\sigma}_+)[\mathbf{b}]^r_0$ can be proved by induction assuming the following equality (cf. [55, p. 205, formula (**)]):

$$b_{i_1} \wedge \cdots \wedge b_{i_r} = \sum_{j=1}^{r} (-1)^{j-1} \sigma_{i_j}(b_0 \wedge b_{i_1} \wedge \ldots \wedge \hat{b}_{i_j} \wedge \cdots \wedge b_{i_r}) \tag{5.55}$$

where $\hat{b}_{i_j}$ means that b_{i_j} has been omitted.

Challenge 5.11.5. Prove formula (5.55) in general or at least for $r = 3$.

5.11.6 (See [132]). Take $p = X^{n+2}$, so that M_p is the span of $(b_0, b_1, \ldots, b_{n+1})$, and let $0 \le m \le n$.

i) Using Newton's multinomial formula, cf. Exercise 4.5.1, find an expression for

$$K(m,n) := \int_{2n} \sigma_1^{2m} \sigma_2^{n-m} [\mathbf{b}]^2_0.$$

ii) Prove that $K(n,n)$ is the nth Catalan number (or, alternatively, the *Plücker degree* of the Grassmannian $G_{2,n+2}$; see Section 5.3).
iii) Prove the recursion formula

$$K(m,n) = K(m+1,n) - K(m,n-1). \tag{5.56}$$

It shows that certain numbers, occurring in a problem of path enumeration studied by Niederhausen (*Catalan traffic at the beach*; see [120]), are top intersection numbers in the cohomology ring of the Grassmannian of lines in complex projective $(n+1)$-space.

5.11.7. Fix $r > 0$ and consider the formal power series

$$\sum_{\boldsymbol{\lambda}\in\mathcal{P}_r} [\mathbf{b}]^r_{\boldsymbol{\lambda}} \cdot \mathbf{t}^{\boldsymbol{\lambda}} \in \bigwedge^r M_0[[t_1, \ldots, t_r]],$$

where if $\boldsymbol{\lambda} \in \mathcal{P}_r$, by $\mathbf{t}^{\boldsymbol{\lambda}}$ we mean $t_1^{\lambda_1} \cdots t_r^{\lambda_r}$. Let $h_j(\mathbf{t}^{-1}) := h_j(t_1^{-1}, \ldots, t_r^{-1})$ be the complete symmetric polynomials in $(t_1^{-1}, \ldots, t_r^{-1})$ (see Exercise 2.8.3). Prove that:

i) $\sigma_+(z) \sum_{\boldsymbol{\lambda}\in\mathcal{P}_r} [\mathbf{b}]^r_{\boldsymbol{\lambda}} \mathbf{t}^{\boldsymbol{\lambda}} = \sum_{\boldsymbol{\lambda}\in\mathcal{P}_r, j\geq 0} [\mathbf{b}]^r_{\boldsymbol{\lambda}} \mathbf{t}^{\boldsymbol{\lambda}} \cdot h_j(\mathbf{t}^{-1}) z^j.$

ii) If $\boldsymbol{\lambda} \in \mathbb{N}^r$, define $\Delta_{\boldsymbol{\lambda}}(H_r)$ as being $\det(h_{\lambda_j - j + i})_{1\leq i,j\leq r}$ if $\boldsymbol{\lambda} \in \mathcal{P}_r$ and 0 otherwise. Show that Pieri formula (5.10) is equivalent to

$$\sigma_+(z) \sum_{\boldsymbol{\lambda}\in\mathcal{P}_r} \Delta_{\boldsymbol{\lambda}}(H_r) \mathbf{t}^{\boldsymbol{\lambda}} = \sum_{\boldsymbol{\lambda}\in\mathbb{N}^r, j\geq 0} \Delta_{\boldsymbol{\lambda}}(H_r) \cdot h_j(\mathbf{t}^{-1}) \mathbf{t}^{\boldsymbol{\lambda}} z^j,$$

where by $\sigma_i \Delta_{\boldsymbol{\lambda}}(H_r)$ we mean the element of B_r defined by the equality $(\sigma_i \Delta_{\boldsymbol{\lambda}}(H_r))[\mathbf{b}]^r_0 = \sigma_i [\mathbf{b}]^r_{\boldsymbol{\lambda}}$, according to the natural B_r-module structure of $\bigwedge^r M_0$.

iii) Try to guess a formula for $\sum_{\boldsymbol{\lambda}\in\mathcal{P}_r} \overline{\sigma}_+(z) [\mathbf{b}]^r_{\boldsymbol{\lambda}} \mathbf{t}^{\boldsymbol{\lambda}}$ similar to i) and prove it.

5.11.8. This exercise presupposes a little familiarity with, e.g. [34, Section14.6]. The notations are as in 5.4.4. Let $\rho : \mathcal{E} \to \mathcal{X}$ be a vector bundle of rank r over an m-dimensional smooth connected variety $\mathcal{X}$ and $\mathbb{P}(\mathcal{E}) := G_1(\mathcal{E})$ the associated projective bundle. Set $M := A_*(\mathbb{P}(E))$ and call $\xi\cap : M \to M$ the map sending $\alpha \in A_*(\mathbb{P}(E))$ to $c_1(\mathcal{S}_1) \cap \alpha$. It is known that M is an $A_*(\mathcal{X})$-module generated by $b_0 := [\mathbb{P}(\mathcal{E})]$ and $b_i := \xi^i \cap b_0$. Consider the trace polynomial operator $\overline{D}^\xi(z) : \bigwedge M \to \bigwedge M[[z]]$ in the sense of Section 4.2. Show that the vocabulary (5.24) is equivalent to saying that the Chern classes $c_i(\mathcal{E})$ of $\rho : \mathcal{E} \to \mathcal{X}$ are the *traces* of the endomorphism $\xi\cap$.

5.11.9. If A is a commutative ring with unit and $P(X) \in A[X]$ an arbitrary monic polynomial of degree n, there is an A-algebra $A_{P;(r,n)}$, unique up to isomorphism, such that P, regarded as element of $A_{P;(r,n)}[X]$, can be written as the product of two monic polynomials $P_r, P_{n-r} \in A_{P;(r,n)}[X]$ of degrees r and $n - r$, respectively, satisfying the following universal property. If A' is any A-algebra in which P decomposes as the product of monic polynomials $Q_r, Q_{n-r} \in A_{P;(r,n)}[X]$, there is a unique algebra homomorphism mapping the coefficients of P_r to those of Q_r and those of P_{n-r} to those of Q_{n-r}. Such distinguished algebra is called in [96] the *universal factorization algebra* of P into the product of monic polynomials of degrees r and $n - r$. Let $n \geq r$. Prove that the ring $B_{r,n}$, which is isomorphic to $\bigwedge^r M_{0,n}$ as a $\mathbb{Z}$-module, is the universal factorization algebra of X^n into the product of two monic polynomials of $B_{r,n}[X]$ of degree r and $n-r$. Hence taking the rth exterior power of a free abelian group of rank n is a geometric procedure to universally decompose the polynomial X^n.

5.11.10. Generalize Exercise 5.11.9. Let $P \in \mathbb{Z}[X]$ be monic of degree n. Let $M_{0,P} := \mathbb{Z}[X]/(P)$. The map $B_r \to \bigwedge^r M_{0,P}$, composition of $\bigwedge^r M_0 \to \bigwedge^r M_{0,P}$ with $B_r \to \bigwedge^r M_0$, factorizes through a $\mathbb{Z}$-algebra $B_{r,P}$. Show that $B_{r,P}$ is the universal decomposition algebra of P into the product of two monic polynomials $P_r Q_{n-r}$ of degree r and $n-r$, respectively. So, in general, taking the exterior power of a free module of rank n is equivalent to universally decompose a monic polynomial of degree n. (Hint: define $P_r = X^r E_r(1/X)$ and $P_r(X)Q_{n-r}(X) = P(X)$).

For more on this subject see [24].

Chapter 6
Decomposable Tensors in Exterior Powers

The geometrical picture presented in this chapter is essentially a reworking, within a finite-dimensional context, of the idea of writing Plücker equations for the infinite Grassmannian parametrizing the solutions of the KP hierarchy, classically due to Sato, Date, Jimbo, Kashiwara and Miwa [18, 19, 133, 134]. All this can be told in a purely algebraic way from the point of view of vertex operators [5, 73, 78, 79].

The purpose is to show how Schubert derivations enable to write a finite-dimensional version of the KP hierarchy by means of a compact formula describing the Grassmann cone of decomposable tensors in an exterior power. That this was possible certainly comes as no surprise. However, while writing this book, we were not aware of a precise reference for the main formula we prove in Theorem 6.4.3, here merely proposed as a different pedagogical point of view. Its proof is based on the following main ingredients. First, we rephrase a well-known criterion (Theorem 6.1.7) to detect decomposable tensors of an exterior power in terms of the vanishing of the residue of a formal power series, as we learned, e.g. in the references [5, 78], where the game is played within an infinite-dimensional context. We will prove it through arguments alternative to [78, Proposition 7.2], in order to stay elementary.

The second ingredient is the pivotal aspect and the main novelty of the chapter, if not of the entire text: a sort of mirror Schubert derivation on $\bigwedge M_0$, denoted by $\sigma_-(z) := \sum_{i\geq 0}\sigma_{-i}z^{-i}$. As anticipated in the introduction, this is the unique HS-derivation $\bigwedge M_0 \to (\bigwedge M_0)[[z^{-1}]]$ extending the (-1)-shift endomorphism of M_0, mapping $b_i \to b_{i-1}$ if $i > 0$ and to 0 otherwise.

This chapter is loosely based on, and inspired by, the material of [44, 45]. Here is how it is organized. Sections 6.2 deal with the HS-derivation $\sigma_-(z)$. Notably, $\sigma_-(z)$ enjoys, unlike $\sigma_+(z)$, a certain *stability* property which permits to define it, in an equivalent way, as a map $B_r \to B_r[z^{-1}]$ (Section 6.2.7).

The approximation of the bosonic counterparts of the vertex operators related to the fermionic vertex algebra appears in Section 6.3. It will be applied in Section 6.4 to prove the main Theorem 6.4.3 and to illustrate a couple of dutiful examples.

L. Gatto, P. Salehyan, *Hasse-Schmidt Derivations on Grassmann Algebras*, IMPA Monographs, DOI 10.1007/978-3-319-31842-4_6

Section 6.5, finally, can be seen as a short appendix regarding the construction of infinite exterior powers as suitable projective limits of finite ones, by invoking the isomorphism $B_r \to \bigwedge^r M_0$.

6.1 A Criterion for Decomposability

Wedging and Contracting 6.1.1. Notation as in 5.6. Recall that any element $m \in M_0$ induces a *wedging homomorphism*

$$\begin{cases} m\wedge : \bigwedge^k M_0 \longrightarrow \bigwedge^{k+1} M_0 \\ \qquad\quad \eta \quad \longmapsto \quad m \wedge \eta. \end{cases}$$

On a dual side, any $\mu \in M_0^\vee := Hom_{\mathbb{Z}}(M_0, \mathbb{Z})$ induces a *contraction homomorphism*

$$\mu\lrcorner : \bigwedge^{r} M_0 \to \bigwedge^{r-1} M_0,$$

which, as specified in Example 3.5.3, is the unique derivation of degree 0 which is trivial on $\mathbb{Z}$ and such that $\mu\lrcorner m = \mu(m)$, for all $m \in M_0$. It can be recursively defined as

$$\mu\lrcorner b_{i_0} \wedge (b_{i_1} \wedge \cdots \wedge b_{i_{r-1}}) = \mu(b_{i_1}) b_{i_1} \wedge \cdots \wedge b_{i_{r-1}} - b_{i_0} \wedge \mu\lrcorner(b_{i_1} \wedge \cdots \wedge b_{i_{r-1}}).$$

Example 6.1.2. If $\mu \in M_0^\vee$ and $b_{i_0} \wedge b_{i_1} \wedge b_{i_2} \in \bigwedge^3 M_0$, one has

$$\mu\lrcorner(b_{i_0} \wedge b_{i_1} \wedge b_{i_2}) = \mu(b_{i_0}) b_{i_1} \wedge b_{i_2} - b_{i_0} \wedge \mu\lrcorner(b_{i_1} \wedge b_{i_2}). \tag{6.1}$$

Now

$$\mu\lrcorner(b_{i_1} \wedge b_{i_2}) = \mu(b_{i_1}) b_{i_2} - b_{i_1} \wedge \mu\lrcorner b_{i_2} = \mu(b_{i_1}) b_{i_2} - \mu(b_{i_2}) b_{i_1}$$

which substituted into (6.1) gives

$$\mu\lrcorner(b_{i_0} \wedge b_{i_1} \wedge b_{i_2}) = \mu(b_{i_0}) b_{i_1} \wedge b_{i_2} - \mu(b_{i_1}) b_{i_1} \wedge b_{i_2} + \mu(b_{i_2}) b_{i_0} \wedge b_{i_1}.$$

Definition 6.1.3. A tensor $\eta \in \bigwedge^r M_0$ is *divisible* by $m \in M_0$ if $\eta \wedge m = 0$; it is *decomposable* if there exist $m_1, \ldots, m_r \in M_0$ such that $\eta = m_1 \wedge \cdots \wedge m_r$.

Remark 6.1.4. It is dutiful to observe that studying decomposability of tensors in $\bigwedge^r M_0$ is equivalent to study decomposability for tensors in $\bigwedge^r M_{0,n}$ for some sufficiently large n. In fact, an arbitrary element of $\bigwedge^r M_0$ is a *finite* linear combination of the form

$$\eta = \sum_{\lambda \in \mathcal{P}_r} a_\lambda [\mathbf{b}]^r_\lambda, \tag{6.2}$$

i.e. there exists $n \in \mathbb{N}$ such that the sum (6.2) coincides with $\sum_{\lambda \in \mathcal{P}_{r,n}} a_\lambda [\mathbf{b}]^r_\lambda$.

6.1.5. All the elements of $\bigwedge M_0$ in degrees 0 and 1 are obviously decomposable. All the elements of $\bigwedge^{n-1} M_0$ are decomposable modulo $\sigma_n M_0 \wedge \bigwedge^{n-2} M_0$ (Exercise 6.7.2). Hence all the tensors of $M_{0,n}$ are decomposable for $n \leq 3$. Thus, indecomposable tensors only occur for exterior powers $\bigwedge^k M_{0,n}$ for $1 < k < n-1$ and $n \geq 4$. It is a simple observation that $\eta \in \bigwedge^r M_0$ is decomposable if the kernel of the map $M_0 \to \bigwedge^{r+1} M_0$ which sends m to $\eta \wedge m$ has rank r.

6.1.6. Let $(\beta_j)_{j\geq 0} \in M_0^\vee$ such that $\beta_j(b_i) = \delta_{ij}$. The vectors $(\beta_j)_{j\geq 0}$ form a basis of $M_0^* \subseteq M_0^\vee$, called the *restricted dual of* M_0. The dual and the restricted dual coincide only if the rank of M_0 is finite. For example, the element $\beta \in M_0^\vee$ such that $\beta(b_j) = 1$, for all $j \geq 0$, cannot be written as a finite linear combination of the β_js; thus it is not an element of the restricted dual M_0^*. The following characterization shall be heavily used in the sequel.

Theorem 6.1.7. *An element $\eta \in \bigwedge^r M_0$ is decomposable if and only if*

$$\sum_{i\geq 0} (b_i \wedge \eta) \otimes (\beta_i \lrcorner \eta) = 0. \tag{6.3}$$

Proof. Note preliminarily that i) expression (6.3) is a finite sum, as η is a finite $\mathbb{Z}$-linear combination $\sum a_{\boldsymbol{\lambda}} [\mathbf{b}]^r_{\boldsymbol{\lambda}}$, and thus $\beta_j \lrcorner \eta = 0$ for all j but finitely many, and ii) that η is *divisible* by m if and only if

$$0 = \eta \wedge m = \eta \wedge \sum_{i=0}^{r-1} \beta_i(m) b_i = \sum_{i=0}^{r-1} (\eta \wedge b_i) \beta_i(m) = 0. \tag{6.4}$$

Suppose now that η is decomposable, i.e. that $\eta = m_0 \wedge \cdots \wedge m_{r-1}$ for some $m_i \in M_0$ $(0 \leq i \leq r-1)$. Then

$$\begin{aligned} \sum_{i\geq 0} (b_i \wedge \eta) \otimes (\beta_i \lrcorner \eta) &= \sum_{i\geq 0} (b_i \wedge \eta) \otimes \beta_i \lrcorner (m_0 \wedge \cdots \wedge m_{r-1}) = \\ &= \sum_{j=0}^{r-1} (-1)^j \sum_{i\geq 0} (b_i \wedge \eta) \beta_i(m_j) m_0 \wedge \cdots \wedge \hat{m}_j \wedge \cdots \wedge m_{r-1} \\ &= 0, \end{aligned}$$

due to (6.4) applied to each sum

$$0 = \sum_{i\geq 0} a_{ij} (b_i \wedge \eta) := \sum_{i\geq 0} (b_i \wedge \eta) \beta_i(m_j), \qquad (0 \leq j \leq r-1),$$

and where the notation '$\hat{m}_j$' means that m_j is omitted. Conversely, assume (6.3) holds and let us prove the claim by induction. If $r = 1$, the property is obviously true. Let us suppose that all $\eta_1 \in \bigwedge^{r-1} M_0$ satisfying equation (6.3) are decomposable, and let $\eta \in \bigwedge^r M_0$ satisfying (6.3) as well. Then η is divisible. If $\eta = 0$, this is clear. If $\eta \neq 0$, then $\beta_j \lrcorner \eta$ cannot vanish for all j, and there must exist $\xi \in (\bigwedge^{r-1} M_0)^\vee$ such that $a_i := \xi(\beta_i \lrcorner \eta) \neq 0$ for some i. Thus

$$\sum_{j\geq 0} a_j(b_j \wedge \eta) = 0$$

is a nontrivial linear relation, implying that η is divisible by 6.4 (remember that $a_j = 0$ for all j but finitely many, as the same holds for $\beta_j \lrcorner m$). Write $\eta = m_1 \wedge \eta_1$, for some $m_1 \in M_0$ and $\eta_1 \in \bigwedge^{r-1} M_0$. Then

$$\begin{aligned} 0 &= \sum_{j\geq 0}(b_j \wedge m_1 \wedge \eta_1) \otimes (\beta_j \lrcorner (m_1 \wedge \eta_1)) \\ &= \sum_{i\geq 0}(b_j \wedge m_1 \wedge \eta_1) \otimes \big(\beta_j(m_1)\eta_1 - m_1 \wedge (\beta_j \lrcorner \eta)\big) \end{aligned}$$

$$= \sum_{i\geq 0}(b_i \wedge m_1 \wedge \eta_1) \otimes (\beta_i(m_1)\eta_1) + \sum_{i\geq 0}(m_1 \wedge b_i \wedge \eta_1) \otimes (m_1 \wedge (\beta_i \lrcorner \eta_1)), \tag{6.5}$$

i.e. finally

$$\sum_{i\geq 0}(m_1 \wedge b_i \wedge \eta_1) \otimes (m_1 \wedge (\beta_i \lrcorner \eta_1)) = 0. \tag{6.6}$$

In fact the first summand of (6.5) is zero, again due to (6.4). By Exercise 6.7.3 equality (6.6) is equivalent to

$$\sum_{i\geq 0}(b_i \wedge \eta_1) \otimes (\beta_i \lrcorner \eta_1) = 0,$$

and then η_1 is decomposable by the inductive hypothesis. Thus η itself is decomposable as claimed. □

Proposition 6.1.8. *The epimorphism $\gamma_n^r : \bigwedge^r M_0 \to \bigwedge^r M_{0,n}$ of* (5.47) *maps decomposable tensors onto decomposable tensors.*

Proof. Suppose $\eta = m_1 \wedge \cdots \wedge m_r \in \bigwedge^r M_0$. Then

$$\gamma_n^r(\eta) = \bigwedge^r \gamma_n^0(\eta) = \gamma_n^0(m_1) \wedge \cdots \wedge \gamma_n^0(m_r),$$

i.e. γ_n^r maps decomposable tensors to decomposable tensors. That a decomposable tensor in $\bigwedge^r M_{0,n}$ is an image of a decomposable tensor of $\bigwedge^r M_0$ is obvious by Remark 6.1.4. □

Corollary 6.1.9. *The tensor* $\eta \in \bigwedge^r M_{0,n}$ *is decomposable if and only if*

$$\sum_{i=0}^{n-1} (b_i \wedge \eta) \otimes (\beta_i \lrcorner \eta) = 0. \tag{6.7}$$

Proof. The inclusion $\bigwedge^r M_{0,n} \subseteq \bigwedge^r M_0$ implies that if η is decomposable, then (6.3) holds. However, since η is a linear combination of $[\mathbf{b}]^r_{\boldsymbol{\lambda}}$, with $\boldsymbol{\lambda} \in \mathcal{P}_{r,n}$, all terms $\beta_j \lrcorner \eta$ vanish for $j \geq n$, whence formula (6.7) follows. Conversely, if $\eta \in \bigwedge^r M_{0,n}$ satisfies (6.7), then

$$0 = \sum_{i=0}^{n-1} (b_i \wedge \eta) \otimes (\beta_i \lrcorner \eta) + \sum_{i \geq n} (b_i \wedge \eta) \otimes (\beta_i \lrcorner \eta),$$

because $\beta_i \lrcorner \eta = 0$ if $i \geq n$. □

6.2 A Schubert Derivation with a Stability Property

6.2.1. Let $\mathbb{Z}[X^{-1}, X] = \bigcup_{n \in \mathbb{N}} \left\{ \sum_{i \geq -n} a_i X^i \;\middle|\; a_i \in \mathbb{Z} \right\}$ be the $\mathbb{Z}$-module of *Laurent polynomials*. As in Section 2.1.6, the obvious epimorphism $\mathbb{Z}[X^{-1}, X] \to \mathbb{Z}[X]$ mapping $\sum_{i \geq -n} a_i X^i$ to its 'holomorphic part' $\sum_{i \geq 0} a_i X^i$ has kernel $X^{-1}\mathbb{Z}[X^{-1}]$. The identification

$$\mathbb{J} : \frac{\mathbb{Z}[X]}{X^{-1}\mathbb{Z}[X^{-1}]} \longrightarrow \mathbb{Z}[X]$$

allows to define, for all $j \geq 0$, the endomorphism $X^{-j} \cdot : \mathbb{Z}[X] \to \mathbb{Z}[X]$ via

$$X^{-j} \cdot \mathfrak{p}(X) := \mathbb{J}\big(X^{-j}\mathfrak{p}(X) + X^{-1}\mathbb{Z}[X^{-1}]\big).$$

In other words $X^{-j} \cdot X^i = X^{i-j}$ if $i \geq j$ and 0 otherwise.

The Schubert derivation in Section 5.1 was denoted by $\sigma_+(z)$ in view of the coming into play of another related *HS*-derivation:

$$\sigma_-(z) : \bigwedge M_0 \to \bigwedge M_0[z^{-1}] \subseteq \bigwedge M_0[[z^{-1}]].$$

It is, by definition, the unique *HS*-derivation such that

$$\sigma_-(z)m = \sum_{j \geq 0} \sigma_{-j} m \cdot z^{-j} := \sum_{j \geq 0} (X^{-j} m) z^{-j}. \tag{6.8}$$

Because of Exercise 6.7.5, it is appropriate to call $\sigma_-(z)$ *Schubert derivation* as well.

Remark 6.2.2. For notational coherence, we should have denoted this further *HS*–derivation by $\sigma_-(z^{-1})$. Nevertheless, we have chosen to write simply $\sigma_-(z)$ to avoid too heavy notation and believe that the subscripts '$\pm$', decorating σ, eliminate any source of confusion.

6.2.3. As in the case of $\sigma_+(z)$, it is also necessary to consider the inverse *HS*-derivation of $\sigma_-(z)$ on $\bigwedge M_0$, namely,

$$\overline{\sigma}_-(z) := \sum_{i\geq 0}(-1)^i\overline{\sigma}_{-i}z^{-i},$$

defined via

$$\overline{\sigma}_-(z)\sigma_-(z) = \sigma_-(z)\overline{\sigma}_-(z) = \mathbb{1}_{\bigwedge M_0} \in \mathrm{End}_{\mathbb{Z}}\Big(\bigwedge M_0\Big)[[z^{-1}]].$$

Observe that for each $\eta \in \bigwedge M_0$, we have $\sigma_-(z)\eta \in \bigwedge M_0[z]$, and 4.2.1 implies that $\overline{\sigma}_-(z)$ is precisely the *trace operator polynomial* associated with the shift endomorphism $X^{-1}\cdot : M_0 \to M_0$ of step -1. Moreover, with respect to the bi-grading

$$\bigwedge M_0 = \bigoplus_{r,w\geq 0}\Big(\bigwedge^r M_0\Big)_w$$

we have:

Proposition 6.2.4. *For all $j \geq 0$, the endomorphisms $\sigma_{-j}, \overline{\sigma}_{-j}$ are bi-homogeneous of degree $(0,-j)$: they map $\bigwedge^r M_0$ to itself, and, when restricted to $\bigwedge^r M_0$, they are homogeneous of degree $-j$ with respect to the weight graduation* (3.2.3):

$$\sigma_{-j}\Big(\bigwedge^r M_0\Big)_w \subseteq \Big(\bigwedge^r M_0\Big)_{w-j} \qquad \text{and} \qquad \overline{\sigma}_{-j}\Big(\bigwedge^r M_0\Big)_w \subseteq \Big(\bigwedge^r M_0\Big)_{w-j}.$$

Proof. Since $\sigma_{-j}, \overline{\sigma}_{-j}$ preserve M_0, an easy induction argument proves them homogeneous of degree 0 with respect to the exterior algebra graduation. Moreover

$$\sigma_{-j}[\mathbf{b}]^r_{\boldsymbol{\lambda}} = \sum_{j_0+\cdots+j_{r-1}=j} b_{\lambda_r-j_0}\wedge\cdots\wedge b_{r-1+\lambda_1-j_{r-1}},$$

a sum of elements of weight $|\boldsymbol{\lambda}|-j$. So σ_{-j} is homogeneous of degree $-j$ with respect to the weight graduation. The same holds for $\overline{\sigma}_{-j}$ by arguing analogously. □

Lemma 6.2.5. For all $j, r \geq 1$,

$$\sigma_{-j}[\mathbf{b}]^r_0 = \overline{\sigma}_{-j}[\mathbf{b}]^r_0 = 0.$$

Proof. We have that $\sigma_{-j}b_0 = 0$, for all $j \geq 1$. Hence we can argue by induction. Suppose now that $\sigma_{-j}[\mathbf{b}]_0^{r-1} = 0$. Then

$$\sigma_{-j}[\mathbf{b}]_0^r = \sum_{k=0}^{j} \sigma_{-j+k}[\mathbf{b}]_0^{r-1} \wedge \sigma_{-k}b_{r-1}$$

and using the inductive hypothesis, this is equal to

$$[\mathbf{b}]_0^{r-1} \wedge \sigma_{-j}b_{r-1} = [\mathbf{b}]_0^{r-1} \wedge b_{r-1-j},$$

which is zero because $r-1-j \leq r-2$ and by the skew-symmetry of the wedge product. The proof for $\overline{\sigma}_{-j}$ is completely analogous and left as an exercise. □

The next result aims to convince the reader that $\sigma_-(z)$ is far from being a useless optional feature, a mere extension of the Schubert derivation. Quite the contrary, it is a denizen of the theory, for crops up in a very natural way even if we were unaware of its existence.

Theorem 6.2.6. *For all* $\eta \in \bigwedge^r M_0$,

$$\sum_{j\geq 0} b_j z^j \wedge \eta = (-1)^r z^r \sigma_+(z)\overline{\sigma}_-(z)(b_0 \wedge \overline{\sigma}_r \eta). \tag{6.9}$$

Proof. First we notice that $\sum_{j\geq 0} b_j z^j \wedge \eta = \sigma_+(z)b_0 \wedge \eta$, by the very definition of the Schubert derivation $\sigma_+(z)$. Integration by parts (4.6) yields

$$\sum_{j\geq 0} b_j z^j \wedge \eta = \sigma_+(z)(b_0 \wedge \overline{\sigma}_+(z)\eta). \tag{6.10}$$

Without loss of generality, we may assume $\eta = [\mathbf{b}]_\lambda^r$. Therefore, recalling that, by Remark 5.1.2 d), $\overline{\sigma}_j[\mathbf{b}]_\lambda^r = 0$ for all $j \geq r+1$, we get:

$$\begin{aligned} b_0 \wedge \overline{\sigma}_+(z)[\mathbf{b}]_\lambda^r &= b_0 \wedge ([\mathbf{b}]_\lambda^r - \overline{\sigma}_1[\mathbf{b}]_\lambda^r z + \cdots + (-1)^r\overline{\sigma}_r[\mathbf{b}]_\lambda^r z^r) \\ &= z^r b_0 \wedge \left(\frac{1}{z^r}[\mathbf{b}]_\lambda^r - \frac{1}{z^{r-1}}\overline{\sigma}_1[\mathbf{b}]_\lambda^r + \cdots + (-1)^r\overline{\sigma}_r[\mathbf{b}]_\lambda^r\right). \end{aligned} \tag{6.11}$$

Now

$$\begin{aligned} b_0 \wedge \overline{\sigma}_i[\mathbf{b}]_\lambda^r &= b_0 \wedge \overline{\sigma}_i(b_{\lambda_r} \wedge b_{1+\lambda_{r-1}} \wedge \cdots \wedge b_{r-1+\lambda_1}) = \\ &= b_0 \wedge \sum b_{\lambda_r+i_1} \wedge b_{1+\lambda_{r-1}+i_2} \wedge \cdots \wedge b_{r-1+\lambda_1+i_r}, \end{aligned} \tag{6.12}$$

where the sum is over all $(i_1, \ldots, i_r)$ such that $0 \leq i_j \leq 1$ and $\sum i_j = i$. Putting $j_\ell = 1 - i_\ell$, so that $0 \leq j_\ell \leq 1$ and $\sum j_\ell = r - i$, formula (6.12) can be rewritten as

$$\sum b_0 \wedge b_{1+\lambda_r-j_1} \wedge b_{2+\lambda_{r-1}-j_2} \wedge \cdots \wedge b_{r+\lambda_1-j_r} \tag{6.13}$$

summing over all $(j_1, \ldots, j_r)$ such that $0 \le j_\ell \le 1$ and $\sum j_\ell = r - i$. Thus, keeping into account that $\overline{\sigma}_i b_0 = 0$ for all $i < 0$, (6.13) is precisely the definition of

$$\overline{\sigma}_{i-r}[\mathbf{b}]^{r+1}_{\lambda} = \overline{\sigma}_{i-r}(b_0 \wedge \overline{\sigma}_r[\mathbf{b}]^r_{\lambda}). \tag{6.14}$$

Plugging the left-hand side of (6.14) into (6.11), in place of $b_0 \wedge \overline{\sigma}_i[\mathbf{b}]^r_{\lambda}$, produces

$$b_0 \wedge \overline{\sigma}_+(z)[\mathbf{b}]^r_{\lambda} = z^r \sum_{j=0}^{r} \frac{(-1)^j \overline{\sigma}_{j-r}}{z^{r-j}}(b_0 \wedge \overline{\sigma}_r[\mathbf{b}]^r_{\lambda}) = (-1)^r z^r \overline{\sigma}_-(z)(b_0 \wedge \overline{\sigma}_r[\mathbf{b}]^r_{\lambda}).$$

Substitution into (6.10) gives (6.9). □

6.2.7. Let $r \ge 1$ be fixed. For all $j \in \mathbb{Z}$, we have $\mathbb{Z}$-module endomorphisms $\sigma_j, \overline{\sigma}_j : B_r \to B_r$ defined as follows:

$$(\sigma_j p)[\mathbf{b}]^r_0 := \sigma_j(p[\mathbf{b}]^r_0) \qquad \text{and} \qquad (\overline{\sigma}_j p)[\mathbf{b}]^r_0 := \overline{\sigma}_j(p[\mathbf{b}]^r_0), \tag{6.15}$$

where p denotes a polynomial in $(e_1, \ldots, e_r)$. This in turn induces homomorphisms $B_r \to B_r[[z^{\pm 1}]]$ given by

$$\sigma_\pm(z)p = \sum_{j \ge 0} (\sigma_{\pm j} p) z^{\pm j} \qquad \text{and} \qquad \overline{\sigma}_\pm(z)p = \sum_{j \ge 0} (\overline{\sigma}_{\pm j} p) z^{\pm j}$$

which in principle depend on the positive integer r.

However, in spite of the ostensive symmetry of (6.15), the behaviours of $\sigma_j, \overline{\sigma}_j$ is quite different, depending on j being negative or positive. Indeed, unlike $\sigma_+(z)$ and $\overline{\sigma}_+(z)$, the *HS*-derivations $\sigma_-(z)$ and $\overline{\sigma}_-(z)$ enjoy nice stability properties, in the sense of the following:

Lemma 6.2.8. *For all integers $j, n \ge 0$ and for all $r \ge 1$, we have*

$$\sigma_-(z)h_n = \sum_{j \ge 0} \frac{h_{n-j}}{z^j}, \tag{6.16}$$

and

$$\overline{\sigma}_-(z)h_n = h_n - \frac{h_{n-1}}{z}. \tag{6.17}$$

Proof. By Definition (6.15), Lemma 6.2.5 and the fact that $\sigma_-(z)$ is a *HS*-derivation,

$$\begin{aligned}(\sigma_-(z)h_n)[\mathbf{b}]^r_0 &= \sigma_-(z)(h_n[\mathbf{b}]^r_0) = \sigma_-(z)[\mathbf{b}]^r_{(n)} = \sigma_-(z)([\mathbf{b}]^{r-1}_0 \wedge b_{r-1+n}) \\ &= \sigma_-(z)[\mathbf{b}]^{r-1}_0 \wedge \sigma_-(z)b_{r-1+n} = [\mathbf{b}]^{r-1}_0 \wedge \sum_{j \ge 0} \sigma_{-j} b_{r-1+n} z^{-j}\end{aligned}$$

Again, by definition of $\sigma_-(z)$ on M_0,

$$[\mathbf{b}]_0^{r-1} \wedge \sum_{j\geq 0} \sigma_{-j} b_{r-1+n} z^{-j} = [\mathbf{b}]_0^{r-1} \wedge b_{r-1+n-j} = \sum_{j\geq 0} [\mathbf{b}]_{(n-j)}^r$$

$$= \left(\sum_{j\geq 0} \frac{h_{n-j}}{z^j} \right) [\mathbf{b}]_0^r$$

and (6.16) follows. We repeat the totally analogous proof for (6.17) due to its exceptional importance for the sequel. We have

$$\begin{aligned}
(\sigma_-(z)h_n)[\mathbf{b}]_0^r &= \overline{\sigma}_-(z)(h_n[\mathbf{b}]_0^r) = \overline{\sigma}_-(z)[\mathbf{b}]_{(n)}^r \\
&= \overline{\sigma}_-(z)\big([\mathbf{b}]_0^{r-1} \wedge b_{r-1+n}\big) \\
&= \overline{\sigma}_-(z)[\mathbf{b}]_0^{r-1} \wedge \overline{\sigma}(z) b_{r-1+n} \\
&= [\mathbf{b}]_0^{r-1} \wedge (b_{r-1+n} + b_{r-1+n-1} z^{-1}) \\
&= [\mathbf{b}]_{(n)}^r - [\mathbf{b}]_{(n-1)}^r z^{-1} = \left(h_n - \frac{h_{n-1}}{z} \right) [\mathbf{b}]_0^r
\end{aligned}$$

which proves (6.17) as well. □

Warning 6.2.9. Although the expression $\sigma_j h_n$ makes perfectly sense in the ring B_r for all $r \geq 1$ and $j > 0$, its value, unlike that of $\sigma_{-j} h_n$, does depend on the integer r. A uniform formula like (6.16) is not available in this case. For example, in B_1 one has

$$\sigma_j h_n = \frac{\sigma_j b_n}{b_0} = \frac{b_{n+j}}{b_0} = h_{n+j}$$

while in B_2

$$\begin{aligned}
\sigma_j h_n &= \frac{\sigma_j (b_0 \wedge b_{1+n})}{b_0 \wedge b_1} = \frac{b_0 \wedge b_{1+n+j} + \sigma_{j-1}(b_1 \wedge b_{1+n})}{b_0 \wedge b_1} = \\
&= h_{n+j} + \frac{\sigma_{j-1}(b_1 \wedge b_{1+n})}{b_0 \wedge b_1} \neq h_{n+j}.
\end{aligned}$$

Proposition 6.2.10. *The operator $\overline{\sigma}_-(z)$ commutes with taking Δ_λ:*

$$\overline{\sigma}_-(z)\Delta_\lambda(H_r) = \Delta_\lambda(\overline{\sigma}_-(z)H_r), \tag{6.18}$$

where $\overline{\sigma}_-(z)H_r$ denotes the sequence $(\overline{\sigma}_-(z)h_j)_{j\in\mathbb{Z}}$.

Proof. By definition

$$(\overline{\sigma}_-(z)\Delta_{\boldsymbol{\lambda}}(H_r))[\mathbf{b}]_0^r = \overline{\sigma}_-(z)(\Delta_{\boldsymbol{\lambda}}(H_r)[\mathbf{b}]_0^r) = \overline{\sigma}_-(z)[\mathbf{b}]_{\boldsymbol{\lambda}}^r$$
$$= \overline{\sigma}_-(z)b_{\lambda_r} \wedge \overline{\sigma}_-(z)b_{1+\lambda_{r-1}} \wedge \cdots \wedge \overline{\sigma}_-(z)b_{r-1+\lambda_1}$$
$$= \left(b_{\lambda_r} - \frac{b_{\lambda_r-1}}{z}\right) \wedge \left(b_{1+\lambda_{r-1}} - \frac{b_{\lambda_{r-1}}}{z}\right) \wedge \cdots \wedge \left(b_{r-1+\lambda_1} - \frac{b_{\lambda_r-2+\lambda_1}}{z}\right)$$
$$= f_0(\sigma_1)b_0 \wedge f_1(\sigma_1)b_0 \wedge \cdots \wedge f_{r-1}(\sigma_1)b_0, \tag{6.19}$$

where

$$f_j(X) = X^{j+\lambda_{r-j}} - \frac{X^{j+\lambda_{r-j}-1}}{z}, \qquad 0 \le j \le r-1.$$

By Lemma 2.6.6

$$\mathrm{Res}\left(\frac{X^{i-1}f_{r-j}(X)}{\mathfrak{p}_r(X)}\right) = h_{\lambda_j-j+i} - \frac{h_{\lambda_j-j+i-1}}{z} = \overline{\sigma}_-(z)h_{\lambda_j-j+i}.$$

By applying Theorem 5.8.1,

$$\overline{\sigma}_-(z)\Delta_{\boldsymbol{\lambda}}(H_r) = \mathrm{Res}\left(\frac{f_{r-1}(X)}{\mathfrak{p}_r(X)}, \ldots, \frac{f_0(X)}{\mathfrak{p}_r(X)}\right)$$
$$= \det(\overline{\sigma}_-(z)h_{\lambda_j-j+i}) = \Delta_{\boldsymbol{\lambda}}(\overline{\sigma}_-(z)H_r)).$$

□

Corollary 6.2.11. *For all* r*-tuples* $(h_{i_1}, \ldots, h_{i_r}) \in B_r^r$,

$$\overline{\sigma}_-(z)(h_{i_1}\cdots h_{i_r}) = \overline{\sigma}_-(z)h_{i_1}\cdots\overline{\sigma}_-(z)h_{i_r}.$$

Proof. For each product $h_{i_1}\cdots h_{i_r}$, there exists $n \ge 0$ such that

$$h_{i_1}\cdots h_{i_r} = \sum_{\boldsymbol{\lambda}\in\mathcal{P}_{r,n}} a_{\boldsymbol{\lambda}}\Delta_{\boldsymbol{\lambda}}(H_r).$$

because Schur polynomials $\Delta_{\boldsymbol{\lambda}}(H_r)$ form a basis of B_r. Therefore

$$\overline{\sigma}_-(z)(h_{i_1}\cdots h_{i_r}) = \sum_{\boldsymbol{\lambda}} a_{\boldsymbol{\lambda}}\overline{\sigma}_-(z)\Delta_{\boldsymbol{\lambda}}(H_r)$$
$$= \sum_{\boldsymbol{\lambda}} a_{\boldsymbol{\lambda}}\Delta_{\boldsymbol{\lambda}}(\overline{\sigma}_-(z)H_r) = \overline{\sigma}_-(z)h_{i_1}\cdots\overline{\sigma}_-(z)h_{i_r},$$

as desired. □

Warning 6.2.12. The $\mathbb{Z}$-module homomorphism $\overline{\sigma}_-(z) : B_r \to B_r((z))$ is not a ring homomorphism. To see this, take $r = 1$. Then

$$\overline{\sigma}_-(z)h_2 = h_2 - \frac{h_1}{z}.$$

On the other hand, the equality $h_2 = h_1^2$ holds in B_1 and

$$\overline{\sigma}_-(z)h_1 \cdot \overline{\sigma}_-(z)h_1 = \left(h_1 - \frac{1}{z}\right)^2 = h_2 - \frac{2h_1}{z} + \frac{1}{z^2} \neq \overline{\sigma}_-(z)h_2 = \overline{\sigma}_-(z)(h_1^2).$$

Proposition 6.2.13. *The operator $\sigma_-(z)$ commutes with taking Δ_λ.*

$$\sigma_-(z)\Big(\Delta_\lambda(H_r)\Big) = \Delta_\lambda\Big(\sigma_-(z)H_r\Big), \tag{6.20}$$

where $\sigma_-(z)H_r$ denotes the sequence $(\sigma_-(z)h_j)_{j\in\mathbb{Z}}$.

Proof.

$$\begin{aligned}(\sigma_-(z)\Delta_\lambda(H_r))[\mathbf{b}]_0^r &= \sigma_-(z)(b_{\lambda_r} \wedge b_{1+\lambda_{r-1}} \wedge \cdots \wedge b_{r-1+\lambda_1})\\ &= \sigma_-(z)b_{\lambda_r} \wedge \cdots \wedge \sigma_-(z)b_{r-1+\lambda_1}\\ &= f_0(\sigma_1)b_0 \wedge f_1(\sigma_1)b_0 \wedge \cdots \wedge f_{r-1}(\sigma_1)b_0,\end{aligned}$$

where

$$f_j(X) = \sum_{p=0}^{j+\lambda_{r-j}} \frac{X^{j+\lambda_{r-j}-p}}{z^p}.$$

By Lemma 2.6.6

$$\mathrm{Res}\left(\frac{X^{i-1}f_{r-j}(X)}{\mathfrak{p}_r(X)}\right) = \sum_{p=0}^{\lambda_j-j+i} \frac{h_{\lambda_j-j+i-p}}{z^p} = \sigma_-(z)h_{\lambda_j-j+i}.$$

In conclusion, by Theorem 5.8.1

$$\sigma_-(z)\Big(\Delta_\lambda(H_r)\Big)[\mathbf{b}]_0^r = \mathrm{Res}\left(\frac{f_{r-1}(X)}{\mathfrak{p}_r(X)}, \frac{f_{r-2}(X)}{\mathfrak{p}_r(X)}, \ldots, \frac{f_0(X)}{\mathfrak{p}_r(X)}\right) = \Delta_\lambda\Big(\sigma_-(z)H_r\Big),$$

as claimed. □

The following is the exact analogue of Corollary 6.2.11, so we will omit the proof.

Corollary 6.2.14. *For all* r*-tuple* $(h_{i_1}, \ldots, h_{i_r}) \in B_r^r$,

$$\sigma_-(z)(h_{i_1} \cdots h_{i_r}) = \sigma_-(z)h_{i_1} \cdots \sigma_-(z)h_{i_r}.$$ □

6.3 On Vertex-Like Operators

6.3.1. We have seen that an arbitrary exterior power $\bigwedge^r M_0$ of a free abelian group M_0 (cf. Section 5.6) can be naturally made into a free B_r-module $\bigwedge^r M_r$ of rank 1 generated by $[\mathbf{b}]_0^r$. The purpose of this section is to study the map $B_r \to B_{r+1}((z))$ given by the unique $\mathbb{Z}$-linear extension of

$$\Gamma_r(z)\Delta_{\boldsymbol{\lambda}}(H_r) = \frac{1_{B_{r+1}} \otimes ((-1)^r \sum_{j\geq 0} b_j z^j \wedge [\mathbf{b}]_{\boldsymbol{\lambda}}^r)}{[\mathbf{b}]_0^{r+1}}. \tag{6.21}$$

Lemma 6.3.2. *For all $r \geq 1$ and all $\boldsymbol{\lambda} \in \mathcal{P}_r$,*

$$b_0 \wedge \overline{\sigma}_r [\mathbf{b}]_{\boldsymbol{\lambda}}^r = \Delta_{\boldsymbol{\lambda}}(H_{r+1})[\mathbf{b}]_0^{r+1}. \tag{6.22}$$

Proof. It is a consequence of the definition of $\overline{\sigma}_r$ and Corollary 5.8.2:

$$\begin{aligned} b_0 \wedge \overline{\sigma}_r [\mathbf{b}]_{\boldsymbol{\lambda}}^r &= b_0 \wedge \overline{\sigma}_r (b_{\lambda_r} \wedge \cdots \wedge b_{r-1+\lambda_1}) \\ &= b_0 \wedge b_{1+\lambda_r} \wedge \cdots \wedge b_{r+\lambda_1} \\ &= \Delta_{\boldsymbol{\lambda}}(H_{r+1}) b_0 \wedge b_1 \wedge \cdots \wedge b_r. \qquad \square \end{aligned}$$

Theorem 6.3.3. *We have*

$$\Gamma_r(z)\Delta_{\boldsymbol{\lambda}}(H_r) = \frac{z^r}{E_{r+1}(z)} \cdot \overline{\sigma}_-(z)\Delta_{\boldsymbol{\lambda}}(H_{r+1}), \tag{6.23}$$

Proof. It is just a matter of applying Theorem 6.2.6 to the numerator of formula (6.21):

$$\begin{aligned} \Gamma_r(z)\Delta_{\boldsymbol{\lambda}}(H_r) &= \frac{1_{B_{r+1}} \otimes ((-1)^r \sum_{j\geq 0} b_j z^j \wedge [\mathbf{b}]_{\boldsymbol{\lambda}}^r)}{[\mathbf{b}]_0^{r+1}} && \text{(definition of } \Gamma_r(z)) \\ &= \frac{z^r \sigma_+(z)\overline{\sigma}_-(z) \cdot (b_0 \wedge \overline{\sigma}_r [\mathbf{b}]_{\boldsymbol{\lambda}}^r)}{[\mathbf{b}]_0^{r+1}} && \text{(Theorem 6.2.6)} \\ &= \frac{z^r \sigma_+(z)\overline{\sigma}_-(z) \cdot \Delta_{\boldsymbol{\lambda}}(H_{r+1})[\mathbf{b}]_0^{r+1}}{[\mathbf{b}]_0^{r+1}} && \text{(Lemma 6.3.2)} \\ &= \frac{z^r}{E_{r+1}(z)} \overline{\sigma}_-(z)\Delta_{\boldsymbol{\lambda}}(H_{r+1}) && \text{(using that } \textstyle\bigwedge^{r+1} M_{r+1} \text{ is an eigenmodule of } \sigma_+(z) \text{ with eigenvalue } H_{r+1}(z) = \frac{1}{E_{r+1}(z)}) \end{aligned}$$

and the claim is proven. $\square$

Corollary 6.3.4. *The following equality holds:*

$$\Gamma_r(z)\Delta_{\boldsymbol{\lambda}}(H_r) = \frac{z^r\Delta_{\boldsymbol{\lambda}}(\overline{\sigma}_-(z)H_{r+1})}{E_{r+1}(z)}.$$

Proof. Just apply Proposition 6.2.10 to the numerator of (6.23). Passing $\sigma_-(z)$ inside $\Delta_{\boldsymbol{\lambda}}$ is allowed because $\boldsymbol{\lambda}$ is a partition of length at most r, so clearly shorter than $r+1$ (cf. Remark 6.3.9 below). □

Corollary 6.3.5. *Let $\boldsymbol{\lambda} \in \mathcal{P}_{r,n}$ so that $[\mathbf{b}]^r_{\boldsymbol{\lambda}} \in \bigwedge^r M_{r,n}$. Then*

$$\sum_{j=0}^{n-1} b_j z^j \wedge [\mathbf{b}]^r_{\boldsymbol{\lambda}} = \frac{1}{E_{r+1}(z)}\Delta_{\boldsymbol{\lambda}}(\overline{\sigma}_-(z)H_{r+1,n})[\mathbf{b}]^{r+1}_0$$

Proof. We use diagram (5.49) for $r+1$. We have

$$\begin{aligned}\sum_{j=0}^{n-1} b_j z^j \wedge [\mathbf{b}]^r_{\boldsymbol{\lambda}} + \sum_{j\geq n} b_j z^j \wedge [\mathbf{b}]^r_{\boldsymbol{\lambda}} &= \sum_{j\geq 0} b_j z^j \wedge [\mathbf{b}]^r_{\boldsymbol{\lambda}} \\ &= \frac{1}{E_{r+1}(z)}\Delta_{\boldsymbol{\lambda}}(\overline{\sigma}_-(z)H_{r+1})[\mathbf{b}]^{r+1}_0. \quad (6.24)\end{aligned}$$

Applying the homomorphism $\gamma^{r+1}_n := \bigwedge^{r+1}\gamma^0_n$ to the first and last side of (6.24) and using Proposition 5.9.7, one obtains

$$\sum_{j=0}^{n-1} b_j z^j \wedge [\mathbf{b}]^r_{\boldsymbol{\lambda}} = \frac{1}{E_{r+1}(z)}\Delta_{\boldsymbol{\lambda}}(\overline{\sigma}_-(z)H_{r+1,n})[\mathbf{b}]^{r+1}_0.$$ □

Definition 6.3.6. For all $r \geq 1$, let $\Gamma^*_r(z) : B_r \to B_{r-1}((z))$ be the unique $\mathbb{Z}$-linear extension of the map

$$\Gamma^*_r(z)\Delta_{\boldsymbol{\lambda}}(H_r) = \frac{1_{B_{r-1}} \otimes ((-1)^{r-1}\boldsymbol{\beta}(z^{-1})\lrcorner[\mathbf{b}]^r_{\boldsymbol{\lambda}})}{[\mathbf{b}]^{r-1}_0}, \quad (6.25)$$

where by $\boldsymbol{\beta}\left(z^{-1}\right)$ we understand the formal power series $\sum_{j\geq 0}\beta_j z^{-j-1}$.

Lemma 6.3.7. *For each $[\mathbf{b}]^r_{\boldsymbol{\lambda}} \in \bigwedge^r M_r$, the following equality holds in $\bigwedge^{r-1} M_{r-1}[z^{-1}]$:*

$$\begin{aligned}&(-1)^{r-1}1_{B_{r-1}} \otimes \left(\boldsymbol{\beta}(z^{-1})\lrcorner[\mathbf{b}]^r_{\boldsymbol{\lambda}}\right) \\ &= \sum_{j\geq 0} z^{-j-1}\beta_j\lrcorner[\mathbf{b}]^r_{\boldsymbol{\lambda}} = \frac{1}{z^r}\begin{vmatrix} z^{-\lambda_1} & z^{-\lambda_2+1} & \dots & z^{-\lambda_r+r-1} \\ h_{\lambda_1+1} & h_{\lambda_2} & \dots & h_{\lambda_r+r-2} \\ \vdots & \vdots & \ddots & \vdots \\ h_{\lambda_1+r-1} & h_{\lambda_2+r-2} & \dots & h_{\lambda_r}\end{vmatrix}[\mathbf{b}]^{r-1}_0. \quad (6.26)\end{aligned}$$

(Sketch of) Proof. Let us first recall that if $\boldsymbol{\lambda}, \boldsymbol{\mu} \in \mathcal{P}_r$, then $\boldsymbol{\lambda} + \boldsymbol{\mu} \in \mathcal{P}_r$ denotes the partition $(\lambda_1 + \mu_1, \ldots, \lambda_r + \mu_r)$. We have:

$$\boldsymbol{\beta}\left(z^{-1}\right) \lrcorner b_{\lambda_r} \wedge b_{-1+\lambda_{r-1}} \wedge \cdots \wedge b_{r-1+\lambda_1} = \sum_{j=0}^{r-1} \frac{1}{z^{j+\lambda_{r-j}+1}} (-1)^j [\mathbf{b}]^{r-1}_{\boldsymbol{\lambda}_j+(1^{j-1})}$$

$$= \frac{1}{z^r} \sum_{j=0}^{r-1} \frac{1}{z^{\lambda_{r-j}+1-(r-j)}} [\mathbf{b}]^{r-1}_{\boldsymbol{\lambda}_j+(1^{j-1})} \tag{6.27}$$

where here by $\boldsymbol{\lambda}_j$ we denote the partition of length at most $r-1$ obtained by removing the j-th part from $\boldsymbol{\lambda} := (\lambda_1, \lambda_2, \ldots, \lambda_r)$. So, for instance $\boldsymbol{\lambda}_1 + (1^0) = (\lambda_2, \ldots, \lambda_r)$, $\boldsymbol{\lambda}_2 + (1) = (\lambda_1 + 1, \lambda_3, \ldots, \lambda_r), \ldots, \boldsymbol{\lambda}_r + (1^{r-1}) = (\lambda_1 + 1, \ldots, \lambda_{r-1} + 1)$. Now

$$[\mathbf{b}]^{r-1}_{\boldsymbol{\lambda}_j+(1^{j-1})} = \Delta_{\boldsymbol{\lambda}_j+(1^{j-1})}(H_{r-1})[\mathbf{b}]^{r-1}_0$$

and so (6.27) is equivalent to:

$$\frac{1}{z^r} \sum_{j=0}^{r-1} \frac{[\mathbf{b}]^{r-1}_{\boldsymbol{\lambda}_j+(1^{j-1})}}{z^{\lambda_{r-j}+1-(r-j)}} = \frac{1}{z^r} \sum_{j=0}^{r-1} (-1)^j \frac{\Delta_{\boldsymbol{\lambda}_j+(1^{j-1})}(H_{r-1})}{z^{\lambda_{r-j}+1-(r-j)}} [\mathbf{b}]^{r-1}_0. \tag{6.28}$$

A quick inspection shows that (6.28) is precisely $(-1)^{r-1}$ times the expansion of the last side of (6.26). □

Theorem 6.3.8. *The following equality holds:*

$$\Gamma_r^*(z)\Delta_{\boldsymbol{\lambda}}(H_r) := \frac{E_{r-1}(z)}{z^r} \Delta_{\boldsymbol{\lambda}}(\sigma_-(z)H_{r-1}) \tag{6.29}$$

$$= \frac{E_{r-1}(z)}{z^r} \begin{vmatrix} \sigma_-(z)h_{\lambda_1} & \sigma_-(z)h_{\lambda_2-1} & \ldots & \sigma_-(z)h_{\lambda_r-r+1} \\ \sigma_-(z)h_{\lambda_1+1} & \sigma_-(z)h_{\lambda_2} & \ldots & \sigma_-(z)h_{\lambda_r+r-2} \\ \vdots & \vdots & \ddots & \vdots \\ \sigma_-(z)h_{\lambda_1+r-1} & \sigma_-(z)h_{\lambda_2+r-2} & \ldots & \sigma_-(z)h_{\lambda_r} \end{vmatrix}. \tag{6.30}$$

Proof. By virtue of Lemma 6.3.7, the proof amounts to manipulate formula (6.26), basing on the definition (2.15) of $u_j \in B_{r-1}[[z]]$ for $j \geqslant 0$

$$u_j = \sum_{n \geqslant 0} h_{n+j} z^n,$$

for $j \geq 0$, where $H_{r-1}(z)E_{r-1}(z) = 1$. Since

$$E_{r-1}(z)u_j = \mathrm{U}_0(u_j) + \mathrm{U}_1(u_j)z + \cdots + \mathrm{U}_{r-1}(u_j)z^{r-1}, \tag{6.31}$$

where each $\mathrm{U}_i(u_j)$ is a B_{r-1}-linear combination of $h_j, h_{j+1}, \ldots, h_{j+i-2}$, the determinant

$$\begin{vmatrix} E_{r-1}(z)u_{\lambda_1+1} & E_{r-1}(z)u_{\lambda_2} & \ldots & E_{r-1}(z)u_{\lambda_r+r-2} \\ h_{\lambda_1+1} & h_{\lambda_2} & \cdots & h_{\lambda_r+r-2} \\ \vdots & \vdots & \ddots & \vdots \\ h_{\lambda_1+r-1} & h_{\lambda_2+r-2} & \cdots & h_{\lambda_r} \end{vmatrix} \tag{6.32}$$

vanishes by skew symmetry. Since

$$\frac{1}{z^\lambda} = \frac{E_{r-1}(z)H_{r-1}(z)}{z^\lambda} = E_{r-1}(z)\left(\sigma_-(z)h_\lambda + zu_{\lambda+1}\right),$$

we have

$$\frac{1}{z^r}\begin{vmatrix} z^{-\lambda_1} & z^{-\lambda_2+1} & \ldots & z^{-\lambda_r+r-1} \\ h_{\lambda_1+1} & h_{\lambda_2} & \cdots & h_{\lambda_r+r-2} \\ \vdots & \vdots & \ddots & \vdots \\ h_{\lambda_1+r-1} & h_{\lambda_2+r-2} & \cdots & h_{\lambda_r} \end{vmatrix} =$$

$$= \frac{E_{r-1}(z)}{z^r}\begin{vmatrix} \sigma_-(z)h_{\lambda_1} + zu_{\lambda_1+1} & \ldots & \sigma_-(z)h_{\lambda_r-r+1} + zu_{\lambda_r-r+2} \\ h_{\lambda_1+1} & \cdots & h_{\lambda_r+r-2} \\ \vdots & \ddots & \vdots \\ h_{\lambda_1+r-1} & \cdots & h_{\lambda_r} \end{vmatrix} =$$

$$= \frac{E_{r-1}(z)}{z^r}\begin{vmatrix} \sigma_-(z)h_{\lambda_1} & \sigma_-(z)h_{\lambda_2-1} & \ldots & \sigma_-(z)h_{\lambda_r-r+1} \\ h_{\lambda_1+1} & h_{\lambda_2} & \cdots & h_{\lambda_r+r-2} \\ \vdots & \vdots & \ddots & \vdots \\ h_{\lambda_1+r-1} & h_{\lambda_2+r-2} & \cdots & h_{\lambda_r} \end{vmatrix}, \tag{6.33}$$

the last equality due to the vanishing of (6.32). To conclude the proof, observe that for all $1 \le i,j \le r$,

$$\sigma_-(z)h_{\lambda_j-j+i} = h_{\lambda_j-j+i} + \sigma_-(z)h_{\lambda_j-j+i-1}$$

and thus, exploiting once again the skew symmetry of the determinant, expression (6.33) is equivalent to

$$= \frac{E_{r-1}(z)}{z^r} \begin{vmatrix} \sigma_-(z)h_{\lambda_1} & \sigma_-(z)h_{\lambda_2-1} & \dots & \sigma_-(z)h_{\lambda_r-r+1} \\ \sigma_-(z)h_{\lambda_1+1} & \sigma_-(z)h_{\lambda_2} & \dots & \sigma_-(z)h_{\lambda_r+r-2} \\ \vdots & \vdots & \ddots & \vdots \\ \sigma_-(z)h_{\lambda_1+r-1} & \sigma_-(z)h_{\lambda_2+r-2} & \dots & \sigma_-(z)h_{\lambda_r} \end{vmatrix},$$

which proves that, according to Definition 6.3.6,

$$\Gamma_r^*(z) = \frac{E_{r-1}(z)}{z^r}\Delta_{\boldsymbol{\lambda}}(\sigma_-(z)H_{r-1}),$$

as desired. □

Remark 6.3.9. It is important to notice that here, in contrast to what happens in the case of $\Gamma_r(z)$ (cf. Theorem 6.3.4), in general

$$\Delta_{\boldsymbol{\lambda}}(\sigma_-(z)H_{r-1}) \neq \sigma_-(z)\Delta_{\boldsymbol{\lambda}}(H_{r-1})$$

For instance, $\Delta_{\boldsymbol{\lambda}}(H_1) = 0$ for all partitions of length 2. Thus

$$\sigma_-(z)\Delta_{(11)}(H_1) = \sigma_-(z)(h_1^2 - h_1^2) = 0 \neq \frac{h_1}{z} = \begin{vmatrix} h_1 + \frac{1}{z} & 1 \\ h_2 & h_1 \end{vmatrix} = \Delta_{(11)}(\sigma_-(z)H_1).$$

However $\Delta_{\boldsymbol{\lambda}}(\sigma_-(z)H_{r-1}) = \sigma_-(z)\Delta_{\boldsymbol{\lambda}}(H_{r-1})$ if $\boldsymbol{\lambda}$ has length at most $r-1$ as a consequence of Proposition 6.2.13.

Corollary 6.3.10. *Let $\boldsymbol{\lambda} \in \mathcal{P}_{r,n}$. Then*

$$\Gamma_r^*(z)\Delta_{\boldsymbol{\lambda}}(H_r)[\mathbf{b}]_0^r = \frac{E_{r-1}(z)}{z^r}\Delta_{\boldsymbol{\lambda}}(\sigma_-(z)H_{r-1,n})[\mathbf{b}]_0^r \tag{6.34}$$

in $\bigwedge^{r-1} M_{0,n}$.

Proof. First of all, by virtue of Theorem (6.3.8), one has

$$\frac{E_{r-1}(z)}{z^r}\Delta_{\boldsymbol{\lambda}}(\sigma_-(z)H_{r-1})[\mathbf{b}]_0^{r-1} = \sum_{j\geq 0}\beta_j(z^{-1})\lrcorner[\mathbf{b}]_{\boldsymbol{\lambda}}^r = \sum_{j=0}^{n-1}\beta_j(z^{-1})\lrcorner[\mathbf{b}]_{\boldsymbol{\lambda}}^r,$$

where in the last equality, we used the fact that if $\boldsymbol{\lambda} \in \mathcal{P}_{r,n}$, then $\beta_j\lrcorner[\mathbf{b}]_{\boldsymbol{\lambda}}^r = 0$ for all $j \geq n-r+1$. Now we observe that $\boldsymbol{\lambda} \in \mathcal{P}_{r,n}$ implies that $\Delta_{\boldsymbol{\lambda}}(\sigma_-(z)H_{r-1})[\mathbf{b}]_0^{r-1}$ belongs in $\bigwedge^{r-1} M_{0,n}$ and equals $\Delta_{\boldsymbol{\lambda}}(\sigma_-(z)H_{r-1,n})[\mathbf{b}]_0^{r-1}$, basically by Proposition 5.9.7. □

6.4 Plücker Equations for Grassmann Cones

Take $p(H_r) \in B_r$ (the evaluation of $p \in \mathbb{Z}[\mathbf{X}]$ at $X_i = h_i$). Then $p(H_r) = \sum a_{\boldsymbol{\lambda}} \Delta_{\boldsymbol{\lambda}}(H_r)$. When does p correspond to a decomposable tensor $p(\sigma_+)[\mathbf{b}]_0^r \in \bigwedge^r M_0$? The trick to ease computations is to use the B_r-module $\bigwedge^r M_r$ constructed out of $\bigwedge^r M_0$.

Theorem 6.4.1. *A polynomial $p(H_r) \in B_r$ corresponds to a decomposable tensor of $\bigwedge^r M_0$ if and only if*

$$\operatorname{Res}_z \Gamma_r(z)p(H_r) \otimes \Gamma_r^*(z)p(H_r) = 0. \tag{6.35}$$

Proof. By Theorem 6.1.7, the tensor $\eta := p(\boldsymbol{\sigma}_+)[\mathbf{b}]_0^r$ is decomposable if and only if the residue at $z = 0$ in the expansion

$$\sum_{i\geq 0}(b_i z^i \wedge p(\boldsymbol{\sigma}_+)[\mathbf{b}]_0^r) \otimes (z^{-i-1}\beta_i \lrcorner p(\boldsymbol{\sigma}_+)[\mathbf{b}]_0^r)$$

vanishes. Now

$$\begin{aligned}&\sum_{i\geq 0}(b_i z^i \wedge p(\boldsymbol{\sigma}_+)[\mathbf{b}]_0^r) \otimes \sum_{j\geq 0}(\beta_j \lrcorner p(\boldsymbol{\sigma}_+)[\mathbf{b}]_0^r) \otimes_{\mathbb{Z}} 1_{B_r}\\ &= (-1)^{2r-1}\Gamma_r(z)p(H_r)[\mathbf{b}]_0^{r+1} \otimes \Gamma_r^*(z)p(H_r)[\mathbf{b}]_0^{r-1}\end{aligned}$$

and so the residue at $z = 0$ of the last side vanishes if and only if (6.35) holds. □

6.4.2. An arbitrary $p(H_r) \in B_r$ is of the form

$$p(H_r) = \sum_{\boldsymbol{\lambda}\in\mathcal{P}_r} a_{\boldsymbol{\lambda}} \Delta_{\boldsymbol{\lambda}}(H_r), \qquad (a_{\boldsymbol{\lambda}} \in \mathbb{Z})$$

where $a_{\boldsymbol{\lambda}} = 0$ for all but finitely many $\boldsymbol{\lambda} \in \mathcal{P}_r$. Substituting in (6.35) the explicit expressions (6.23) and (6.29) of $\Gamma_r(z)$ and $\Gamma_r^*(z)$, respectively, we have proven, as promised in the introduction, the following:

Theorem 6.4.3. *The $\mathbb{Z}$-linear combination $\sum_{\boldsymbol{\lambda}} a_{\boldsymbol{\lambda}} \Delta_{\boldsymbol{\lambda}}(H_r) \in B_r$ corresponds to a decomposable tensor in $\bigwedge^r M_0$ if and only if*

$$\operatorname{Res}_z \sum_{\boldsymbol{\lambda},\boldsymbol{\mu}\in\mathcal{P}_r} a_{\boldsymbol{\lambda}} a_{\boldsymbol{\mu}} E_{r-1}(z)\Delta_{\boldsymbol{\lambda}}(\sigma_-(z)H_{r-1}) \otimes \frac{\overline{\sigma}_-(z)\Delta_{\boldsymbol{\mu}}(H_{r+1})}{E_{r+1}(z)} = 0. \tag{6.36}$$

□

Corollary 6.4.4. *Let $p(H_r) := \sum_{\boldsymbol{\lambda}\in\mathcal{P}_{r,n}} a_{\boldsymbol{\lambda}} \Delta_{\boldsymbol{\lambda}}(H_r)$. The polynomial $p(H_r)$ corresponds to a decomposable tensor if and only if*

$$\operatorname{Res}_{z=0} \sum_{\boldsymbol{\lambda},\boldsymbol{\mu}\in\mathcal{P}_{r,n}} a_{\boldsymbol{\lambda}} a_{\boldsymbol{\mu}} E_{r-1}(z)\Delta_{\boldsymbol{\lambda}}(\sigma_-(z)H_{r-1,n}) \otimes \frac{\overline{\sigma}_-(z)\Delta_{\boldsymbol{\mu}}(H_{r+1,n})}{E_{r+1}(z)} = 0. \tag{6.37}$$

Proof. In fact $\eta := p(H_r)[\mathbf{b}]^r_0 \in \bigwedge^r M_{r,n}$ *by hypothesis. Then we may apply Corollaries 6.3.5 and 6.3.10.* □

This formula can be written in a more intelligible form once one identifies the tensor product of polynomial rings with a bigger polynomial ring (cf. 1.4.4)

$$B_{r-1} \otimes B_{r+1} = \mathbb{Z}[e'_1, \dots, e'_{r-1}, e''_1, \dots, e''_{r+1}].$$

Let $E'_{r-1}(z) = 1 - e'_1 z + \dots + (-1)^{r-1} e'_{r-1} z^r$ and $E''_{r+1}(z) = 1 - e''_1 z + \dots + (-1)^{r+1} e''_{r+1} z^{r+1}$. Similarly, let

$$H'_{r-1}(z) = \sum_{n\geq 0} h'_n z^n \qquad \text{and} \qquad H''_{r-1}(z) = \sum_{n\geq 0} h''_n z^n$$

be the inverses of $E'_{r-1}(z)$ and $E''_{r+1}(z)$ in $B'_r[[z]]$ and $B''_r[[z]]$, respectively. Formula (6.36) then reads

$$\mathrm{Res}_z \frac{E'_{r-1}(z)}{E''_{r+1}(z)} \sum_{\lambda,\mu} a_\lambda a_\mu \Delta_\lambda(\sigma_-(z) H'_{r-1}) \cdot \overline{\sigma}_-(z) \Delta_\mu(H''_{r+1}) = 0.$$

Corollary 6.4.5. *The polynomial*

$$\sum_{\lambda \in \mathcal{P}_{r,n}} a_\lambda \Delta_\lambda(H_{r,n}) \in B_{r,n}$$

corresponds to a decomposable tensor in $\bigwedge^r M_{0,n}$ *if and only if*

$$\mathrm{Res}_z \frac{E'_{r-1}(z)}{E''_{r+1}(z)} \sum_{\lambda,\mu \in \mathcal{P}_{r,n}} a_\lambda a_\mu \Delta_\lambda(\sigma_-(z) H'_{r-1,n}) \cdot \overline{\sigma}_-(z) \Delta_\mu(H''_{r+1,n}) = 0.$$

Proof. Let $\eta = \sum_{\lambda \in \mathcal{P}_{r,n}} a_\lambda [\mathbf{b}]^r_\lambda \in \bigwedge^r M_{0,n}$ be decomposable. Then $\Delta_\lambda(H_r)[\mathbf{b}]^r_0 = \Delta_\lambda(H_{r,n})[\mathbf{b}]^r_0$, and because of the inclusion $\bigwedge^r M_{0,n} \subseteq \bigwedge^r M_0$, we have

$$\begin{aligned} 0 &= \mathrm{Res}_z \frac{E'_{r-1}(z)}{E''_{r+1}(z)} \sum_{\lambda,\mu} a_\lambda a_\mu \Delta_\lambda(\sigma_-(z) H'_{r-1}) \cdot \overline{\sigma}_-(z) \Delta_\mu(H''_{r+1}) \\ &= \mathrm{Res}_z \frac{E'_{r-1}(z)}{E''_{r+1}(z)} \sum_{\lambda,\mu} a_\lambda a_\mu \Delta_\lambda(\sigma_-(z) H'_{r-1,n}) \cdot \overline{\sigma}_-(z) \Delta_\mu(H''_{r+1,n}). \ \square \end{aligned}$$

Example 6.4.6. Let

$$p(H_2) := a_0 + a_1 h_1 + a_2 h_2 + a_{11} \Delta_{(11)}(H_2) + a_{21} \Delta_{(21)}(H_2) + a_{22} \Delta_{(22)}(H_2) \in B_2.$$

The goal is to determine conditions on the coefficients $(a_{\boldsymbol{\lambda}} \mid \boldsymbol{\lambda} \in \mathcal{P}_{2,4})$ ensuring that $p(H_2)[\mathbf{b}]_0^2$ is decomposable in $\bigwedge^2 M_0$. To this purpose we use the B_2-module structure of $\bigwedge^2 M_2 := B_2 \otimes \bigwedge^2 M_0$. Since $p(H_2)[\mathbf{b}]_0^2 \in \bigwedge^2 M_{2,4}$, the B_2-module structure of $\bigwedge^2 M_2$ can be factorizes through that of $B_{2,4}$. Let us compute

$$\Gamma_2^*(z)p(H_{2,4}) \in B_{1,4}((z))$$

according to the recipe (6.34). We have

$$\begin{aligned} p(\sigma_-(z)H_{1,4}) &= a_0 + a_1\sigma_-(z)h_1 + a_2\sigma_-(z)h_1^2 \\ &\quad + a_{11}\Delta_{(11)}(\sigma_-(z)H_{1,4}) + a_{21}\Delta_{(21)}(\sigma_-(z)H_{1,4}) \\ &\quad + \Delta_{(22)}(\sigma_-(z)H_{1,4}) = \\ &= \sum_{\boldsymbol{\lambda}\in\mathcal{P}_{2,4}} a_{\lambda_1,\lambda_2} \det\left(\sum_{k=0}^{\lambda_j-j+i} h_{\lambda_j-j+i-k}z^{-k}\right)_{1\le i,j\le 2} \end{aligned}$$

and

$$\begin{aligned} p(\overline{\sigma}_-(z)H_{3,4}) &= \sum_{\boldsymbol{\lambda}} a_{\boldsymbol{\lambda}}\Delta_{\boldsymbol{\lambda}}(\overline{\sigma}_-(z)H_{3,4}) = \\ &= \sum_{\boldsymbol{\lambda}\in\mathcal{P}_{2,4}} a_{(\lambda_1,\lambda_2)} \det(h_{\lambda_j-j+i-1} - h_{\lambda_j-j+i}z^{-1})_{1\le i,j\le 2} \end{aligned}$$

from which

$$\Gamma_2^*(z)p(H_{2,4}) = \frac{1-e_1z}{z^2}\sum_{\boldsymbol{\lambda}\in\mathcal{P}_{2,4}} a_{\lambda_1,\lambda_2} \det\left(\sum_{k=0}^{\lambda_j-j+i} h_{\lambda_j-j+i-k}z^{-k}\right)_{1\le i,j\le 2}$$

in $B_{1,4}((z))$and

$$\Gamma_2(z)p(H_{2,4}) = \frac{z^2}{E_3(z)}\sum_{\boldsymbol{\lambda}} a_{(\lambda_1,\lambda_2)} \det(h_{\lambda_j-j+i-1} - h_{\lambda_j-j+i}z^{-1})_{1\le i,j\le 2}$$

in $B_{3,4}((z))$. Now observe that:

$$B_{3,4} = \frac{\mathbb{Z}[e_1,e_2,e_3]}{(h_2,h_3,h_4)} \cong \frac{\mathbb{Z}[y]}{(y^4)},$$

where we put $y = e_1 + (h_2,h_3,h_4) = h_1 + (h_2,h_3,h_4)$. Indeed, the relation $h_2 = 0$ implies that $e_2 = e_1^2$. Moreover $h_3 - e_1h_2 + e_2h_1 + e_3 = 0$ together with $h_2 = h_3 = 0$ yields $e_3 = h_1e_2 = e_1^3$. In addition $h_4 - e_1h_3 + e_2h_2 - e_3h_1 = 0$, whence $e_3h_1 = e_1^4 = 0 \mod (h_2,h_3,h_4)$.

$$B_{1,4} = \frac{\mathbb{Z}[x]}{(h_4)} = \frac{\mathbb{Z}[x]}{(x^4)}$$

where $x = e_1 + (h_4)$. In fact the relation

$$(1 - e_1 z)(1 + h_1 z + h_2 z^2 + h_3 z^3) = 1,$$

holding in $B_{1,4}$, says that $h_1 = e_1$, $h_2 = e_1^2$, $h_3 = e_1^3$ and $e_1^4 = 0$. So we have

$$\Gamma_2^*(z)p(H_{2,4}) = \frac{1 - xz}{z^2}\left(a_0 + a_1\left(x + \frac{1}{z}\right) + a_2\left(x^2 + \frac{x}{z} + \frac{1}{z^2}\right) + \right.$$

$$\left. + a_{11}\frac{x}{z} + a_{21}\left(\frac{x}{z^2} + \frac{x^2}{z}\right) + a_{22}\frac{x^2}{z^2}\right)$$

and

$$\Gamma_2(z)p(H_{2,4}) = z^2(1 + yz)\left[a_0 + a_1\left(y - \frac{1}{z}\right) - a_2\frac{y}{z} + a_{11}\left(y^2 - \frac{y}{z} + \frac{1}{z^2}\right)\right.$$

$$\left. + a_{21}\left(\frac{y}{z^2} - \frac{y^2}{z}\right) + a_{22}\frac{y^2}{z^2}\right].$$

Using these expressions in Corollary 6.4.5, and using a little CoCoA software [1], eventually gives

$$\mathrm{Res}_z(\Gamma_2(z)p(H_{2,4})) \cdot (\Gamma_2^*(z)p(H_{2,4}))$$

$$= (-a_{11}a_2 + a_1a_{21} - a_0a_{22})x^3 + (a_{11}a_2 - a_1a_{21} + a_0a_{22})x^2y -$$

$$(a_{11}a_2 - a_1a_{21} + a_0a_{22})xy^2 + (a_{11}a_2 - a_1a_{21} + a_0a_{22})y^3$$

$$= (a_{11}a_2 - a_1a_{21} + a_0a_{22})(y^3 - y^2x + x^2y - x^3)$$

which is identically zero if and only if the Plücker equation

$$a_{11}a_2 - a_1a_{21} + a_0a_{22} = 0 \tag{6.38}$$

holds.

Remark 6.4.7. If $V = M_{0,4} \otimes_{\mathbb{Z}} \mathbb{C} = \mathbb{C}^4$ and

$$\sum_{2\geq\lambda_1\geq\lambda_2\geq 0} a_{(\lambda_1,\lambda_2)} b_{\lambda_2} \wedge b_{1+\lambda_1} \in \bigwedge^2 \mathbb{C}^4,$$

equation (6.38) defines the *Klein quadric* in $\mathbb{P}^5$, whose zero locus corresponds to points of the Grassmannian $G_{2,4}$ (cf. formula (5.20) in 5.3.2).

Example 6.4.8. In a similar way to that of Example 6.4.6, one can find the locus of polynomials of $B_2 := \mathbb{Z}[e_1, e_2]$:

$$p(H_2) := \sum_{\lambda \in \mathcal{P}_{2,5}} a_\lambda \Delta_\lambda(H_2)$$

corresponding to decomposable tensors in $\bigwedge^2 M_0$. As before, to do the computations, one can use the $B_{2,5}$-module $\bigwedge^2 M_{2,5}$. In $B_{1,5}$ we put $h'_1 = x$ and in $B_{3,5}$ we set $h''_1 = y_1$ and $h''_2 = y_2$. Thus the equation of decomposable tensors is given by the vanishing of the residue of the formal power series

$$(1 - xz)(1 + y_1 z + y_2 z^2) \sum_{\lambda,\mu \in \mathcal{P}_{2,5}} \Delta_\lambda(\sigma_-(z)H'_{1,5})\Delta_\lambda(\overline{\sigma}_-(z)H''_{3,5}).$$

Reproducing the computations of Example 6.4.6, one finds

$$(a_{11}a_{20} - a_{10}a_{21} + a_{00}a_{22})P_3(x,\mathbf{y}) + (a_{11}a_{30} - a_{10}a_{31} + a_{00}a_{32})P_4(x,\mathbf{y})$$

$$+ (a_{21}a_{30} - a_{20}a_{31} + a_{00}a_{33})P_5(x,\mathbf{y}) + (a_{22}a_{30} - a_{20}a_{32} + a_{10}a_{33})P_6(x,\mathbf{y})$$

$$+ (a_{22}a_{31} - a_{21}a_{32} + a_{11}a_{33})P_7(x,\mathbf{y}) = 0$$

where we have set $\mathbf{y} = (y_1, y_2)$ and

$$P_3(x,\mathbf{y}) := y_1^3 - 2y_1y_2 - xy_1^2 + xy_2 + x^2y_1 - x^3;$$

$$P_4(x,\mathbf{y}) := y_1^2y_2 - y_2^2 - y_1y_2x + y_2x^2 - x^4;$$

$$P_5(x,\mathbf{y}) := y_1y_2^2 - y_2^2x + y_2x^3 - y_1x^4;$$

$$P_6(x,\mathbf{y}) := y_2^3 - y_2^2x^2 + y_1y_2x^3 - y_1^2x^4 + y_2x^4;$$

$$P_7(x,\mathbf{y}) := y_2^3x - y_1y_2^2x^2 + y_1^2y_2x^3 - x^3y_2^2 - 2x^4y_1y_2.$$

The above long expression vanishes if and only if the coefficients of the forms of degrees $3, 4, 5, 6$ *and* 7 in (x, y_1, y_2) vanish. One easily recognizes in such coefficients the five 4×4 Pfaffians of the 5×5 skew-symmetric matrix

$$\begin{pmatrix} 0 & a_{00} & a_{10} & a_{20} & a_{30} \\ -a_{00} & 0 & a_{11} & a_{21} & a_{31} \\ -a_{10} & -a_{11} & 0 & a_{22} & a_{32} \\ -a_{20} & -a_{21} & -a_{22} & 0 & a_{33} \\ -a_{30} & -a_{31} & -a_{32} & -a_{33} & 0 \end{pmatrix}$$

as it should rightly be (cf., for instance, [116]).

6.5 On the Infinite Exterior Power I

There is a first naïve way to figure out infinite exterior powers. We have seen that there is a $\mathbb{Z}$-module isomorphism $B_r \to \bigwedge^r M_0$ mapping $\Delta_{\boldsymbol{\lambda}}(H_r) \mapsto [\mathbf{b}]^r_{\boldsymbol{\lambda}}$. If we replace the finite sequence of indeterminates $(e_1, e_2, \ldots, e_r)$ with an infinite sequence $(e_1, e_2, \ldots)$, it still makes sense to consider the polynomial ring

$$B_\infty = \mathbb{Z}[e_1, e_2, \ldots]$$

in infinitely many indeterminates. The latter is the projective limit of B_r in the category of graded rings [109] and can be constructed explicitly as follows. Recall that each B_r is graded by the weight:

$$(B_r)_w = \bigoplus_{|\boldsymbol{\lambda}|=w} \mathbb{Z}\Delta_{\boldsymbol{\lambda}}(H_r)$$

and there is a $\mathbb{Z}$-module isomorphism $(B_r)_w \to (\bigwedge^r M)_w$ mapping $\Delta_{\boldsymbol{\lambda}}(H_r) \longmapsto \Delta_{\boldsymbol{\lambda}}(H_r)[\mathbf{b}]^r_0$. For all $s \geq r$, there is a diagram of $\mathbb{Z}$-modules

$$\begin{array}{ccc} (\bigwedge^s M_0)_w & \xrightarrow{\phi_{rs}} & (\bigwedge^r M_0)_w \\ \downarrow & & \downarrow \\ (B_s)_w & \xrightarrow{\beta_{rs}} & (B_r)_w \end{array} \tag{6.39}$$

whose horizontal arrows are the epimorphisms mapping

$$\Delta_{\boldsymbol{\lambda}}(H_s) \longmapsto \Delta_{\boldsymbol{\lambda}}(H_r) \qquad \text{and} \qquad [\mathbf{b}]^s_{\boldsymbol{\lambda}} \mapsto [\mathbf{b}]^r_{\boldsymbol{\lambda}} \tag{6.40}$$

if $\ell(\boldsymbol{\lambda}) \leq r$ and 0 otherwise. Either of (6.40) has an obvious section. The former is given by $\Delta_{\boldsymbol{\lambda}}(H_r) \mapsto \Delta_{\boldsymbol{\lambda}}(H_s)$ and the latter by $[\mathbf{b}]^r_{\boldsymbol{\lambda}} \mapsto [\mathbf{b}]^s_{\boldsymbol{\lambda}}$.

A diagram like (6.39) factorizes through $(B_t)_w \to (\bigwedge^t M_0)_w$ for all $s \geq t \geq r \geq 1$:

$$\begin{array}{ccccc} (\bigwedge^s M_0)_w & \xrightarrow{\phi_{ts}} & (\bigwedge^t M_0)_w & \xrightarrow{\phi_{rt}} & (\bigwedge^r M_0)_w \\ \downarrow & & \downarrow & & \downarrow \\ (B_s)_w & \xrightarrow{\beta_{ts}} & (B_t)_w & \xrightarrow{\beta_{rt}} & (B_r)_w \end{array}$$

Hence the data $((B_s)_w \to (\bigwedge^s M_0)_w, \phi_{rs}, \beta_{rs})$ is an inverse system ($\phi_{ts}\phi_{sr} = \phi_{tr}$ and $\beta_{ts}\beta_{sr} = \beta_{tr}$, for all $s \geq t \geq r$), and one can then take the inverse limit $(B_\infty)_w \longrightarrow (\bigwedge^\infty M_0)_w$, where we may set:

$$\bigwedge^\infty M_0 := \lim_{\leftarrow}(\bigwedge^r M_0)_w := \lim_{\leftarrow}(B_r)_w \cdot [\mathbf{b}]^r_0$$

One then defines

$$\bigwedge^{\infty} M_0 := \bigoplus_{w\geq 0}(\bigwedge^{\infty} M_0)_w = \mathbb{Z}[\mathbf{b}]^{\infty}_{\lambda}$$

where $[\mathbf{b}]^{\infty}_0 := b_0 \wedge b_1 \wedge b_2 \wedge \cdots$ and $[\mathbf{b}]^{\infty}_{\lambda}$ stands for $\Delta_{\lambda}(H_{\infty})[\mathbf{b}]^{\infty}_0$. To see what one should mean by decomposable element of $\bigwedge^{\infty} M_0$ the reader is referred to the recent [46].

In Chapter 7, will be identified the module $M_r := M_0 \otimes_{\mathrm{ev}_{\sigma_+}} B_r$, as defined in Section 5.7, with the $\mathbb{Z}$-module of generic linear recurrence sequences of order r. This will offer a more direct approach to "discover" approximations of the boson-fermion correspondence, phrased via a formalism involving infinite exterior powers related to those outlined above. The corresponding vertex operators will be computed and shown to coincide with those introduced in Section 1.2.1.

6.6 Notes and References

- It goes without saying that Theorem 6.4.3, on detecting decomposable tensors in exterior powers, does not solve an open problem. General criteria for decomposability are mentioned in almost every algebraic geometry textbook, like [55, 64], and one could use the elegant criterion of [58, p. 64–65]. See also the comprehensive exposition by M. Marcus [111], where the reader may find a catalogue of alternative methods. More can be found in the rich, although not exhaustive, bibliography we put at the end of the text. The explicit equivalent version we needed in this book can be found in [5, Sect. 4] and [78, Proposition 7.2], applied to infinite exterior powers of an infinite vector space V, seen as representation spaces of the Clifford algebra sketchily described in Example 3.3.2 (see also the introduction to [79]). We preferred to use the latter, after 'downgrading' from infinite to finite exterior powers.
- We cannot refrain from citing Kasman's beautiful *glimpses* [81, Ch. 10], a presentation both deep and recommended to beginners for its amenability. Among other things, Kasman discusses the following astonishing example [81, Section 10.4]: in the vector space V spanned by four linearly independent solutions (g_1, g_2, g_2, g_4) of the Hirota's bilinear form (1.18) of the *KP* equation, an arbitrary linear combination $\sum_{i<j} a_{ij} Wr(g_i, g_j)$ is itself a KP solution if and only if $\sum_{i<j} a_{ij} g_i \wedge g_j$ is a decomposable tensor in $\bigwedge^2 V$, where $Wr(g_i.g_j)$ denotes the Wronskian of the pair (g_i, g_j) with respect to the time parameter.
- The idea of working out the KP equation in a finite-dimensional context is well present in Kasman and Gekhtman's investigations and philosophy [52]. We also point out the interesting and somewhat mysterious result of Kasman, Pedings, Reiszl and Shiota [83], whereby decomposability can be checked using just rank

6 quadrics. Put otherwise, a Grassmannian in its Plücker embedding is, at least set theoretically, cut out by Klein-like quadrics.

- We emphasize that the approximate operators Γ_r and Γ_r^* computed in Section 6.3 were expressed using special generators of the ring B_r: we employed e_i (which can be interpreted as elementary symmetric polynomials) or h_i (which can be seen as complete symmetric polynomials). The reason is that formulas become nicer and more tractable, than those expressed in terms of power sum symmetric polynomials (see Section 7.4). Our choice is also sanctioned by an important work by I. Frenkel, Penkov and Serganova [31] and Jing and Rozhkovskaya [71].

6.7 Exercises

6.7.1. Prove by hands that in $\bigwedge^3 M_0$

$$a_3 \cdot b_0 \wedge b_1 \wedge b_2 + a_2 \cdot b_0 \wedge b_1 \wedge b_3 + a_1 \cdot b_0 \wedge b_2 \wedge b_3 + a_0 \cdot b_1 \wedge b_2 \wedge b_3$$

is decomposable.

6.7.2. Let M_r be the free B_r-module of rank r as in Section 5.7. Prove that all the elements of $\bigwedge^{r-1} M_r$ are decomposable and find an explicit decomposition. Prove that each element $\eta \in \bigwedge^{r-1} M_0$ is decomposable modulo $\sigma_r M_0 \wedge \bigwedge^{r-2} M_0$.

6.7.3. Let $\eta \in \bigwedge^r M_0$ and $0 \neq m \in M_0$. Prove that the equality

$$\sum_{i \geq 0} (m \wedge b_i \wedge \eta) \otimes (m \wedge (\beta_i \lrcorner \eta)) = 0$$

implies $\sum_{i \geq 0} (b_i \wedge \eta) \otimes (\beta_i \lrcorner \eta) = 0$. This has been used in the final part of the proof of Theorem 6.1.7.

6.7.4. Let M_0 be as in Section 5.6, and consider $\bigwedge^2 M_0$ as B_2-module. Compute by hand

$$\sum_{i \geq 0} b_i z^i \wedge \eta \quad \text{and} \quad \sum_{j \geq 0} z^{-j-1} \beta_j \lrcorner \eta.$$

for $\eta := b_2 \wedge b_5$. To get nice formulas, use the B_3-module structure of M_0 in the first case and the B_1 structure in the second (hint: try with $\eta = b_0 \wedge b_1$ first).

6.7.5. Let $\boldsymbol{\lambda} \in \mathcal{P}_{r,n}$ and denote by $n - \boldsymbol{\lambda}$ the partition $(n - \lambda_r, \ldots, n - \lambda_1)$. Define the involution $\phi : \bigwedge^r M_{0,n} \to \bigwedge^r M_{0,n}$ as $\phi([\mathbf{b}]^r_{\boldsymbol{\lambda}}) = [\mathbf{b}]^r_{n-\boldsymbol{\lambda}}$. Then

$$\sigma_{-j}[\mathbf{b}]^r_{\boldsymbol{\lambda}} = \phi(\sigma_j [\mathbf{b}]^r_{n-\boldsymbol{\lambda}}) \mod \sigma_n M_0 \wedge \bigwedge^{r-1} M_0.$$

In other words $\sigma_-(z)$ behaves on $M_{0,n}$ exactly as $\sigma_+(z)$ does, up to reversing the order of the basis of $\bigwedge^r M_0$ and renaming by z^{-1} the formal variable z. Thus, in all the respect, $\sigma_-(z)$ is a Schubert derivation ('mirror') in the sense of Definition 5.1.1.

6.7.6. Can you imitate Exercise 5.11.7, replacing $\sigma_+(z)$ by $\sigma_-(z)$ in the statement (and t_j^{-1} with t_j)? Make a few examples involving partitions of length at most two, to feel how trickier the situation becomes.

6.7.7. Let $M_{0,n}$ be the free abelian group generated by $(b_0, b_1, \ldots, b_{n-1})$ and let $M_{0,n}^\vee$ its dual. Define an isomorphism $\phi : \bigwedge M_{0,n} \to \bigwedge M_{0,n}^\vee$ which maps $\bigwedge^r M_{0,n}$ isomorphically onto $\bigwedge^{n-r} M_{0,n}^\vee$, for all $0 \le r \le n$, and such that

$$\phi(\beta \lrcorner \eta) = \beta \wedge \phi(\eta).$$

Using such isomorphism define a suitable Schubert derivation on $\bigwedge M_0^*$ (analogous to $\sigma_+(z)$ on $\bigwedge M_0$, but essentially an avatar of $\sigma_-(z)$) to provide an alternative and shorter proof of (6.29) using integration by parts as in 6.2.6.

6.7.8. For all $e_i \in B_r$, $1 \le i \le r$, show that

$$\overline{\sigma}_-(z)e_i = \frac{(-1)^i E_i(z)}{z^i} \qquad \textit{and} \qquad \sigma_-(z)E_i(z) = (-1)^i e_i z^i.$$

Conclude that

$$\Gamma_r(z)e_i = \frac{(-1)^i E_i(z)}{z^i E_{r+1}(z)}$$

and find a formula for $\Gamma_r^*(z)E_i(z)$ as well.

6.7.9. Let z, w be two formal variables. Using the B_{r+1} and B_{r+2} module structures of $\bigwedge^{r+1} M_0$ and $\bigwedge^{r+2} M_0$, respectively, find a nice formula expressing

$$\sum_{i\ge 0} b_i z^i \wedge \sum_{j\ge 0} b_j w^j \wedge \eta$$

in a compact way, for all $\eta \in \bigwedge^r M_0$. Can you generalize it to an arbitrary number $z_1, \ldots, z_n$ of formal variables? (Hint: use the definition of $\sigma_+(z)$ and integration by parts (4.6)).

6.7.10. Let (β_i) be the basis of the restricted dual M_0^*, dual of $(b_j)_{j\ge 0}$ $(\beta_j(b_i) = \delta_{ij})$. With the same notation of 6.7.9, find nice expressions for

$$\sum_{i,j\ge 0} \beta_i \lrcorner (\beta_j \lrcorner \eta) z^{-i} w^{-j}, \quad \sum_{i,j\ge 0} \beta_i \lrcorner (b_j \wedge \eta) z^{-i} w^j \quad \text{and} \quad \sum_{i,j\ge 0} b_i \wedge (\beta_j \lrcorner \eta) z^i w^{-j}$$

using the appropriate B_r-module structure.

6.7.11. The decomposability condition (6.36) applied to $\sum_{\lambda\in\mathcal{P}_{2,n}} a_\lambda [\mathbf{b}]_\lambda^2$ gives the equation of the Pfaffians of all of the 4×4 minors of an $n\times n$ skew-symmetric matrix, $n\geq 4$, which in turn are the equations of the quadrics cutting $G_{2,n}$ in its own Plücker embedding. Can you find an equation similar to (6.36), encoding all the 6×6 Pfaffians of an $n\times n$ skew-symmetric matrix, with $n\geq 6$?

Chapter 7
Vertex Operators via Generic LRS

This closing part of the book is concerned with the identification of the free abelian group M_0, studied in Chapters 5 and 6, with the $\mathbb{Z}$-module spanned by generic LRSs of finite order. The latter will be embedded in the larger free abelian group $\mathbb{U}_r := \bigoplus_{i\in\mathbb{Z}} \mathbb{Z}\cdot u_i$, where $(u_i)_{i\in\mathbb{Z}}$ is the fundamental sequence of formal power series introduced in Section 2.3.1. Certain abelian subgroups F_i^r of semi-infinite exterior products of elements of $\mathbb{U}_r$ are interpreted as approximations of fermionic Fock spaces in the sense of [78]. The index r keeps track that everything is constructed by means of formal power series with B_r-coefficients, while the index i is called *charge*, in compliance with the terminology of standard reference texts on the subject. The vertex operators of the Prologue re-enter the game in the last section of this chapter, where the results of Chapter 6 are recast in terms of the natural basis of the LRSs of order r, encountered in Chapter 2.

The fermionic module of zero charge, F_0^r, plays a special role in the theory. It is in fact canonically isomorphic to the *bosonic* space B_r via the *universal Cauchy formula* (2.24). The $\mathbb{Z}$-module isomorphism $B_r \to F_0^r$ will be called the *boson–fermion correspondence* of order r, for it equips F_0^r with a structure of free B_r-module of rank 1 generated by the *vacuum vector* $\Phi_0^r := u_0 \wedge u_{-1} \wedge \cdots$. It will be used to equip the direct sum $F^r := \bigoplus_{i\in\mathbb{Z}} F_i^r$ with a structure of free $B_r[\ell, \ell^{-1}]$-module freely generated by Φ_0^r, where ℓ is a further indeterminate.

A word of warning: the Schubert derivations on $\bigwedge \mathbb{U}_r$ will be denoted by $\mathcal{D}_{\pm 1}(z)$ (instead of $\sigma_{\pm}(z)$ as in Chapters 5 and 6), to emphasize that it is 'induced' by the endomorphism D of $B_r[[t]]$ which, modulo conjugation by a formal Laplace transform, is essentially the derivative of a formal power series as explained in Section 2.5. An essential ingredient of this chapter consists in extending the Schubert derivations of $\bigwedge \mathbb{U}_r$ to an endomorphism of F^r (Section 7.5) regarded as a subgroup of $\bigwedge^{\infty/2} \mathbb{U}_r$, the *semi-infinite* wedge power of $\mathbb{U}_r$. The task is not difficult but slightly tricky.

In Section 7.4 we discuss the representation of the $(2r + 2)$-dimensional Heisenberg algebra [148] on B_r and F_0^r (bosonic and fermionic).

L. Gatto, P. Salehyan, *Hasse-Schmidt Derivations on Grassmann Algebras*, IMPA Monographs, DOI 10.1007/978-3-319-31842-4_7

The promised expression of the vertex operators is worked out in Section 7.8, based on the respective approximating maps $B_r[\ell^{-1}, \ell] \to B_r[\ell^{-1}, \ell]((z))$ introduced in Section 7.7.

The philosophy underlying the book in general, and this chapter in particular, is not, or at least not only, to prove old results from a different point of view. Rather, we wish to detect vestiges of the representation theory of the Heisenberg Lie algebra or of the Lie algebra $\mathfrak{gl}(\infty)$ of infinite matrices, in elementary situations, such as Schubert calculus for finite-dimensional Grassmannians. This is a perspective which is also endorsed in [3, 71, 119].

7.1 Bosonic and Fermionic Fock Spaces of Finite Order

We assume known all the definitions and notation of Chapter 2. The polynomial ring $\mathbb{Z}[e_1, \ldots, e_r]$ was there named B_r to be reminiscent of the role it plays in the bosonic representation, so to speak, of the $(2r+2)$-dimensional Heisenberg algebra in the sense of Section 7.4 (see [148, Definition 1.7]). The following definition will constitute part of the jargon adopted in this chapter.

Definition 7.1.1. The *r-order bosonic Fock space* is the polynomial ring $B_r = \mathbb{Z}[e_1, \ldots, e_r]$.

7.1.2. For all $r \geq 1$, let

$$\mathbb{U}_r := \bigoplus_{i \in \mathbb{Z}} \mathbb{Z} u_i,$$

where u_i are as in 2.3.1. For notational commodity, we think of the module $\mathbb{U}_r$ as equipped of the inner product $(\,,\,)$ for which $(u_i)_{i\in\mathbb{Z}}$ is an orthogonal basis, i.e. $(u_i, u_j) = \delta_{i,j}$. We also agree, since now, to denote by $(u_i^*)_{i\in\mathbb{Z}}$ the elements of the dual $\mathbb{U}_r^\vee$ of $\mathbb{U}_r$, defined by $u_i^*(u_j) = (u_i, u_j) = \delta_{ij}$, for all $i, j \in \mathbb{Z}$. Recall that the elements u_j^* do not span all the dual $\mathbb{U}_r^\vee$, but only the submodule $\mathbb{U}_r^*$ of it, known as the *restricted dual* (see Section 6.1.6).

We have a filtration $\mathbb{U}_{r,i+1} \subset \mathbb{U}_{r,i} \subset \mathbb{U}_{r,i-1} \subset \cdots$, $(i \in \mathbb{Z})$, where

$$\mathbb{U}_{r,i} := \bigoplus_{j>i} \mathbb{Z} \cdot u_j.$$

In particular, for all $i \in \mathbb{Z}$, we have the $(\,,\,)$-orthogonal direct sum decomposition

$$\mathbb{U}_r = \mathbb{U}_{r,i} \oplus \mathbb{U}_{r,i}^\perp,$$

where

$$\mathbb{U}_{r,i}^\perp = \bigoplus_{j\leq i} \mathbb{Z} \cdot u_j. \tag{7.1}$$

As $u_j \in B_r[[t]]$, for all $j \in \mathbb{Z}$, the universal bilinear map $B_r \times \mathbb{U}_r \to B_r \otimes_{\mathbb{Z}} \mathbb{U}_r$ is just the multiplication:

$$p(e_1, \ldots, e_r)u_j = \sum_{n\geq 0}(p(e_1, \ldots, e_r)h_{n+j})t^n. \tag{7.2}$$

Proposition 7.1.3. *For all $i \leq 0$, $B_r \otimes \mathbb{U}_{r,i}^{\perp}$ is a free B_r-module of infinite rank.*

Proof. This is a consequence of Proposition 2.3.3 or, even easier, the fact that $u_{-i} = t^i u_0$ for $i \geq 0$, which are clearly linearly independent over B_r. □

For $i = -r$, equality (7.1) reads $\mathbb{U}_r = \mathbb{U}_{r,-r} \oplus \mathbb{U}_{r,-r}^{\perp}$. Tensoring both sides by B_r,

$$\begin{aligned} B_r \otimes_{\mathbb{Z}} \mathbb{U}_r &= B_r \otimes \mathbb{U}_{r,-r} \oplus B_r \otimes \mathbb{U}_{r,-r}^{\perp} = K_r \oplus \left(B_r \otimes \mathbb{U}_{r,-r}^{\perp}\right) \\ &= B_r \otimes \mathbb{U}_{r,0}^{\perp} = \bigoplus_{j\geq 0} B_r u_{-j}, \end{aligned}$$

because $B_r \otimes \mathbb{U}_{r,-r}$ is precisely the rank r free B_r-module K_r of the generic LRS.

Notation 7.1.4. For each $(i, k) \in \mathbb{Z}\times\mathbb{N}$ and $\boldsymbol{\lambda} \in \mathcal{P}_r$, define, in a purely conventional way:

i) $[\mathbf{u}]_{i+\boldsymbol{\lambda}}^r := u_{i+\lambda_1} \wedge u_{i-1+\lambda_2} \wedge \cdots \wedge u_{i-r+1+\lambda_r}$;
ii) $\Phi_i^r := u_i \wedge u_{i-1} \wedge \cdots \wedge u_{i-r+1} \wedge u_{i-r} \wedge \cdots$;
iii) $\Phi_{i+\boldsymbol{\lambda}}^r := [\mathbf{u}]_{i+\boldsymbol{\lambda}}^r \wedge \Phi_{i-r}^r$.

By formally extending the associativity and the skew symmetry of the wedge product to infinite exterior powers, we have $u_i \wedge \Phi_i^r = (u_i \wedge u_i) \wedge \Phi_{i-1}^r = 0$, for all $i \in \mathbb{Z}$.

Definition 7.1.5. The r-order *fermionic Fock space* of *charge* $i \in \mathbb{Z}$ is the abelian group

$$F_i^r := \bigwedge^r \mathbb{U}_{r,i} \wedge \Phi_{i-r}^r = \bigoplus_{\boldsymbol{\lambda}\in\mathcal{P}_r} \mathbb{Z}\Phi_{i+\boldsymbol{\lambda}}^r \subsetneq \bigoplus_{\boldsymbol{\lambda}\in\mathcal{P}} \mathbb{Z}\Phi_{i+\boldsymbol{\lambda}}^r,$$

i.e. the set of all finite $\mathbb{Z}$-linear combinations of expressions of the type

$$[\mathbf{u}]_{i+\boldsymbol{\lambda}}^r \wedge \Phi_{i-r}^r := u_{i+\lambda_1} \wedge \cdots \wedge u_{i-r+1+\lambda_r} \wedge \Phi_{i-r}^r, \quad \boldsymbol{\lambda} \in \mathcal{P}_r.$$

The superscript r in the notation emphasizes the fact that the coefficients of the formal power series u_i are taken in B_r. Notice that each element of F_i^r is a fake infinite exterior power, as we have embedded the honest $\bigwedge^r \mathbb{U}_{r,i-r}$ into an *infinite* exterior power by wedging with the fixed formal expression Φ_{i-r}^r. As $i \neq j$ implies $F_i^r \cap F_j^r = \{\mathbf{0}\}$, one can form the direct sum

$$F^r = \bigoplus_{i\geq 0} F_i^r$$

which is an abelian subgroup of $\bigwedge^{\infty/2}\mathbb{U}_r$, the semi-infinite wedge power of $\mathbb{U}_r$, whose notation is here borrowed from many references, as, e.g. [84, p. 1086], [121, Sect. 1.3].

Remark 7.1.6. Definition 7.1.5 might seem slightly artificial. Why not defining F_i^r as the free abelian group generated by $\Phi^r_{i+\boldsymbol{\lambda}}$ with no restriction on the length of the partition $\boldsymbol{\lambda}$? Even more, why adding the formal piece Φ^r_{i-r} to work formally with infinite exterior powers instead of with finite ones? The reason will be clear in Section 7.4 when we will face the fermionic representation of the finite-dimensional Heisenberg algebra. We want to stress, however, that here most of the issue is purely formal and/or aesthetic, and the difference between the two definitions would be lost in the case $r=\infty$, which this chapter is really concerned with.

7.2 The Finite Order Boson–Fermion Correspondence

Let us begin with a dutiful remark. By the way it has been defined, each fermionic Fock spaces F_i^r can be equipped with two in principle different B_r module structures: an *abstract* one and a *natural one*. The former relies on the fact that there is an obvious abstract free abelian group isomorphism $B_r \to F_i^r$, mapping $\Delta_{\boldsymbol{\lambda}}(H_r) \mapsto \Phi^r_{i+\boldsymbol{\lambda}}$. This enables to induce on F_i^r a structure of free B_r-module of rank 1 generated by Φ_i^r. On the other hand, one may consider a B_r-module structure on F_i^r, by just taking the tensor product $B_r \otimes F_i^r$, compatible with the module structure induced by (7.2). The issue of this section is to compare the abstract with the natural B_r-module structure of F_i^r. It is easy to realize that the behaviour of $B_r \otimes F_i^r$ substantially changes depending on i being positive, negative or zero. See below.

7.2.1. For all $i>0$, the module $B_r \otimes F_i^r$ is trivial. From Section 2.1.6 recall the endomorphism D of $B_r[[t]]$ defined by

$$D\sum_{n\geq 0} a_n t^n = \sum_{n\geq 0} a_{n+1}t^n, \qquad (a_n \in B_r) \tag{7.3}$$

that maps $u_i \mapsto u_{i+1}$. Now $u_{-r+1+j} \in K_r := \ker \mathfrak{p}_r(D)$ for all $j \geq 0$ (Proposition 2.3.6), i.e. it is a generic LRS of order r. By Corollary 2.3.9 the B_r-module K_r has rank r. Thus, in the category of B_r-modules, $\bigwedge^{r+1} K_r = 0$, that is, for all $\boldsymbol{\lambda} \in \mathcal{P}_r$,

$$\begin{aligned} 1_{B_r} \otimes \Phi^r_{i+\boldsymbol{\lambda}} &= 1_{B_r} \otimes ([\mathbf{u}]^r_{i+\boldsymbol{\lambda}} \wedge \Phi^r_{i-k}) \\ &= 1_{B_r} \otimes (u_{i+\lambda_1} \wedge \cdots \wedge u_{i-r+1+\lambda_r} \wedge u_{i-r+\lambda_{r+1}} \wedge \Phi^r_{i-r-1}) = 0 \end{aligned} \tag{7.4}$$

in that the expression begins with the wedge product of at least $r+1$ (linearly dependent) generic LRS.

7.2.2. The behaviour of $B_r \otimes F^r_{-i}$, $i > 0$, is instead quite nasty. We limit ourselves to show an example with $r = 1$, leaving the reader to guess how the combinatorics gets more complicated as r and i increase. Consider the $\mathbb{Z}$-module F^1_{-2} whose basis is $\left(\Phi^1_{-2+(\lambda)} := u_\lambda \wedge \Phi^1_{-3}\right)_{\lambda \geq -2}$. It can be written as follows:

$$F^1_{-2} := \bigoplus_{\lambda \geq -2} \mathbb{Z}(u_{2+\lambda} \wedge \Phi^1_{-3}) \oplus \mathbb{Z}(u_{-1} \wedge \Phi^1_{-3}) \oplus \mathbb{Z}(u_{-2} \wedge \Phi^1_{-3}).$$

Taking the tensor product with $B_1 = \mathbb{Z}[e_1] = \mathbb{Z}[h_1]$ and remembering that, in this case, $u_{2+\lambda} = h_1^{2+\lambda} u_0$, we see that $B_1 \otimes F^1_{-2}$ is a free B_1-module of rank 3:

$$B_1 \otimes F^1_{-2} = B_1 \Phi^1_{-2+(2)} \oplus B_1 \Phi^1_{-2+(1)} \oplus B_1 \Phi^1_{-2}.$$

In general, $B_1 \otimes F^1_{-i}$ is a free B_1-module of rank $i + 1$.

In the case $i = 0$, the situation is as nice as one hopes, because the natural B_r-module structure of F^r_0 does coincide with the abstract one.

Proposition 7.2.3. *The abelian group $B_r \otimes F^r_0$ has a natural structure of free module of rank 1 generated by Φ^r_0 such that*

$$\Phi^r_{0+\lambda} = \Delta_\lambda(H_r)\Phi^r_0, \tag{7.5}$$

which then coincides with the abstract $\mathbb{Z}$-module isomorphism $B_r \to F^r_0$ mapping $\Delta_\lambda(H_r) \mapsto \Phi^r_{0+\lambda}$ (the boson–fermion correspondence of order r in charge 0).

Proof. To directly prove (7.5), observe that i) $\mathfrak{p}_r(D)u_{-r+1+j} = 0$ for all $j \geq 0$ and ii) that $u_{-i+1+\lambda_i} = D_{r-i+\lambda_i} u_{-r+1}$. Therefore

$$\begin{aligned}
\Phi^r_{0+\lambda} &= u_{\lambda_1} \wedge \cdots \wedge u_{-r+1+\lambda_r} \wedge \Phi^r_{-r} = D_{r-1-\lambda_1} u_{-r+1} \wedge \cdots \wedge D_{\lambda_r} u_{-r+1} \wedge \Phi^r_{-r} \\
&= \mathrm{Res}_X \left(\frac{X^{r-1-\lambda_1}}{\mathfrak{p}_r(X)}, \ldots, \frac{X^{\lambda_r}}{\mathfrak{p}_r(X)} \right) u_0 \wedge \cdots \wedge u_{-r+1} \wedge \Phi^r_{-r} \\
&= \Delta_\lambda(H_r)[\mathbf{u}]^r_0 \wedge \Phi^r_{-r} = \Delta_\lambda(H_r)\Phi^r_0
\end{aligned}$$

having applied Theorem 5.8.1 and Corollary 5.8.2. □

Remark 7.2.4. Define the *Wronskian module*:

$$\mathcal{W}(\mathbf{u}_r) := \bigoplus_{\lambda \in \mathcal{P}_r} \mathbb{Z} \cdot W_{\mathbf{X}^\lambda}(\mathbf{u}_r).$$

By Theorem 2.6.7, it is a free $\mathbb{Z}$-module, because of the $\mathbb{Z}$-module isomorphism

$$B_r \to \mathcal{W}(\mathbf{u}_r) \tag{7.6}$$

mapping $\Delta_{\boldsymbol{\lambda}}(H_r) \mapsto W_{\mathbf{X}^{\boldsymbol{\lambda}}}(\mathbf{u}_r)$. In turn, the map $W_{X^{\boldsymbol{\lambda}}}(\mathbf{u}_r) \mapsto [\mathbf{u}]^r_{\boldsymbol{\lambda}}$ defines an obvious homomorphism of free abelian groups $\Psi : \mathcal{W}(\mathbf{u}_r) \to \bigwedge^r K_r$ which is a B_r-module isomorphism as well. In fact

$$\begin{aligned}\Psi(\Delta_{\boldsymbol{\lambda}}(H_r)W_{\mathbf{X}}(\mathbf{u}_r)) &= \Psi(W_{\mathbf{X}^{\boldsymbol{\lambda}}}(\mathbf{u}_r)) = u_{\lambda_1} \wedge u_{-1+\lambda_2} \wedge \cdots \wedge u_{-r+1+\lambda_r} \\ &= \Delta_{\boldsymbol{\lambda}}(H_r)u_0 \wedge u_{-1} \wedge \cdots \wedge u_{-r+1} \\ &= \Delta_{\boldsymbol{\lambda}}(H_r)\Psi(W_{\mathbf{X}}(\mathbf{u}_r)).\end{aligned}$$

There is in sum a canonical identification of the generalized Wronskians with the monomials $[\mathbf{u}]^r_{\boldsymbol{\lambda}} \in \bigwedge^r K_r$. However, contrarily to $\mathcal{W}(\mathbf{u}_r)$, the module $\bigwedge^r K_r$ can be defined also for $r = \infty$. In all the respects, the elements of $\bigwedge^r K_r$ can be considered as generalized Wronskians, and the map (7.6), which as remarked in 2.6.8 is a generalization of the Liouville's theorem, is yet another realization of the finite order boson–fermion correspondence we dealt with in this section.

7.3 On the Infinite Exterior Power II

Although up to now we have just dealt with *fake* infinite exterior powers (for instance, F^r_0 can be treated in all the respects as $\bigwedge^r \mathbb{U}_{r,-r}$), it is dutiful to spend a few words on what one should really mean for an 'infinite exterior power of a module of infinite rank'. We do not want, here, to be too technical for two reasons: the former is that such an explanation goes well beyond the scopes of the book, whose focus is on derivations on Grassmann algebras and not on representation theory; secondly, the story we are going to summarize below is superbly told in many excellent references like [32, p. 85], [73, p. 90ff], [74, p. 50], [72, p. 219], [78, p. 52], [79, p. 191–192].

Let $\mathbb{U}^*_r := \bigoplus_{i \in \mathbb{Z}} \mathbb{Z}u^*_i$ be the *restricted dual* of $\mathbb{U}_r$, as in Section 7.1.2, where $u^*_i \in \mathbb{U}^\vee_r$ are such that $u^*_i(u_j) = \delta_{ij}$. Roughly speaking, we can consider an irreducible module $\overline{F}^r$ over the Clifford algebra $\mathcal{C}(\mathbb{U}_r \oplus \mathbb{U}^*_r)$ constructed with respect to the canonical bilinear symmetric form of $\mathbb{U}_r \oplus \mathbb{U}^*_r$ (see Section 3.3.2). It is generated by a vector $|0\rangle_r$ such that $u_i|0\rangle_r = 0$ for all $i \le 0$ and $u^*_i|0\rangle_r = 0$ for all $i > 0$. One then defines the module $\overline{F}^r$ to be the $\mathbb{Z}$-module generated by all the expressions of the form

$$u_{i_1} \cdots u_{i_h} u^*_{j_1} \cdots u^*_{j_k} |0\rangle_r.$$

Define the *vacuum* of charge $i \in \mathbb{Z}$ as

$$|i\rangle_r = \begin{cases} u_i u_{i-1} \cdots u_1 |0\rangle_r, & \textit{if} \;\; i > 0 \\ u^*_{i+1} u^*_{i+2} \cdots u^*_0 |0\rangle_r, & \textit{if} \;\; i < 0 \end{cases}$$

and $\Phi^r_{i+\boldsymbol{\lambda}} = u_{i+\lambda_1}\cdots u_{i-r+1+\lambda_r}u^*_{i-r+1}\cdots u^*_i|i\rangle_r$, for all $\boldsymbol{\lambda} \in \mathcal{P}_r$. Thus, if $\boldsymbol{\lambda} = (0)$ and $i > 0$,

$$\begin{aligned}\Phi^r_i = u_iu^*_i|i\rangle_r &= u_iu^*_iu_iu_{i-1}\cdots u_1|0\rangle_r = u_i(1-u_iu^*_i)u_{i-1}\cdots u_1|0\rangle_r\\ &= u_iu_{i-1}\cdots u_1|0\rangle_r - u_iu^*_iu_{i-1}\cdots u_1|0\rangle_r\\ &= u_iu_{i-1}\cdots u_1|0\rangle_r - u_iu_{i-1}\cdots u_1u^*_i|0\rangle_r\\ &= u_iu_{i-1}\cdots u_1|0\rangle_r\\ &= |i\rangle_r.\end{aligned}$$

The case $i < 0$ is analogous. One can then define $\overline{F}^r_i := \bigoplus_{\boldsymbol{\lambda}\in\mathcal{P}}\mathbb{Z}\Phi^r_{i+\boldsymbol{\lambda}}$, which is of course larger of our previously defined F^r_i (cf. Definition 7.1.5), in that we are not bounding the length of the partitions. Writing $\Phi^r_{i+\boldsymbol{\lambda}}$ as $[\mathbf{u}]^r_{i+\boldsymbol{\lambda}} \wedge |i-r\rangle_r$, the representation of $\mathcal{C}(\mathbb{U}_r\oplus\mathbb{U}^*_r)$ on $\overline{F}^r_0$ can be seen as an extension of the usual wedging and contracting isomorphisms, namely,

$$u_j \wedge \Phi^r_{i+\boldsymbol{\lambda}} := u_ju_{i+\lambda_1}\cdots u_{i-r+1+\lambda_r}u^*_{i-r+1}\cdots u^*_i|0\rangle_r,$$

and

$$u^*_j \lrcorner \Phi^r_{i+\boldsymbol{\lambda}} := u^*_ju_{i+\lambda_1}\cdots u_{i-r+1+\lambda_r}u^*_{i-r+1}\cdots u^*_i|0\rangle_r.$$

7.4 On the Finite-Dimensional Heisenberg Algebra

7.4.1. By mixing both [148, Definition 1.7] and [78], let

$$\mathcal{H}_r := \bigoplus_{-r\le n\le r}\mathbb{Z}\cdot d_n \oplus \mathbb{Z}\cdot\hbar,$$

be the *Heisenberg Lie algebra* of rank $2r+2$, with commutation relations

$$[d_n,\hbar] = 0, \qquad [d_m,d_n] = m\delta_{m,-n}\hbar, \tag{7.7}$$

i.e. both d_0 and $\hbar$ are central elements. If $r=\infty$, one obtains precisely the definition as in [78, p. 12]. The first of (7.7) says that $\mathcal{H}_r$ is a one-dimensional central extension of the trivial Lie algebra of rank $2r+1$ generated by $(d_i)_{-r\le i\le r}$.

This section aims to sketch the construction of both the *bosonic* and the *fermionic* representation of $\mathcal{H}_r$. By the former we it mean representing $\mathcal{H}_r$ on $\overline{B}_r := B_r \otimes_{\mathbb{Z}} \mathbb{Q} = \mathbb{Q}[e_1,\dots,e_r]$, by the latter representing it on $F^r_0 \otimes_{\mathbb{Z}} \mathbb{Q}$. It is clear that each representation induces a fermionic one via the boson–fermion correspondence (7.5).

Bosonic Representation of $\mathcal{H}_r$ 7.4.2. In the ring $\overline{B}_r$, consider the sequence of elements $\mathbf{x}_r := (x_1, x_2, \ldots)$ implicitly defined by the equation

$$E_r(z)\exp(\sum_{i\geq 1} x_i z^i) = 1. \tag{7.8}$$

Equation (7.8) implies the following recursions (see Exercise 7.10.3):

$$jx_j - e_1 \cdot (j-1)x_{j-1} + \cdots + (-1)^{j-1}e_{j-1}x_1 + (-1)^j je_j = 0, \quad (1 \leq j \leq r) \tag{7.9}$$

and

$$(r+i)x_{r+i} + \sum_{j=1}^{r}(-1)^j e_j(r+i-j)x_{r+i-j} = 0, \quad (i \geq 1). \tag{7.10}$$

They prove in fact that each x_j is a homogeneous polynomial expression in $x_1, \ldots, x_r$ (with $\overline{B}_r$-coefficients). A simple exercise shows that the expressions of $x_1, \ldots, x_r$ as polynomials in $(e_1, \ldots, e_r)$ are invertible, i.e. each e_j can be expressed as a polynomial in $(x_1, \ldots, x_j)$ of degree j. In particular the map $\mathbb{Q}[T_1, T_2, \ldots] \to \overline{B}_r$, defined by $T_i \mapsto x_i$, is surjective, and we denote by $\mathbb{Q}[\mathbf{x}_r]$ its image, to signify that the ring $\overline{B}_r$ is indeed generated by $(x_1, \ldots, x_r)$ as a $\mathbb{Q}$-algebra. For the sake of notational brevity, let

$$\partial_j := \frac{\partial}{\partial x_j}, \qquad (1 \leq j \leq r)$$

and make $\mathbb{Q}[\mathbf{x}_r]$ into an $\mathcal{H}_r$-module by defining

$$d_j p = \partial_j p, \qquad d_{-j}p := ix_j p, \qquad d_0 p := \mu p, \qquad \hbar p := p, \tag{7.11}$$

for all $p \in \mathbb{Q}[\mathbf{x}_r]$ and $j > 0$, where $\mu \in \mathbb{Q}$, and by simplicity we set $\hbar = 1$. As remarked in [78], where one works over the complex numbers and with $r = \infty$, the value of $\mu \in \mathbb{C}$ is important from the point of view of classifying representations, but we shall not be concerned with such aspects. Because of the equalities

$$\left[\partial_i, jx_j\right] = i\delta_{ij}, \qquad (i, j > 0),$$

equations (7.11) define a representation of $\mathcal{H}_r$ on $\overline{B}_r$.

Fermionic Representation 7.4.3. The representation (7.11) induces a *fermionic* one, defined by

$$d_i \Phi^r_{0+\boldsymbol{\lambda}} := (d_i \Delta_{\boldsymbol{\lambda}}(H_r))\Phi^r_0.$$

For instance,

$$d_1\Phi^r_{0+(1)} = d_1(h_1\Phi^r_0) := (\partial_1 h_1)\Phi^r_0 = \partial_1 h_1 \Phi^r_0 = \Phi^r_0,$$

while

$$\begin{aligned} d_{-1}\Phi^r_{0+(1)} &= d_{-1}(h_1\Phi^r_0) := (d_{-1}h_1)\Phi^r_0 = (x_1h_1)\Phi^r_0 \\ &= h_1^2\Phi^r_0 = (h_1^2 - h_2)\Phi^r_0 + h_2\Phi^r_0 = \Phi^r_{0+(1,1)} + \Phi^r_{0+(2)}. \end{aligned}$$

Clearly $d_0\Phi^r_{0+\lambda} = \mu\Phi^r_{0+\lambda}$. We have so obtained a representation of $\mathcal{H}_r$ on $F^r \otimes_{\mathbb{Z}} \mathbb{Q}$. Our task then reduces to make more explicit the form of the operators d_i in the fermionic case. To this purpose we shall heavily rely on the boson–fermion correspondence 7.2.3.

7.4.4. Let us start by introducing the notion of the *trace* of an endomorphism of $\overline{B}_r \otimes_{\mathbb{Z}} \mathbb{U}_r$. It is a relative of the coefficient of the first degree of the trace polynomial operator of Section 4.2. Here, we prefer to define it directly to avoid too many preliminaries that would bring us far from the scope of the present exposition. Here we think of $\overline{B}_r[[t]]$ as the completion of $\mathbb{U}_r \otimes_{\mathbb{Z}} \overline{B}_r$ in the sense of Exercise 7.10.1. We say that the *trace* of $\phi \in \mathrm{End}_{\mathbb{Q}}(\overline{B}_r[[t]])$ is defined on $\overline{B}_r \otimes_{\mathbb{Z}} F^r_0$ if and only if there exists $n \in \mathbb{Z}$ such that

$$1_{B_r} \otimes (\phi(u_i) \wedge \Phi^r_{i-1}) = 0, \tag{7.12}$$

for all $i \le n$. In this case one defines the trace $\hat{\phi}$ of ϕ on $\overline{B}_r \otimes_{\mathbb{Z}} F^r_0$ as

$$\begin{aligned} \hat{\phi}(\Phi^r_{0+\lambda}) &= \hat{\phi}([\mathbf{u}]^r_{0+\lambda} \wedge \Phi^r_{-r}) \\ &:= \hat{\phi}([\mathbf{u}]^r_{0+\lambda}) \wedge \Phi^r_{-r} + [\mathbf{u}]^r_{0+\lambda} \wedge \hat{\phi}(\Phi^r_{-r}), \end{aligned} \tag{7.13}$$

where

$$\hat{\phi}([\mathbf{u}]^r_{0+\lambda}) := 1_{\overline{B}_r} \otimes \sum \phi^{i_1}(u_{\lambda_1}) \wedge \phi^{i_2}(u_{-1+\lambda_2}) \wedge \cdots \wedge \phi^{i_r}(u_{-r+1+\lambda_r}),$$

summed over all $(i_1, \ldots, i_r) \in \{0, 1\}^r$ such that $\sum_j i_j = 1$, and

$$\widehat{\phi}(\Phi^r_{-r}) = 1_{\overline{B}_r} \otimes \sum \phi^{i_{r+1}}(u_{-r}) \wedge \phi^{i_{r+2}}(u_{-r-1}) \wedge \phi^{i_{r+3}}(u_{-r-2}) \wedge \cdots,$$

summed over all the sequences of the set $\{0, 1\}$ supported at just one term, which is finite because of condition (7.12). Considering the expression with $\overline{B}_r$-coefficients, the right-hand side of (7.13) is a finite sum, and that's why we say that it is defined. Recall the endomorphism D of $\mathbb{U}_r$ mapping $u_j \mapsto u_{j+1}$, as in (2.6).

Lemma 7.4.5. *The trace $\widehat{D^i}$ of D^i is defined on $\overline{B}_r \otimes_{\mathbb{Z}} F^r_0$, for all $i \ge 1$.*

Proof. Since $D^i u_{j-i-k} = u_{j-k}$ for all $k \ge 0$, it follows that

$$\begin{aligned} \widehat{D^i}(\Phi^r_j) &= \widehat{D^i}(u_j \wedge \cdots \wedge u_{j-i+1}) \wedge \Phi^r_{j-i} + u_i \wedge \cdots \wedge u_{i-j+1} \wedge \widehat{D^i}\Phi^r_{i-j} \\ &= \widehat{D^i}(u_j \wedge \cdots \wedge u_{j-i+1}) \wedge \Phi^r_{j-i}. \end{aligned}$$

In fact, in all the summands obtained by expanding the expression $u_i \wedge \cdots \wedge u_{i-j+1} \wedge \widehat{D^i}\Phi^r_{i-j}$, according to the definition of $\widehat{D^i}$, a pair u_h and u_k of exterior factors occurs with $h = k$, causing its vanishing by skew symmetry. □

7.4.6. Notice now that the equality

$$\sum_{n\geq 0} h_n z^n = \exp(\sum_{i\geq 1} x_i z^i), \tag{7.14}$$

obtained from (7.8), exhibits h_j as an explicit homogeneous polynomial of degree j in $x_1, \ldots, x_r$. Thus, it makes sense to speak of the partial derivatives of h_j with respect to $x_1, \ldots, x_r$, and then each u_j, $j \in \mathbb{Z}$, of the basis of $\mathbb{U}_r$ can be itself regarded as a function of $(x_1, \ldots, x_r)$. (Cf. [78, p. 59] and Section 1.4.3 where the h_is are called S_i. We are rather using Macdonald notation [109] for the complete symmetric polynomials.)

Example 7.4.7. Let $r \geq 3$. Then

$$h_1 = x_1, \qquad h_2 = \frac{x_1^2}{2} + x_2, \qquad h_3 = \frac{x_1^3}{3!} + x_1x_2 + x_3.$$

Hence $\partial_1 h_1 = \partial_2 h_2 = \partial_3 h_3 = 1$, $\partial_1 h_2 = x_1$, $\partial_1 h_3 = h_2$, $\partial_2 h_3 = h_1$ and $\partial_i h_1 = \partial_i h_2 = \partial_i h_3 = 0$, for all $4 \leq i \leq r$.

Lemma 7.4.8. *For each $i \in \mathbb{Z}$ and $1 \leq j \leq r$, the equality $\partial_i u_j := \sum_{n\geq 0}(\partial_i h_{n+j})t^n$. defines a $\mathbb{Q}$-endomorphism of $B_r[[t]]$, and the following formula holds:*

$$\frac{\partial u_i}{\partial x_k} = u_{i-k} + \sum_{j\geq 1} \frac{\partial x_{r+j}}{\partial x_k} \cdot u_{i-r-j}, \tag{7.15}$$

Proof. Equation 7.14 implies

$$\sum_{n\geq 0} \frac{\partial h_n}{\partial x_k} z^n = \left(z^k + \frac{\partial x_{r+1}}{\partial x_k} z^{r+1} + \frac{\partial x_{r+2}}{\partial x_k} z^{r+2} + \cdots\right) \sum_{n\geq 0} h_n z^n. \tag{7.16}$$

Equating the coefficients of z^{n+i} ($i \in \mathbb{Z}$) on both sides of (7.16), one obtains

$$\begin{aligned}\frac{\partial h_{n+i}}{\partial x_k} &= h_{n+i-k} + \frac{\partial x_{r+1}}{\partial x_k} h_{n+i-r-1} + \frac{\partial x_{r+2}}{\partial x_k} h_{n+i-r-2} + \cdots = \\ &= h_{n+i-k} + \sum_{j\geq 1} \frac{\partial x_{r+j}}{\partial x_k} h_{n+i-r-j}.\end{aligned}$$

Thus

$$\frac{\partial u_i}{\partial x_k} = \sum_{n\geq 0} \frac{\partial h_{n+i}}{\partial x_k} t^n = \sum_{n\geq 0} \left(h_{n+i-k} + \sum_{j\geq 1} \frac{\partial x_{r+j}}{\partial x_k} h_{n+i-r-j} \right) t^n$$

$$\sum_{n\geq 0} h_{n+i-k} z^n + \sum_{j\geq 1} \frac{\partial x_{r+j}}{\partial x_k} \sum_{n\geq 0} h_{n+i-r-j} z^n = u_{i-k} + \sum_{j\geq 1} \frac{\partial x_{r+j}}{\partial x_k} u_{i-r-j},$$

which proves (7.15). □

Corollary 7.4.9. *For all $j \in \mathbb{Z}$ and $1 \le i \le r$, the trace $\widehat{\partial}_i$ of ∂_i is defined on $\overline{B}_r \otimes F_0^r$. Indeed*

$$\widehat{\partial}_i(\mathbb{1}_{\overline{B}_r} \otimes \Phi_j^r) = 0. \tag{7.17}$$

Proof. To prove (7.17), observe that by definition

$$\begin{aligned}\widehat{\partial}_i(1_{\overline{B}_r}\otimes)\Phi_j^r &= 1_{\overline{B}_r} \otimes (\partial_i u_j \wedge u_{j-1} \wedge u_{j-2} \wedge \cdots + u_j \wedge \partial_i u_{j-1} \wedge u_{j-2} \wedge \cdots \\ &\quad + u_j \wedge u_{j-1} \wedge \partial_i u_{j-2} \wedge \cdots + \cdots).\end{aligned} \tag{7.18}$$

Due to (7.15), each $\partial_i u_j$ is equal to an infinite $\overline{B}_r$-linear combination of u_k, with $k < j$, and then all the summands vanish by skew symmetry. □

Remark 7.4.10. Notice that if $r = \infty$, meaning that $\overline{B}_\infty = \mathbb{Q}[e_1, e_2, \ldots]$ is a ring in infinitely many indeterminates, and setting:

$$H_\infty(z) := \sum_{n\ge 0} h_n z^n = \exp(\mathbf{x}_\infty(z)) = \frac{1}{E_\infty(z)}$$

with $x_1, x_2, \ldots$ all algebraically independent, then Lemma 7.4.8 gives

$$\partial_i u_j = u_{i-j}$$

because $\partial_i x_j = \delta_{ij}$ in this case. It follows that, for $r = \infty$, both D_j and ∂_i are well-defined *shift operators* on $\mathbb{U}_\infty$, reproducing exactly the situation studied in [78, p. 32]. The traces $\widehat{\partial}_i$, like $\widehat{D^j}$, are defined over the rationals. Therefore, tensoring the vector space $F_0 \otimes \mathbb{Q} := F_0^\infty \otimes \mathbb{Q}$ by B_∞ is not necessary to let the manipulations as in 7.4.9 making sense.

Theorem 7.4.11. *For all $1 \le i \le r$, the map $d_i \mapsto \widehat{\partial}_i$, $d_{-i} \mapsto \widehat{D^i}$ is the fermionic representation of $\mathcal{H}_r$, i.e.:*

i) $\widehat{D^i}(\Phi_{0+\lambda}^r) = ix_i \Delta_\lambda(H_r) \cdot \Phi_0^r$;
ii) $\widehat{\partial}_i(\Phi_{0+\lambda}^r) = \partial_i \Delta_\lambda(H_r) \cdot \Phi_0^r$;
iii) $[\widehat{\partial}_i, \widehat{D^i}] = i\delta_{ij}$.

Before sketching the proof of Theorem 7.4.11, let us see how things work through the following:

Example 7.4.12. The equalities below hold in the ring B_r, for all $r \ge 2$:

$$h_1 = e_1 = x_1, \qquad h_2 = \frac{x_1^2}{2} + x_2, \qquad e_2 = \frac{x_1^2}{2} - x_2.$$

If $r = 2$, the relation

$$h_{n+2} = e_1 h_{n+1} - e_2 h_n,$$

gives in particular $h_3 = e_1h_2 - e_2h_1 = 2x_1x_2$ and the relation $3x_3 = e_1 \cdot 2x_2 - e_2x_1$ yields the equality

$$x_3 = x_1x_2 - \frac{x_1^3}{3!}$$

Hence

$$\Delta_{(2,1)}(H_2) = h_1h_2 - h_3 = e_2h_1 = \frac{x_1^3}{2} - x_1x_2.$$

Therefore, in the *bosonic* picture, one has, e.g.

$$d_2(\Delta_{(2,1)}(H_2)) = \partial_2(\Delta_{(2,1)}(H_2)) = \partial_2\left(\frac{x_1^3}{2} - x_1x_2\right) = -x_1 = -h_1 \tag{*}$$

and

$$d_{-2}\Delta_{(2,1)}(H_2) = 2x_2\Delta_{(2,1)}(H_2) = x_1^3x_2 + 2x_1x_2^2. \tag{**}$$

We want to perform the corresponding computations within the *fermionic* picture. Under the boson–fermion correspondence, $\Delta_{(2,1)}(H_2)$ is mapped to $\Phi^2_{0+(2,1)}$. We have

$$\begin{aligned} \widehat{\partial_2}\Phi^2_{0+(2,1)} &= \widehat{\partial_2}(u_2 \wedge u_0 \wedge \Phi^2_{-2}) \\ &= \partial_2 u_2 \wedge u_0 \wedge \Phi^2_{-2} + u_2 \wedge \partial_2 u_0 \wedge \Phi^2_{-2} + u_2 \wedge u_0 \wedge \widehat{\partial_2}\Phi^2_{-2} \\ &= \partial_2 u_2 \wedge u_0 \wedge \Phi^2_{-2} + u_2 \wedge \partial_2 u_0 \wedge \Phi^2_{-2}, \end{aligned} \tag{7.19}$$

where we used $u_2 \wedge u_0 \wedge \widehat{\partial_2}\Phi^2_{-2} = 0$, due to formula (7.17). Now using formula (7.15),

$$\partial_2 u_2 = u_0 + \frac{\partial x_3}{\partial x_2}u_{-1} + \frac{\partial x_4}{\partial x_2}u_{-2} + \cdots \tag{7.20}$$

and

$$\partial_2 u_0 = u_{-2} + \frac{\partial x_3}{\partial x_2}u_{-3} + \frac{\partial x_4}{\partial x_2}u_{-4} + \cdots. \tag{7.21}$$

Using the skew symmetry of the exterior product after substituting (7.20) and (7.21) into (7.19) one obtains (*):

$$\begin{aligned} \widehat{\partial_2}\Phi^2_{0+(2,1)} &= \left(u_0 + \frac{\partial x_3}{\partial x_2}u_{-1} + \frac{\partial x_4}{\partial x_2}u_{-2} + \cdots\right) \wedge u_0 \wedge u_{-2} \wedge u_{-3} \wedge \cdots + \\ &+ u_2 \wedge \left(u_{-2} + \frac{\partial x_3}{\partial x_2}u_{-3} + \frac{\partial x_4}{\partial x_2}u_{-4} + \cdots\right) \wedge u_{-2} \wedge u_{-3} \wedge \cdots = \\ &= \frac{\partial x_3}{\partial x_2}u_{-1} \wedge u_0 \wedge \Phi^2_{-2} = -\frac{\partial x_3}{\partial x_2}\Phi^2_0 = -x_1\Phi^2_0 = -h_1\Phi^2_0, \end{aligned}$$

because of the equality $x_1 = h_1$. Notice on the other hand that for all $r \geq 3$,

$$\frac{\partial}{\partial x_2}(h_1 h_2 - h_3) = h_1 - h_1 = 0,$$

and in this case, in fact, $\widehat{\partial_2} \Phi'_{0+(2,1)} = 0$. Moreover

$$\begin{aligned}\widehat{D^2} \Phi^2_{0+(2,1)} &= \widehat{D^2} u_2 \wedge u_0 \wedge \Phi^2_{-2} + u_2 \wedge \widehat{D^2} u_0 \wedge \Phi^2_{-2} \\ &= u_4 \wedge u_0 \wedge \Phi^2_{-2} = \Delta_{(4,1)}(H_2)\Phi^2_0 = (h_1 h_4 - h_5)\Phi^2_0.\end{aligned}$$

whence

$$\frac{\widehat{D^2} \Phi^2_{0+(2,1)}}{\Phi^2_0} = h_1 h_4 - h_5 = e_2 h_3 = e_2(e_1 h_2 - e_2 h_1) = x_1^3 x_2 + 2 x_1 x_2^2$$

which coincides with (**) as expected.

The rest of this section shall be devoted to sketch the proof of Theorem 7.4.11.

Notation 7.4.13. Let n,k be positive integers. Following Section 2.2.2, denote by $(n, 1^k)$ the partition $(1^k n^1)$. By convention set $(n, 1^0) := (n)$.

The following result will be assumed without proof by referring the reader to [44, Lemma 3.11].

Lemma 7.4.14. *Let r,n be positive integers. Then, for each non-negative integer k,*

$$\Delta_{(n,1^k)}(H_r) + \sum_{j=1}^{n-1} (-1)^j e_j \Delta_{(n-j,1^k)}(H_r) + (-1)^n e_{n+k} = 0, \tag{7.22}$$

In particular if $n + k > r$, then $e_{n+k} = 0$, and formula (7.22) reduces to

$$\Delta_{(n,1^k)}(H_r) + \sum_{j=1}^{n-1} (-1)^j e_j \Delta_{(n-j,1^k)}(H_r) = 0. \tag{7.23}$$

□

- **Proof of Theorem 7.4.11, i).** For all $1 \leq j \leq r$, put $\phi = D^j$ in 7.13. Then

$$\widehat{D^j} \Phi^r_0 = 1_{\overline{B}_r} \otimes (u_j \wedge \Phi^r_{-1} + u_0 \wedge u_{j-1} \wedge \Phi^r_{-2} + \cdots + u_0 \wedge u_{-1} \wedge \cdots \wedge u_1 \wedge \Phi_{-j}),$$

as a consequence of the proof of Lemma 7.4.5. Therefore

$$\begin{aligned}\widehat{D^j} \Phi^r_0 &= \Delta_{(j)}(H_r)\Phi^r_0 \quad \Delta_{(j-1,1)}(H_r)\Phi^r_0 + \Delta_{(j-2,1^2)}(H_r)\Phi^r_0 + \cdots + (-1)^j \Delta_{(1^j)}(H_r)\Phi^r_0 \\ &= \left(\Delta_{(j)}(H_r) + \sum_{i=1}^{j-1} (-1)^i \Delta_{(j-i,1^i)}(H_r) \right) \Phi^r_0.\end{aligned} \tag{7.24}$$

In particular

$$\Delta_\lambda(H_r)\widehat{D^j}\Phi_0^r = \widehat{D^j}\Delta_\lambda(H_r)\Phi_0^r,$$

because each $\Delta_{(j-i,1^i)}(H_r)$ commutes with any other Schur polynomials ($\overline{B}_r$ is a commutative ring). This means that

$$\Delta_{(j)}(H_r) + \sum_{i=1}^{j-1}(-1)^i\Delta_{(j-i,1^i)}(H_r)$$

is an eigenvalue of $\widehat{D^j}$ whose eigenspace is the entire $F_0^r \otimes \mathbb{Q}$. We contend that

$$\begin{cases} \widehat{D^1} = x_1 \\ \widehat{D^j} + \sum_{k=1}^{j-1}(-1)^k e_k\widehat{D^{j-k}} + (-1)^j je_j = 0, \qquad 2 \le j \le r. \end{cases} \tag{7.25}$$

Expressions (7.25) would prove the claim because they are the same recursion relations (7.9) for the sequence $(x_1, 2x_2, 3x_3, \ldots, rx_r)$ and with the same initial conditions. The first of (7.25) is clear. In fact $\widehat{D^1} = \Delta_1(H_r) = h_1 = x_1$. The issue is then to prove formula (7.25). First of all notice that for each $1 \le i \le j-1$, the equality

$$\sum_{k=0}^{j-i-1}(-1)^k e_k\Delta_{(j-i-k,1^{i+k})}(H_r) + (-1)^{j-i}e_j = 0 \tag{7.26}$$

holds because of Lemma (7.4.14) applied to $n = j - i$. Then

$$\begin{aligned} &\widehat{D^j} + \sum_{1\le k\le j-1}(-1)^k e_k\widehat{D^{j-k}} + (-1)^j je_j = \\ &= \sum_{k=0}^{j-1}(-1)^k e_k\left(\Delta_{(j-k)}(H_r) + \sum_{i=1}^{j-k-1}(-1)^i\Delta_{(j-k-i,1^i)}(H_r)\right) + (-1)^j je_j = \\ &= \sum_{i=0}^{j-1}(-1)^i\left(\sum_{k=0}^{j-1-i}(-1)^k e_k\Delta_{(j-i-k,1^{i+k})}(H_r) + (-1)^{j-i}e_j\right) = 0 \end{aligned}$$

where the vanishing is due to (7.26). Hence formula (7.25) follows and part i) of Theorem 7.4.11 is proven. □

- The **Proof of Theorem 7.4.11 part ii**) is boring but not difficult. It amounts to a patient computation of the derivative with respect to x_i, $1 \le i \le r$, of the Schur determinant $\Delta_\lambda(H_r)$ in order to compare with the expression of $\hat{\partial}_i\Phi_{0+\lambda}^r$. This procedure works because to compute the derivative of a determinant is sufficient to apply Leibniz rule with respect to the columns. We refer to [44, p. 177] for the complete details.

- **Proof of Theorem 7.4.11, iii).** The last part is now easy. In fact

$$[\widehat{\partial}_i, \widehat{D^j}]\Phi^r_{0+\lambda} = \widehat{\partial}_i(\widehat{D^j}\Phi^r_{0+\lambda}) - \widehat{D^j}(\widehat{\partial}_i\Phi^r_{0+\lambda})$$

Now

$$\widehat{\partial}_i(\widehat{D^j}\Phi^r_{0+\lambda}) = \widehat{\partial}_i(jx_j\Delta_\lambda(H_r)\Phi^r_0).$$

By item ii) this is equal to

$$\partial_i(jx_j\Delta_\lambda)\Phi^r_0 = j\delta_{ij}\Phi_{0+\lambda}b + jx_j\widehat{\partial}_i\Phi^r_{0+\lambda}.$$

On the other hand,

$$\widehat{D^j}(\widehat{\partial}_i\Phi^r_{0+\lambda}) = jx_j\hat{\partial}_i\Phi^r_{0+\lambda},$$

and so $[\widehat{\partial}_i, \widehat{D^j}] = i\delta_{ij}$ as required. □

7.5 Schubert Derivations on the Fermionic Modules

7.5.1. By restriction of the coefficients to the ring of the integers, the B_r-endomorphism (7.3) permits to construct the unique *HS*-derivation (see Proposition 4.1.13)

$$\mathcal{D}_+(z) := \sum_{j\geq 0} D_jz^j : \bigwedge \mathbb{U}_r \to \bigwedge \mathbb{U}_r[[z]],$$

such that $D_ju_i = D^ju_i = u_{i+j}$. Similarly, let us denote by

$$\mathcal{D}_-(z) := \sum_{j\geq 0} D_{-j}z^{-j} : \bigwedge \mathbb{U}_r \to \bigwedge \mathbb{U}_r[[z^{-1}]],$$

the unique *HS*-derivation on $\bigwedge \mathbb{U}_r$ such that $D_{-j}u_i = u_{i-j}$. In agreement with our general notational convention, let $\overline{\mathcal{D}}_+(z)$ be the inverse of $\mathcal{D}_+(z)$ in $\mathrm{End}_{\mathbb{Z}}(\bigwedge \mathbb{U}_r)[[z]]$ and $\overline{\mathcal{D}}_-(z)$ be the inverse of $\mathcal{D}_-(z)$ in $\mathrm{End}_{\mathbb{Z}}(\bigwedge \mathbb{U}_r)[[z^{-1}]]$. Notice that $\overline{\mathcal{D}}_\pm(z)$ are precisely the *characteristic polynomial operators* associated with the shift endomorphism of step ± 1. We point out that, for $j \neq 0$, D_{-j} behaves quite differently from D_j. For instance, D_1 is B_r-linear by construction (it is precisely the B_r-endomorphism (2.6)), while D_{-1} is not. To see this, notice that $u_{i-1} \in K_r := \ker p_r(D)$ for all $i \geq 1$ and that

$$u_{i-1} = D_{-1}(u_i) = D_{-1}(\sum_{j=0}^{r-1} \mathrm{U}_j(u_i)u_{-j}) \neq \sum_{j=0}^{r-1} \mathrm{U}_j(u_i)u_{-j-1} \notin \ker \mathfrak{p}_r(D).$$

7.5.2. Our next task is to extend the *HS*-derivations $\mathcal{D}_\pm(z)$ and $\overline{\mathcal{D}}_\pm(z)$ to F^r. It suffices to see how do they extend to F_i^r for fixed i. Naively, the idea is to require that $D_{\pm j}$ satisfy Leibniz rules against the infinite exterior products of the abelian group F_i^r. Doing the job for $D_{\pm 1}$ seems promising:

$$D_1\Phi_i^r = D_1u_i \wedge \Phi_{i-1}^r + u_i \wedge D_1u_{i-1} \wedge \Phi_{i-2}^r + u_i \wedge u_{i-1} \wedge D_1u_{i-2} \wedge \Phi_{i-3}^r + \cdots .$$

All the summands, but the first, vanish, eventually giving $D_1\Phi_i^r = u_{i+1} \wedge \Phi_{i-1}^r$. The issue is however trickier than expected, as the following example shows.

Example 7.5.3. Let us directly attempt to extend to F_0^r the Leibniz rule enjoyed by $D_2 : \bigwedge \mathbb{U}_r \to \bigwedge \mathbb{U}_r$. A first output would be

$$\begin{aligned} D_2(\Phi_0^r) &= D_2u_0 \wedge \Phi_{-1}^r + D_1u_0 \wedge D_1\Phi_{-1}^r + u_0 \wedge D_2\Phi_{-1}^r \\ &= u_2 \wedge \Phi_{-1}^r + u_1 \wedge u_0 \wedge \Phi_{-2}^r + u_0 \wedge D_2(u_{-1} \wedge \Phi_{-2}^r) \\ &= u_2 \wedge \Phi_{-1}^r + u_1 \wedge u_0 \wedge \Phi_{-2}^r + u_0 \wedge (u_1 \wedge \Phi_{-2}^r \\ &\quad + u_0 \wedge \Phi_{-3}^r + u_{-1} \wedge D_2\Phi_{-2}^r) \\ &= u_2 \wedge \Phi_{-1}^r + u_0 \wedge u_{-1} \wedge D_2\Phi_{-2}^r. \end{aligned}$$

That is, determining $D_2\Phi_0^r$ amounts to know $D_2\Phi_{-2}^r$. Going further, one gets

$$D_2(\Phi_0^r) = u_2 \wedge \Phi_{-1}^r + u_0 \wedge u_{-1} \wedge u_{-2} \wedge u_{-3} \wedge D_2\Phi_{-4}^r.$$

In general

$$D_2(\Phi_0^r) = u_2 \wedge \Phi_{-1}^r + [\mathbf{u}]_{0+(0^{2n})}^r \wedge D_2\Phi_{-2n}^r,$$

where, abusing notation, we have set $[\mathbf{u}]_{0+(0^{2n})}^r := u_0 \wedge u_{-1} \wedge \cdots \wedge u_{-2n}$. As it is apparent, the process does not stop.

7.5.4. To circumvent the trouble highlighted in Example 7.5.3, we first extend $\overline{\mathcal{D}}_\pm(z)$ to *HS*-derivations of $F_i^r \to F_i^r[[z^{\pm 1}]]$, and then we define $\mathcal{D}_\pm(z)$ as their inverse. To this purpose, recall that for all $r \geq 1$, all $\boldsymbol{\lambda} \in \mathcal{P}_r$ and all $j \in \mathbb{N}^*$, the $\mathbb{Z}$-endomorphisms $\overline{D}_{\pm j} : \bigwedge^r \mathbb{U}_{r,i-r} \to \bigwedge^r \mathbb{U}_{r,i-r}$ are defined by

$$\overline{D}_{\pm j}[\mathbf{u}]_{i+\boldsymbol{\lambda}}^r = \sum_{\mathbf{a}} u_{i+\lambda_1 \pm \mathbf{a}(1)} \wedge u_{i-1+\lambda_2 \pm \mathbf{a}(2)} \wedge \cdots \wedge u_{i-r+1+\lambda_r \pm \mathbf{a}(r)}$$

the sum being taken over all $\mathbf{a} : \{1, \ldots, r\} \to \{0, 1\}$ such that $\sum_{n=1}^{r} \mathbf{a}(n) = j$. In particular $\overline{D}_{\pm j}[\mathbf{u}]^r_{i+\lambda} = 0$ for all $j \geq r$. Denote by $\{0, 1\}^{\mathbb{N}}$ the set of all the sequences $\mathbf{a} : \mathbb{N} \to \{0, 1\}$ and let $|\mathbf{a}| = \sum_{n\geq 0} \mathbf{a}(n) \in \mathbb{N} \cup \{\infty\}$. If $|\mathbf{a}| < \infty$, then '**a**' takes only finitely many non-zero values.

Definition 7.5.5. For all $i \in \mathbb{Z}$ and all $j \geq 0$, set

$$\overline{D}_{\pm j}\Phi^r_{i+\lambda} = \sum_{\mathbf{a}\in\{0,1\}^{\mathbb{N}},\ |\mathbf{a}|=j} u_{j+\lambda_1\pm\mathbf{a}(1)} \wedge u_{j-1+\lambda_2\pm\mathbf{a}(2)} \wedge u_{j-2+\lambda_3\pm\mathbf{a}(3)} \wedge \cdots ,$$

where the sum is over all $\mathbf{a} \in \{0, 1\}^{\mathbb{N}}$ such that $\sum_{n=0}^{r-1} \mathbf{a}(n) = j$.

Proposition 7.5.6. *For all $r > 0$ and all $0 < j \leq r$, we have*

$$\overline{D}_{\pm j}\Phi^r_{i+\lambda} = \overline{D}_{\pm j}([\mathbf{u}]^r_{i+\lambda} \wedge \Phi^r_{i-r}) = \sum_{h=0}^{j} \overline{D}_{\pm j\mp h}[\mathbf{u}]^r_{i+\lambda} \wedge \overline{D}_{\pm h}\Phi^r_{i-r} \tag{7.27}$$

Proof. Look initially at the right-hand side of (7.27). The first summand $\overline{D}_{\pm j}[\mathbf{u}]^r_{i+\lambda} \wedge \Phi^r_{i-r}$ (corresponding to $h = 0$) is itself the sum of all the terms obtained by adding (resp. subtracting) 1 to the index of j distinct exterior factors of $[\mathbf{u}]^r_{i+\lambda}$, in all possible ways; $\overline{D}_{\pm j\mp 1}[\mathbf{u}]^r_{i+\lambda} \wedge \overline{D}_{\pm 1}\Phi^r_{i-r}$ is the sum of all the terms obtained by adding (resp. subtracting) 1 to the index of $(j-1)$ distinct exterior factors of $[\mathbf{u}]^r_{i+\lambda}$ and 1 to the index of an exterior factor of Φ^r_{i-r}, in all possible ways; ..., $[\mathbf{u}]^r_{i+\lambda} \wedge \overline{D}_j\Phi^r_{i-r}$ adds (resp. subtracts) 1 to the index of j distinct exterior factors of Φ^r_{i-r}, in all possible ways. It follows that the sum occurring on the right-hand side of (7.27) is precisely the definition of the left-hand side, namely, the sum of semi-infinite exterior monomials obtained by adding (resp. subtracting) 1 to the index of j distinct exterior factors of $\Phi^r_{i+\lambda}$, in all possible ways. □

Proposition 7.5.7. *For all $i \in \mathbb{Z}$ and all $j > 0$:*

i) $\overline{D}_{-j}\Phi^r_i = 0$;
ii) $\overline{D}_j\Phi^r_i = \Phi^r_{i+(1^j)} = u_{i+1} \wedge \cdots \wedge u_{i-j+2} \wedge \Phi^r_{i-j}$,

where according to 2.2.3, (1^j) *denotes the partitions having j parts equal to* 1.

Proof. To prove item i), recall that:

$$\overline{D}_{-j}\Phi^r_i = \sum_{\mathbf{a}\in\{0,1\}^{\mathbb{N}},\ |a|=j} u_{j-\mathbf{a}(1)} \wedge u_{j-1-\mathbf{a}(2)} \wedge u_{j-2-\mathbf{a}(3)} \wedge \cdots , \tag{7.28}$$

Since there is at least one $h \in \mathbb{N}^*$ such that $\mathbf{a}(h) = 1$ and $\mathbf{a}(h + 1) = 0$, the expression $\overline{D}_{-j}\Phi^r_i$ must contain the monomial $u_{i-h} \wedge u_{i-h} (= 0)$, causing the vanishing of the right-hand side of (7.28).

As for item ii),

$$\begin{aligned}\overline{D}_j\Phi_i^r &= \sum_{a\in\{0,1\}^{\mathbb{N}},\,|a|=j} u_{i+a(0)}\wedge u_{i-1+\mathbf{a}(1)}\wedge\cdots\wedge u_{i-j+1+\mathbf{a}(j-1)}\wedge u_{i-j+\mathbf{a}(j)}\wedge\cdots\\ &= \overline{D}_j(u_i\wedge\cdots\wedge u_{i-j+1})\wedge\Phi_{i-j}^r+\sum_{k=1}^{j}\overline{D}_{j-k}(u_i\wedge\cdots\wedge u_{i-j+1})\wedge\overline{D}_k\Phi_{i-j}^r,\end{aligned}\tag{7.29}$$

having used Proposition 7.5.6 with $\eta = u_i \wedge \cdots \wedge u_{i-j+1}$ and $\theta = \Phi_{i-j}^r$. For any monomial occurring in the sum $\sum_{k=1}^{j}\overline{D}_{j-k}(u_i\wedge\cdots\wedge u_{i-j+1})\wedge\overline{D}_k\Phi_{i-j}^r$, there is at least one $h \geq 1$ such that $0 = \mathbf{a}(h-1) < \mathbf{a}(h) = 1$. The only exception is $\overline{D}_j(u_i\wedge\cdots\wedge u_{i-j+1})\wedge\Phi_{i-j}$ already taken into account. It follows that

$$u_{i-k+1+\mathbf{a}(k-1)} = u_{i-k+1} = u_{i-k+\mathbf{a}(k)}$$

and then that summand vanishes. Item ii) is so proven. □

Hence the formal power series

$$\overline{D}_+(z)\Phi_i^r = \sum_{j\geq 0}\Phi_{i+(1^j)}^r z^j \in F_i^r[[z]]$$

is well defined.

More generally, consider the maps $\overline{\mathcal{D}}_\pm(z) : F_i^r \to F_i^r[[z^{\pm 1}]]$ defined by

$$\overline{\mathcal{D}}_\pm(z)\Phi_{i+\boldsymbol{\lambda}}^r = \sum_{i\geq 0}\overline{D}_{\pm j}\Phi_{i+\boldsymbol{\lambda}}^r\cdot z^{\pm j}, \qquad (\boldsymbol{\lambda}\in\mathcal{P}_r).$$

Corollary 7.5.8. *For all $j \geq 0$ and all $i \in \mathbb{Z}$,*

$$u_i\wedge u_{i-1}\wedge\cdots\wedge u_{i-j+1}\wedge\overline{D}_j\Phi_{i-j}^r = 0.$$

Proof. It is a consequence of the proof of 7.5.7, item ii). □

Proposition 7.5.9. *We have*

$$\overline{\mathcal{D}}_\pm(z)\Phi_{i+\boldsymbol{\lambda}} = \overline{\mathcal{D}}_\pm(z)[\mathbf{u}]_{i+\boldsymbol{\lambda}}^r\wedge\overline{\mathcal{D}}_\pm(z)\Phi_{i-r}. \tag{7.30}$$

Proof. In fact by Proposition 7.5.6,

$$\overline{D}_{\pm j}([\mathbf{u}]_{i+\boldsymbol{\lambda}}^r\wedge\Phi_{i-r}^r) = \sum_{k=0}^{j}\overline{D}_{\pm j\mp k}([\mathbf{u}]_{i+\boldsymbol{\lambda}}^r)\wedge\overline{D}_{\pm k}\Phi_{i-r}^r,$$

and this is equivalent to (7.30). □

Corollary 7.5.10. *For all $i \in \mathbb{Z}$ and $\boldsymbol{\lambda} \in \mathcal{P}_r$:*

i) $\overline{\mathcal{D}}_-(z)\Phi_i^r = \Phi_i^r$;
ii) $\overline{\mathcal{D}}_-(z)\Phi_{i+\boldsymbol{\lambda}}^r = \overline{\mathcal{D}}_-(z)[\mathbf{u}]_{i+\boldsymbol{\lambda}}^r \wedge \Phi_{i-r}^r$.

Proof. Since $\overline{\mathcal{D}}_-(z) = 1 + \sum_{j\geq 0}(-1)^j\overline{D}_{-j}z^{-j}$ and $\overline{D}_{-j}$ vanishes on Φ_i for all $j \geq 1$, item i) follows; item ii) is a consequence of (7.30) and item i). □

Remark 7.5.11. We leave to the reader's care to check that indeed $\mathcal{D}_-(z)$ and $\overline{\mathcal{D}}_-(z)$ map F_i^r to $F_i^r[z^{-1}]$ (i.e. the image of any element of F_r^i is not a formal power series but only a polynomial in z^{-1}).

Finally:

Definition 7.5.12. The (extended) Schubert derivation $\mathcal{D}_\pm(z) : F_i^r \to F_i^r[[z]]$ is the inverse of $\overline{\mathcal{D}}_\pm(z)$ in $\mathrm{End}_\mathbb{Z}(F_i^r)[[z^{\pm 1}]]$.

Proposition 7.5.13. *For all $i \in \mathbb{Z}$ and $\boldsymbol{\lambda} \in \mathcal{P}_r$,*

$$\mathcal{D}_\pm(z)\Phi_{i+\boldsymbol{\lambda}} = \mathcal{D}_\pm(z)[\mathbf{u}]_{i+\boldsymbol{\lambda}}^r \wedge \mathcal{D}_\pm(z)\Phi_{i-r}.$$

Proof. It works as in 4.1.7 iii). Let us make the check for $\mathcal{D}_+(z)$, because that for $\mathcal{D}_-(z)$ is totally analogous. We have

$$\begin{aligned}\mathcal{D}_+(z)([\mathbf{u}]_{i+\boldsymbol{\lambda}}^r \wedge \Phi_{i-r}) &= \mathcal{D}_+(z)(\overline{\mathcal{D}}_+(z)\mathcal{D}_+(z)[\mathbf{u}]_{i+\boldsymbol{\lambda}}^r \wedge \overline{\mathcal{D}}_+(z)\mathcal{D}_+(z)\Phi_{i-r})\\ &= \mathcal{D}_+(z)(\overline{\mathcal{D}}_+(z)\big(\mathcal{D}_+(z)[\mathbf{u}]_{i+\boldsymbol{\lambda}}^r \wedge \mathcal{D}_+(z)\Phi_{i-r}\big)\\ &= \mathcal{D}_+(z)[\mathbf{u}]_{i+\boldsymbol{\lambda}}^r \wedge \mathcal{D}_+(z)\Phi_{i-r},\end{aligned}$$

as desired. □

Corollary 7.5.14. *Integration by parts holds:*

$$\mathcal{D}_\pm(z)[\mathbf{u}]_{i+\boldsymbol{\lambda}}^r \wedge \Phi_{i-r}^r = \mathcal{D}_\pm(z)\left([\mathbf{u}]_{i+\boldsymbol{\lambda}}^r \wedge \overline{\mathcal{D}}_\pm(z)\Phi_{i-r}^r\right). \tag{7.31}$$

Proof. It is an obvious consequence of Proposition 7.5.13. Notice that, in particular,

$$\mathcal{D}_-(z)[\mathbf{u}]_{i+\boldsymbol{\lambda}}^r \wedge \Phi_{i-r}^r = \mathcal{D}_-(z)\left([\mathbf{u}]_{i+\boldsymbol{\lambda}}^r \wedge \Phi_{i-r}^r\right)$$

because of 7.5.10, i).

Lemma 7.5.15. Let $(i_1, \ldots, i_r)$ be any finite sequence of non-negative integers. Then

$$D_{i_1}\cdots D_{i_r}\Phi_i^r = D_{i_1}\cdots D_{i_r}(u_i \wedge u_{i-1} \wedge \cdots \wedge u_{i-r+1}) \wedge \Phi_{i-r}^r \tag{7.32}$$

Proof. Let $(z_1, \ldots, z_r)$ be indeterminates over $\mathrm{End}_\mathbb{Z}(F_i^r)$. All the possible expressions of the type

$$D_{i_1}\cdots D_{i_r}$$

occur as coefficients of the product $\mathcal{D}_+(z_r)\cdots\mathcal{D}_+(z_1)$. Iterating integration by parts (7.31),

$$\begin{aligned}
&\mathcal{D}_+(z_r)\cdots\mathcal{D}_+(z_1)(u_i\wedge\cdots\wedge u_{i-r+1})\wedge\Phi^r_{i-r}\\
&=\mathcal{D}_+(z_r)\cdots\mathcal{D}_+(z_1)(u_i\wedge\cdots\wedge u_{i-r+1}\wedge\overline{\mathcal{D}}_+(z_r)\cdots\overline{\mathcal{D}}_+(z_1)\Phi_{i-r})\\
&=\mathcal{D}_+(z_r)\cdots\mathcal{D}_+(z_1)(u_i\wedge\cdots\wedge u_{i-r+1}\wedge\Phi_{i-r}),
\end{aligned}\tag{7.33}$$

where in the last equality, we have repeatedly applied Corollary 7.5.8, which in turn, equating the coefficient of the powers of $z_1^{i_1}\cdot\ldots\cdot z_r^{i_r}$ of either side of (7.33), proves (7.32). □

Let $\mathbf{T}:=(T_0,T_1,T_2,\ldots)$ be a sequence of indeterminates, and let

$$\Delta_{\boldsymbol\lambda}(\mathcal{D}_+):=\det(T_{i_j-j+i})_{T_i=D_i}$$

Corollary 7.5.16. *For all $\boldsymbol\lambda\in\mathcal{P}_r$,*

$$\Delta_{\boldsymbol\lambda}(\mathcal{D}_+)\Phi_i=\Delta_{\boldsymbol\lambda}(\mathcal{D}_+)(u_i\wedge u_{i-1}\wedge\cdots\wedge u_{i-r+1})\wedge\Phi^r_{i-r}.$$

Proof. In fact all Schur polynomials associated with partitions of length at most r are integral linear combinations of monomials of the form $D_{i_1}\cdots D_{i_r}$. □

We are now in the position to prove another version of the *boson–fermion* correspondence, which works uniformly for all the values of the *charge*:

Proposition 7.5.17. *For all $i\in\mathbb{Z}$ and $\boldsymbol\lambda\in\mathcal{P}_r$,*

$$\Phi^r_{i+\boldsymbol\lambda}=\Delta_{\boldsymbol\lambda}(\mathcal{D}_+)\Phi^r_i.$$

Proof. The proof works as follows. The obvious isomorphism of free abelian groups $M_0\mapsto\bigoplus_{k\geq i}\mathbb{Z}u_k$ given by $b_j\mapsto u_{j+i}$ (notation as in Section 5.6) maps $[\mathbf{b}]^r_{\boldsymbol\lambda}\mapsto[\mathbf{u}]^r_{i+\boldsymbol\lambda}$ and $\sigma_k b_j\mapsto D_k u_{j+i}$. Through such an isomorphism, $\Delta_{\boldsymbol\lambda}(\boldsymbol\sigma_+)[\mathbf{b}]^r_0$ maps to both $\Delta_{\boldsymbol\lambda}(\mathcal{D}_+)[\mathbf{u}]^r_i$ and $[\mathbf{u}]^r_{i+\boldsymbol\lambda}$ because $\Delta_{\boldsymbol\lambda}(\boldsymbol\sigma_+)[\mathbf{b}]^r_0=[\mathbf{b}]^r_{\boldsymbol\lambda}$, by Corollary 5.8.2. Thus

$$\Delta_{\boldsymbol\lambda}(\mathcal{D}_+)\Phi^r_i=\Delta_{\boldsymbol\lambda}(\mathcal{D}_+)[\mathbf{u}]^r_{i+\boldsymbol\lambda}\wedge\Phi^r_{i-r}=\Phi^r_{i+\boldsymbol\lambda}$$

as desired. □

Proposition 7.5.18. *The module $B_r\otimes_{\mathbb{Z}}F^r_0$ is an eigenmodule of $\Delta_{\boldsymbol\lambda}(\mathcal{D}_+)$ with respect to the eigenvalue $\Delta_{\boldsymbol\lambda}(H_r)$.*

Proof. Since the space $B_r\otimes_{\mathbb{Z}}F^r_0$ is a free B_r-module of rank 1, it suffices to check that Φ^r_0 is an eigenvalue. We have $\Delta_{\boldsymbol\lambda}(\mathcal{D}_+)\Phi^r_0=\Phi^r_{0+\boldsymbol\lambda}$, because of Proposition 7.5.17, which is in turn equal to $\Delta_{\boldsymbol\lambda}(H_r)\Phi^r_0$, by Proposition (7.2.3). □

Corollary 7.5.19. *The following equalities hold:*

$$\mathcal{D}_+(z)\Phi^r_{0+\boldsymbol{\lambda}} = \frac{1}{E_r(z)}\Phi^r_{0+\boldsymbol{\lambda}} \quad \text{and} \quad \overline{\mathcal{D}}_+(z)\Phi^r_{0+\boldsymbol{\lambda}} = E_r(z)\Phi^r_{0+\boldsymbol{\lambda}}.$$

Proof. It is clearly sufficient to prove the claim for $\boldsymbol{\lambda} = 0$, which easily descends from (7.5.18). □

7.5.20. For all $r > 0$ and $\boldsymbol{\lambda} \in \mathcal{P}_r$, define

$$\overline{\mathcal{D}}_-(z)\Delta_{\boldsymbol{\lambda}}(H_r) := \frac{\overline{\mathcal{D}}_-(z)\Phi^r_{0+\boldsymbol{\lambda}}}{\Phi^r_0} \in B_r[z^{-1}] \tag{7.34}$$

and

$$\mathcal{D}_-(z)\Delta_{\boldsymbol{\lambda}}(H_r) := \frac{\mathcal{D}_-(z)\Phi^r_{0+\boldsymbol{\lambda}}}{\Phi^r_0} \in B_r[z^{-1}]. \tag{7.35}$$

Theorem 7.5.21 (Stability Properties). *The equalities*

$$\overline{\mathcal{D}}_-(z)h_n = h_n - \frac{h_{n-1}}{z} \tag{7.36}$$

and

$$\mathcal{D}_-(z)h_n = \sum_{i=0}^{n} \frac{h_{n-i}}{z^i} \tag{7.37}$$

hold in B_r for all $r \geq 1$.

Proof. It works analogously to 6.2.8. For all $r \geq 1$, the following equalities hold true:

$$\begin{aligned}\overline{\mathcal{D}}_-(z)\Phi^r_{0+(n)} &= \overline{\mathcal{D}}_-(z)u_n \wedge \Phi^r_{-1} = \overline{\mathcal{D}}_-(z)u_n \wedge \overline{\mathcal{D}}_-(z)\Phi^r_{-1} \\ &= \left(u_n - \frac{u_{n-1}}{z}\right) \wedge \Phi^r_{-1} = \left(h_n - \frac{h_{n-1}}{z}\right)\Phi^r_0\end{aligned}$$

proving (7.36). Similarly

$$\begin{aligned}\mathcal{D}_-(z)\Phi^r_{0+(n)} &= \mathcal{D}_-(z)u_n \wedge \Phi^r_{-1} = \mathcal{D}_-(z)u_n \wedge \overline{\mathcal{D}}_-(z)\Phi^r_{-1} \\ &= \left(\sum_{j\geq 0} \frac{u_{n-j}}{z^j}\right) \wedge \Phi^r_{-1} = \left(h_n - \frac{h_{n-1}}{z}\right)\Phi^r_0 \\ &= \sum_{j\geq 0} \frac{1}{z^j}(u_{n-j} \wedge \Phi^r_{-1}) = \sum_{j\geq 0} \frac{1}{z^j} h_{n-j}\Phi^r_0\end{aligned}$$

and (7.36) is proven as well. □

Theorem 7.5.22. *The homomorphisms $\mathcal{D}_-(z), \overline{\mathcal{D}}_-(z)$ commute with taking the Schur determinants, i.e. for all $1 \leq s \leq r$ and all $\boldsymbol{\lambda}$ of length* s*, the following equalities hold:*

$$\mathcal{D}_-(z)\Delta_{\boldsymbol{\lambda}}(H_r) = \Delta_{\boldsymbol{\lambda}}(\mathcal{D}_-(z)H_r), \qquad \overline{\mathcal{D}}_-(z)\Delta_{\boldsymbol{\lambda}}(H_r) = \Delta_{\boldsymbol{\lambda}}(\overline{\mathcal{D}}_-(z)H_r).$$

Proof. It is the same as that of 6.2.10 and 6.2.13, up to renaming σ_- and $\overline{\sigma}_-$ by $\mathcal{D}_-$ and $\overline{\mathcal{D}}_-$, respectively. □

Corollary 7.5.23. *For all $1 \leq s \leq r$ and all $(h_{i_1}, \ldots, h_{i_s}) \in B_r^s$,*

$$\overline{\mathcal{D}}_-(z)(h_{i_1} \cdots h_{i_s}) = \overline{\mathcal{D}}_-(z)h_{i_1} \cdots \overline{\mathcal{D}}_-(z)h_{i_s} \tag{7.38}$$

and

$$\mathcal{D}_-(z)(h_{i_1} \cdots h_{i_s}) = \mathcal{D}_-(z)h_{i_1} \cdots \mathcal{D}_-(z)h_{i_s}. \tag{7.39}$$

Proof. It works the same as that of Corollaries 6.2.11 and 6.2.14.

7.6 Extending the Boson–Fermion Correspondence

We have seen in Section 7.2 that F_0^r possesses a canonical structure of free B_r-module of rank 1 because K_r, the module of generic LRS, is a free B_r-module of rank r. It is a quite privileged structure because in this way F_0^r is canonically an eigenmodule of $\Delta_{\boldsymbol{\lambda}}(\mathcal{D}_+) \in \mathrm{End}_{B_r}(B_r \otimes_{\mathbb{Z}} F^r)$. We want to equip all F_i^r with a B_r-module structure in some sense compatible with that of F_0^r. To do that, we first notice that for all $i \in \mathbb{Z}$, the $\mathbb{Z}$-module isomorphism $D_i : \mathbb{U}_r \to \mathbb{U}_r$ induces a determinant $\mathbb{Z}$-homomorphism

$$\ell^i : F_j^r \to F_{j+i}^r$$

mapping $\Phi_{j+\boldsymbol{\lambda}}^r \mapsto \ell^i \Phi_{j+\boldsymbol{\lambda}}^r = \Phi_{j+i+\boldsymbol{\lambda}}^r$. In other words

$$\ell^i \Phi_{j+\boldsymbol{\lambda}}^r = \ell^i[\mathbf{u}]_j^r \wedge \ell^i \Phi_{j-r} = [\mathbf{u}]_{i+j}^r \wedge \Phi_{j+i-r},$$

where

$$\begin{aligned} \ell^i(u_{j-r} \wedge u_{j-r-1} \wedge u_{j-r-2} \wedge \cdots) &= D_i u_{j-r} \wedge D_i u_{j-r-1} \wedge D_i u_{j-r-2} \wedge \cdots \\ &= u_{i+j-r} \wedge u_{i+j-r-1} \wedge u_{i+j-r-2} \wedge \cdots \end{aligned}$$

Strongly abusing notation, but for the sake of notational brevity, we denote by

$$B_r(\ell) := B_r[\ell^{-1}, \ell].$$

Thanks to the maps ℓ^i we can endow $F^r := \bigoplus_{i\in\mathbb{Z}} F_i^r$ with a structure of free $B_r(\ell)$-module of rank 1 generated by Φ_0^r, basing on the following equalities:

$$\Delta_\lambda(\mathcal{D}_+)\Phi_i^r = \Phi_{i+\lambda}^r = \ell^i\Phi_{0+\lambda}^r = \ell^i\Delta_\lambda(H_r)\Phi_0^r.$$

We shall also write

$$\frac{1_{B_r(\ell)} \otimes \Phi_{i+\lambda}^r}{\Phi_0^r} := \ell^i\Delta_\lambda(H_r)$$

to mean that $\ell^i\Delta_\lambda(H_r)$ is the unique element of $B_r(\ell)$ which 'translates' Φ_0^r to $\Phi_{i+\lambda}^r$. We are so given of an extension of the boson–fermion correspondence (7.5), $B_r \to F_0^r$, as a map $B_r(\ell) \to F^r$.

7.6.1. Recall the notation of 7.1.2 and define operators

$$\mathbb{X}_r(z)\wedge : F_i^r \to F_{i+1}^r[[z^{-1}, z]] \qquad \text{and} \qquad \mathbb{X}_r^*(z)\lrcorner : F_i^r \to F_{i-1}^r[[z^{-1}, z]] \tag{7.40}$$

as follows:

$$\mathbb{X}_r(z) \wedge \Phi_{i+\lambda}^r = \sum_{j\in\mathbb{Z}} z^j u_j \wedge \Phi_{i+\lambda}^r \tag{7.41}$$

and

$$\mathbb{X}_r^*(z)\lrcorner\Phi_{i+\lambda}^r = \sum_{j\in\mathbb{Z}} z^{-j-1}u_j^*\lrcorner\Phi_{i+\lambda}^r. \tag{7.42}$$

For $r = \infty$, expression (7.41) is the same as the first of [78, formula (5.21)] (up to the notation), while (7.41) is the second of [78, formula (5.21)] divided by z.

Definition 7.6.2. Define operators $\Gamma_r(z), \Gamma_r^*(z) : B_r(\ell) \to B_r(\ell)[[z]]$ through the equalities

$$\Gamma_r(z)\ell^i\Delta_\lambda(H_r) = \frac{1_{B_r(\ell)} \otimes (\mathbb{X}_r(z) \wedge \Phi_{i+\lambda}^r)}{\Phi_0^r}, \tag{7.43}$$

and

$$\Gamma_r^*(z)\ell^i\Delta_\lambda(H_r) = \frac{1_{B_r(\ell)} \otimes (\mathbb{X}_r^*(z)\lrcorner\Phi_{i+\lambda}^r)}{\Phi_0^r}. \tag{7.44}$$

They will be said vertex operators *truncated* to the order r.

7.6.3. The goal is now to compute explicitly $\Gamma_r(z)$ and $\Gamma_r^*(z)$ of Definition 7.6.2. They are related to those studied in Section 6.3, there called *vertex-like operator*.

Their mild difference disappears as $r \to \infty$, accounting for the abuse of notation according which they are indicated by the same symbol.

7.6.4. To compute $\Gamma_r(z)$ and $\Gamma^*(z)$, we shall exploit the B_r-module structure of F_0^r. To this purpose observe that

$$\begin{aligned}\mathbb{X}_r(z) \wedge \Phi^r_{i+\lambda} &= \sum_{j\in\mathbb{Z}} z^j u_j \wedge \Phi^r_{i+\lambda} = z^{i+1}\sum_{j\in\mathbb{Z}} z^{j-i-1} D_{i+1}u_{j-i-1} \wedge \ell^{i+1}\Phi^r_{-1+\lambda}\\ &= z^{i+1}\sum_{j\in\mathbb{Z}} z^{j-i-1}\ell^{i+1}(u_{j-i-1} \wedge \Phi^r_{-1+\lambda})\\ &= \ell^{i+1}z^{i+1}\sum_{j\in\mathbb{Z}} u_j z^j \wedge \Phi^r_{-1+\lambda}.\end{aligned}$$

Therefore, determining the expression of $\Gamma_r(z)$ amounts to work out the formal power series $\sum_{j\in\mathbb{Z}} z^j u_j \wedge \Phi^r_{-1+\lambda}$.

7.6.5. The computation of $\Gamma_r^*(z)$ can also be reduced to a special case because a straightforward computation shows that (see Exercise 7.10.4) the following equality holds:

$$\mathbb{X}_r^*(z) \lrcorner \Phi^r_{i+\lambda} = \ell^{i-1}z^{-i+1}\mathbb{X}_r^*(z) \lrcorner \Phi^r_{1+\lambda}. \tag{7.45}$$

It will then suffice to analyze the formal power series $\mathbb{X}_r^*(z) \lrcorner \Phi^r_{1+\lambda}$.

7.7 Computing Truncated Vertex Operators

The following is the analogue of Theorem 6.2.6.

Lemma 7.7.1. *For all* $\boldsymbol{\lambda} \in \mathcal{P}_r$*:*

i) $u_{-r} \wedge \overline{\mathcal{D}}_+(z)\Phi^r_{-1+\lambda} = (-1)^r\overline{\mathcal{D}}_+(z)[\mathbf{u}]^r_{-1+\lambda} \wedge \Phi^r_{-r}$;
ii) $u_{-r} \wedge \overline{D}_j\Phi^r_{-1+\lambda} = (-1)^r\overline{D}_j[\mathbf{u}]^r_{-1+\lambda} \wedge \Phi^r_{-r}$, *for all* $0 \le j \le r$;
iii) $u_{-r} \wedge \overline{D}_j\Phi^r_{-1+\lambda} = (-1)^r\overline{D}_{j-r}\Phi^r_{0+\lambda}$, *for all* $0 \le j \le r$.

Proof. To prove (i) we recall that $\overline{\mathcal{D}}_+(z)\Phi^r_{-r-1} = \Phi^r_{-r-1} + \sum_{i\ge1}(-1)^i\Phi^r_{-r-1+(1^i)}z^i$, whence

$$u_{-r} \wedge \overline{\mathcal{D}}_+(z)\Phi^r_{-r-1} = u_{-r} \wedge \Phi^r_{-r-1} = \Phi^r_{-r}.$$

Therefore

$$\begin{aligned}u_{-r} \wedge \overline{\mathcal{D}}_+(z)\Phi^r_{-1+\lambda} &= u_{-r} \wedge \overline{\mathcal{D}}_+(z)\big([\mathbf{u}]^r_{-1+\lambda} \wedge \Phi^r_{-r-1}\big)\\ &= u_{-r} \wedge \overline{\mathcal{D}}_+(z)[\mathbf{u}]^r_{-1+\lambda} \wedge \overline{\mathcal{D}}_+(z)\Phi^r_{-r-1}\\ &= (-1)^r\overline{\mathcal{D}}_+(z)[\mathbf{u}]^r_{-1+\lambda} \wedge u_{-r} \wedge \Phi^r_{-r-1}\\ &= (-1)^r\overline{\mathcal{D}}_+(z)[\mathbf{u}]^r_{-1+\lambda} \wedge \Phi^r_{-r}.\end{aligned}$$

Item ii) is a simple consequence of i), after writing $\overline{\mathcal{D}}_+(z) = \sum_{k\geq 0}(-1)^k\overline{D}_k z^k$. To prove iii) we observe that

$$\overline{D}_j[\mathbf{u}]^r_{-1+\boldsymbol{\lambda}} = \overline{D}_j(u_{-1+\lambda_1} \wedge \cdots \wedge u_{-r+\lambda_r}) = \sum_{\mathbf{a}} u_{-1+\lambda_1+\mathbf{a}(1)} \wedge \cdots \wedge u_{-r+\lambda_r+\mathbf{a}(r)}$$

summing over all the sequences $\mathbf{a} : \{1, \ldots, r\} \to \{0, 1\}$ such that $\sum_{k=1}^r \mathbf{a}(k) = j$. Defining $\mathbf{c}(k) = 1 - \mathbf{a}(k)$, the last side is equal to

$$\sum_{\mathbf{c}} u_{\lambda_1-\mathbf{c}(1)} \wedge \cdots \wedge u_{-r+1+\lambda_r-\mathbf{c}(r)},$$

where the sum is over all the sequences $\mathbf{c} : \{1, \ldots, r\} \to \{0, 1\}$ such that $\sum_{k=0}^r \mathbf{c}(k) = r - j$, i.e. precisely the definition of $\overline{D}_{j-r}[\mathbf{u}]^r_{0+\boldsymbol{\lambda}}$. Plugging it into ii) gives

$$u_{-r} \wedge \overline{D}_j\Phi^r_{-1+\boldsymbol{\lambda}} = (-1)^r\overline{D}_j[\mathbf{u}]^r_{-1+\boldsymbol{\lambda}} \wedge \Phi^r_{-r} = (-1)^r\overline{D}_{j-r}[\mathbf{u}]^r_{0+\boldsymbol{\lambda}} \wedge \Phi^r_{-r}$$

and since $-r \leq j - r \leq 0$, this is equal to

$$(-1)^r\overline{D}_{j-r}([\mathbf{u}]^r_{0+\boldsymbol{\lambda}} \wedge \Phi^r_{-r}) = (-1)^r\overline{D}_{j-r}\Phi^r_{0+\boldsymbol{\lambda}},$$

as desired. □

Corollary 7.7.2. *For all $\boldsymbol{\lambda} \in \mathcal{P}_r$,*

$$\frac{1}{z^r}\left(u_{-r} \wedge \overline{\mathcal{D}}_+(z)\Phi^r_{-1+\boldsymbol{\lambda}}\right) = \overline{\mathcal{D}}_-(z)\Phi^r_{0+\boldsymbol{\lambda}}.$$

Proof. In fact, basing on Lemma 7.7.1,

$$\begin{aligned} u_{-r} \wedge \overline{\mathcal{D}}_+(z)\Phi^r_{-1+\boldsymbol{\lambda}} &= u_{-r} \wedge \sum_{j\geq 0}(-1)^j z^j\overline{D}_j\Phi^r_{-1+\boldsymbol{\lambda}} = \sum_{j\geq 0}(-1)^{j+r}z^j\overline{D}_{j-r}\Phi^r_{0+\boldsymbol{\lambda}} \\ &= z^r\sum_{j\geq 0}(-1)^{j+r}z^{j-r}\overline{D}_{j-r}\Phi^r_{0+\boldsymbol{\lambda}} = z^r\overline{\mathcal{D}}_-(z)\Phi^r_{0+\boldsymbol{\lambda}}, \end{aligned}$$

which proves the claim. □

Theorem 7.7.3. *The following equalities hold:*

$$\Gamma_r(z)\ell^i\Delta_{\boldsymbol{\lambda}}(H_r) = \frac{1_{B_r} \otimes (\mathbb{X}_r(z) \wedge \Phi^r_{i+\boldsymbol{\lambda}})}{\Phi^r_0} = \frac{\ell^{i+1}z^{i+1}}{E_r(z)}\overline{\mathcal{D}}_-(z)\Delta_{\boldsymbol{\lambda}}(H_r).$$

Proof. By definition

$$\mathbb{X}_r(z) \wedge \Phi^r_{-1+\boldsymbol{\lambda}} = \sum_{j\in\mathbb{Z}} z^j u_j \wedge u_{-1+\lambda_1} \wedge \cdots \wedge u_{-r+\lambda_r} \wedge u_{-r+1} \wedge \cdots$$

which is equal to

$$\sum_{j\geq -r} z^j u_j \wedge \Phi^r_{-1+\lambda} = \frac{1}{z^r}\mathcal{D}_+(z)u_{-r} \wedge \Phi^r_{-1+\lambda}. \tag{7.46}$$

Since $\mathcal{D}_+(z)$ behaves as a derivation on the exterior algebra (cf. Proposition 7.5.13), the last side of (7.46) can be written as

$$\mathcal{D}_+(z)(u_{-r} \wedge \overline{\mathcal{D}}_+(z)\Phi^r_{-1+\lambda}),$$

which, invoking Corollaries 7.7.2 and 7.5.18, is equal to

$$\mathcal{D}_+(z)(1_{B_r} \otimes \overline{\mathcal{D}}_-(z)\Phi^r_{0+\lambda}) = E_r(z)\overline{\mathcal{D}}_-(z)\Delta_\lambda(H_r)\Phi^r_0.$$

Thus, finally

$$\Gamma_r(z)\ell^i\Delta_\lambda(H_r) = \frac{1_{B_r(\ell)} \otimes (\mathbb{X}_r(z) \wedge \Phi^r_{i+\lambda})}{\Phi^r_0} = \ell^{i+1}z^{i+1}\frac{1_{B_r} \otimes (\mathbb{X}_r(z) \wedge \Phi^r_{-1+\lambda})}{\Phi^r_0}$$

$$= \frac{\ell^{i+1}z^{i+1}}{E_r(z)}\overline{\mathcal{D}}_-(z)\Delta_\lambda(H_r). \qquad \square$$

Corollary 7.7.4. *We have*

$$\Gamma_r(z)\ell^i\Delta_\lambda(H_r) = \frac{\ell^{i+1}z^{i+1}}{E_r(z)}\Delta_\lambda(\overline{\mathcal{D}}_-(z)H_r).$$

Proof. Due to Proposition 6.2.10, according which $\overline{\mathcal{D}}_-(z)$ (there called $\overline{\sigma}_-(z)$) commutes with taking Δ_λ. $\square$

Let us prove now the following analogue of Theorem 6.3.8.

Theorem 7.7.5. *Let $\Gamma^*_r(z)$ as in formula (7.44). Then, for all $i \in \mathbb{Z}$,*

$$\Gamma^*_r(z)\ell^i\Delta_\lambda(H_r) = \ell^{i-1}z^{-i-1}E_r(z)\mathcal{D}_-(z)\Delta_\lambda(H_r). \tag{7.47}$$

Notice that because of Proposition 6.2.13, formula (7.47) is equivalent to

$$\Gamma^*_r(z)\ell^i\Delta_\lambda(H_r) = \ell^{i-1}z^{-i}E_r(z)\Delta_\lambda(\mathcal{D}_-(z)H_r). \tag{7.48}$$

The proof of Theorem 7.7.5 will be split into the proofs of some lemmas, analogous to 6.3.7.

Lemma 7.7.6. *Let $\boldsymbol{\lambda} := (\lambda_1, \ldots, \lambda_r)$ be any partition of length at most* r. *Then*

$$\frac{(z^2\mathbb{X}_r^*(z)\lrcorner\Phi^r_{1+\boldsymbol{\lambda}})\otimes 1_{B_r}}{\Phi^r_0} = \begin{vmatrix} z^{-\lambda_1} & z^{1-\lambda_2} & \ldots & z^{r-1-\lambda_r} \\ h_{\lambda_1+1} & h_{\lambda_2} & \ldots & h_{\lambda_r+r-2} \\ \vdots & \vdots & \ddots & \vdots \\ h_{\lambda_1+r-1} & h_{\lambda_2+r-2} & \ldots & h_{\lambda_r} \end{vmatrix}$$

$$+ (-1)^r z^r \overline{D}_r \Delta_{\boldsymbol{\lambda}}(H_r). \tag{7.49}$$

Proof. With the same notation as in the proof of Lemma 6.3.7, equality (7.49) is straightforward. In fact:

$$z^2 \cdot \mathbb{X}_r^*(z)\lrcorner\Phi^r_{1+\boldsymbol{\lambda}} = z^2\mathbb{X}_r^*(z)\lrcorner(u_{1+\lambda_1}\wedge u_{\lambda_2}\wedge\cdots\wedge u_{-r+1+\lambda_r}\wedge u_{-r}\wedge\Phi^r_{-r+1})$$

$$\begin{aligned} &= z^{-\lambda_1}\Delta_{(\lambda_2,\ldots,\lambda_r)}(H_r) - z^{1-\lambda_2}\Delta_{(\lambda_1+1,\lambda_3,\ldots,\lambda_r)}(H_r) \\ &+ \cdots + (-1)^{j-1}\Delta_{\boldsymbol{\lambda}_j+(1^{j-1})} + \cdots \\ &+ \cdots + (-1)^{r-1}z^{r-1+\lambda_r}\Delta_{(\lambda_1,\ldots,\lambda_{r-1})}(H_r) + \\ &+ (-1)^r z^r\Delta_{(\lambda_1+1,\ldots,\lambda_r+1)}(H_r), \end{aligned}$$

from which

$$= \sum_{j=1}^{r}(-1)^{j-1}z^{j-1+\lambda_j}\Delta_{\boldsymbol{\lambda}_j+(1^{j-1})}(H_r) + (-1)^r z^r\Delta_{(\lambda_1+1,\ldots,\lambda_r+1)}(H_r). \tag{7.50}$$

The first summand of (7.50) is precisely the determinant occurring in (7.49), while the second determinant $\Delta_{(\lambda_1+1,\ldots,\lambda_r+1)}(H_r)$ is precisely $\overline{D}_r\Delta_{\boldsymbol{\lambda}}(H_r)$. We remark that no power of z with exponent bigger than r can occur in the expansion above, because its coefficient would be the Schur determinant associated with H_r and with a partition of length bigger than r that vanishes. □

The proposition below is the analogous of Theorem 6.3.8. The same kind of proof gets rid of $(-1)^r z^r\overline{D}_r\Delta_{\boldsymbol{\lambda}}(H_r)$, the additional summand occurring in formula (7.49).

Lemma 7.7.7. *We have*

$$\frac{(z^2\mathbb{X}_r^*(z)\lrcorner\Phi^r_{1+\boldsymbol{\lambda}})\otimes 1_{B_r}}{\Phi^r_0} = E_r(z)\Delta_{\boldsymbol{\lambda}}(\mathcal{D}_-(z)H_r). \tag{7.51}$$

Proof. As in the proof of Lemma 6.3.7, one first observes that

$$E_r(z)u_j = \mathrm{U}_0(u_j) + \mathrm{U}_1(u_j)z + \cdots + \mathrm{U}_{r-1}(u_j)z^{r-1}, \tag{7.52}$$

where each $\mathrm{U}_i(u_j)$ is equal to h_{j+i} plus a B_r-linear combination of h_{j+i-1}, $h_{j+i-2}, \ldots, h_i$. Then

$$\begin{aligned}\frac{1}{z^{\lambda_i-i+1}} = \frac{E_r(z)H_r(z)}{z^{\lambda_i-i+1}} &= E_r(z)\left(\mathcal{D}_-(z)h_{\lambda_i-i+1} + zu_{\lambda_i-i+1}\right)\\ &= E_r(z)\mathcal{D}_-(z)h_{\lambda_i-i+1} + zE_r(z)u_{\lambda_i-i+1}\\ &= E_r(z)\mathcal{D}_-(z)h_{\lambda_i-i+1} + \sum_{j=1}^{r}\mathrm{U}_{j-1}(u_{\lambda_i-i+1})z^j.\end{aligned}$$

Substituting into the displayed determinant on the right-hand side of (7.49) and using skew symmetry, one obtains

$$= E_r(z)\begin{vmatrix}\mathcal{D}_-(z)h_{\lambda_1} + h_{\lambda_1+r}z^r & \cdots & \mathcal{D}_-(z)h_{\lambda_r-r+1} + h_{\lambda_r+1}z^r\\ h_{\lambda_1+1} & \cdots & h_{\lambda_r+r-2}\\ \vdots & \ddots & \vdots\\ h_{\lambda_1+r-1} & \cdots & h_{\lambda_r}\end{vmatrix}$$

$$= E_r(z)\begin{vmatrix}\mathcal{D}_-(z)h_{\lambda_1} & \cdots & \mathcal{D}_-(z)h_{\lambda_r-r+1}\\ h_{\lambda_1+1} & \cdots & h_{\lambda_r+r-2}\\ \vdots & \ddots & \vdots\\ h_{\lambda_1+r-1} & \cdots & h_{\lambda_r}\end{vmatrix} + (-1)^{r-1}z^r\overline{D}_r\Delta_{\boldsymbol{\lambda}}(H_r). \tag{7.53}$$

Thus the right-hand side of (7.49) is equal to

$$E_r(z)\begin{vmatrix}\mathcal{D}_-(z)h_{\lambda_1} & \cdots & \mathcal{D}_-(z)h_{\lambda_r-r+1}\\ h_{\lambda_1+1} & \cdots & h_{\lambda_r+r-2}\\ \vdots & \ddots & \vdots\\ h_{\lambda_1+r-1} & \cdots & h_{\lambda_r}\end{vmatrix},$$

due to the cancellation of $(-1)^{r-1}z^r\overline{D}_r\Delta_{\boldsymbol{\lambda}}(H_r)$ with the summand $(-1)^r z^r\overline{D}_r\Delta_{\boldsymbol{\lambda}}(H_r)$. To conclude the proof, observe that for all $1 \le i,j \le r$,

$$\mathcal{D}_-(z)h_{\lambda_j-j+i} = h_{\lambda_j-j+i} + \mathcal{D}_-(z)h_{\lambda_j-j+i-1}$$

and thus, exploiting once again the skew symmetry of the determinant, expression (7.53) is equivalent to $E_r(z)\Delta_{\boldsymbol{\lambda}}(\mathcal{D}_-(z)H_r)$, as desired. □

Proof of Theorem 7.7.5. By definition and using Exercise 7.10.4,

$$\Gamma_r^*(z)\ell^i\Delta_{\boldsymbol{\lambda}}(H_r) = \frac{1_{B_r(\ell)}\otimes(\mathbb{X}_r^*(z)\lrcorner\Phi^r_{i+\boldsymbol{\lambda}})}{\Phi^r_0} = \ell^{i-1}z^{-i+1}\frac{1_{B_r}\otimes(\mathbb{X}_r^*(z)\lrcorner\Phi^r_{1+\boldsymbol{\lambda}})}{\Phi^r_0}.$$

Thus

$$\Gamma_r^*(z)\ell^i\Delta_\lambda(H_r) = \ell^{i-1}z^{-i-1}\frac{(z^2\mathbb{X}_r^*(z)\lrcorner\Phi_{1+\lambda}^r)\otimes 1_{B_r}}{\Phi_0^r},$$

which, by Lemma 7.7.7, gives

$$\begin{aligned}\Gamma_r^*(z)\ell^i\Delta_\lambda(H_r) &= \ell^{i-1}z^{-i-1}E_r(z)\Delta_\lambda(\mathcal{D}_-(z)H_r)\\ &= \ell^{i-1}z^{-i-1}E_r(z)\mathcal{D}_-(z)\Delta_\lambda(H_r),\end{aligned}$$

where in the last equality, we used Proposition 6.2.13. □

7.8 Vertex Operators in the Classical Boson–Fermion Correspondence

This final section is devoted to deduce the expression of the vertex operators introduced in Section 1.2.1, as the limit for $r \to \infty$ of the truncations $\Gamma_r(z)$ and $\Gamma_r^*(z)$, formerly computed in Section 7.7. See also the Preprint [46].

7.8.1. As in Section 6.5, let us denote by B_∞ the polynomial ring $\mathbb{Z}[e_1, e_2, \ldots]$ in infinitely many indeterminates. Let $E_\infty(z) = 1 + \sum_{j\geq 1}(-1)^j e_j z^j$ and $H_\infty(t) = \sum_{n\in\mathbb{Z}} h_n t^n$ defined by $H_\infty(z)\sum_{j\geq 0} E_\infty(z) = 1$ in the ring $B_\infty[[z]]$. In this case the sequence $(h_1, h_2, \ldots)$ is a sequence of algebraically independent indeterminates over $\mathbb{Z}$, and the same holds for $B_\infty := \mathbb{Z}[h_1, h_2, \ldots]$. Let us pass to rational coefficients and, as in Sections 1.2 and 7.4, denote by B the ring $B_\infty \otimes_{\mathbb{Z}} \mathbb{Q}$. Thus

$$B := \mathbb{Q}[e_1, e_2, \ldots] = \mathbb{Q}[h_1, h_2, \ldots,] = \mathbb{Q}[x_1, x_2, \ldots],$$

where the sequence $\mathbf{x}_\infty := (x_1, x_2, \ldots)$ is defined as in (7.8) for $r = \infty$:

$$\exp(\sum_{i\geq 1} x_i z^i) = \frac{1}{E_\infty(z)} = \sum_{n\geq 0} h_n z^n. \tag{7.54}$$

Then, as in Section 7.4, each h_n can be regarded as a function of $(x_1, x_2, \ldots)$.

Lemma 7.8.2. *The $\mathbb{Q}$-linear extensions of the Schubert derivations $B \to B[z^{-1}] \subseteq B[[z^{-1}]]$ of Section 7.5.20, denoted by the same symbol $\overline{\mathcal{D}}_-(z), \mathcal{D}_-(z)$, are ring homomorphisms.*

Proof. Naively the proof amounts to argue that for $r = \infty$, Theorem 7.5.22 and Corollary 7.5.23 hold for arbitrarily large $s \geq 1$. Since the ring B is generated as a $\mathbb{Q}$-algebra by $H_\infty := (h_1, h_2, \ldots)$, it suffices to prove that for all $s \geq 1$ and each $1 \leq i_1 < i_2 < \cdots < i_s$,

$$\overline{\mathcal{D}}_-(z)(h_{i_1}\cdots h_{i_s}) = \overline{\mathcal{D}}_-(z)h_{i_1}\cdots\overline{\mathcal{D}}_-(z)h_{i_s} \tag{7.55}$$

and similarly

$$\mathcal{D}_-(z)(h_{i_1}\cdots h_{i_s}) = \mathcal{D}_-(z)h_{i_1}\cdots\mathcal{D}_-(z)h_{i_s}. \tag{7.56}$$

Let $w = i_1 + \cdots + i_s$ and $r > s$. Then in $(B_r)_w$ we may apply Corollaries 6.2.11 and 6.2.14, i.e. (7.55) and (7.56) hold for all $r > s$. Hence they hold in $B_w := \bigoplus_{|\boldsymbol{\lambda}|=w}\mathbb{Z}\Delta_{\boldsymbol{\lambda}}(H_\infty)$ as well, because the latter is the projective limit $\varprojlim \mathbb{Q}[H_r]_w$, taken with respect to the projection maps $\mathbb{Q}[H_r]_w \to \mathbb{Q}[H_s]_w$ (6.40), defined for all $r \geq s$. □

Lemma 7.8.3. *For each $(n,j) \in \mathbb{Z}\times\mathbb{N}^*$,*

$$\frac{\partial h_n}{\partial x_j} = \frac{\partial^j h_n}{\partial x_1^j} = h_{n-j}. \tag{7.57}$$

Proof. In fact

$$\begin{aligned}\sum_{n\geq 0}\frac{\partial h_n}{\partial x_j}z^n &= \frac{\partial}{\partial x_j}\sum_{n\geq 0}h_n z^n = \frac{\partial}{\partial x_j}\exp(\sum_{i\geq 0}x_i z^i) = z^j\exp(\sum_{i\geq 0}x_i z^i)\\ &= \sum_{n\geq 0}h_{n-j}z^n,\end{aligned}$$

whence (7.57), by equating the coefficients of the same power of z. □

We can finally prove the following.

Lemma 7.8.4.

$$\overline{\mathcal{D}}_-(z) = \exp\left(-\sum_{i\geq 1}\frac{1}{iz^i}\frac{\partial}{\partial x_i}\right) \tag{7.58}$$

and

$$\mathcal{D}_-(z) = \exp\left(\sum_{i\geq 1}\frac{1}{iz^i}\frac{\partial}{\partial x_i}\right). \tag{7.59}$$

Proof. By (7.36) we have

$$\overline{\mathcal{D}}_-(z)h_n = h_n - \frac{h_{n-1}}{z} = \left(1 - \frac{1}{z}\frac{\partial}{\partial x_1}\right)h_n. \tag{7.60}$$

Writing the right-hand side of (7.60) as the exponential of its logarithm and using the first equality in (7.57),

$$\overline{\mathcal{D}}_-(z)h_n = \exp\left(-\sum_{i\geq 1}\frac{1}{iz^i}\frac{\partial^i}{\partial x_1^i}\right)h_n = \exp\left(-\sum_{i\geq 1}\frac{1}{iz^i}\frac{\partial}{\partial x_i}\right)h_n.$$

The right-hand side of (7.58) is a ring homomorphism $B \to B[[z^{-1}]]$, by Proposition 3.4.2, as it is the exponential of $-\sum_{i\geq 1} z^{-i}\partial_i/i$ – a first order differential operator. Since $H_\infty := (h_1, h_2, \ldots)$ generate B as a $\mathbb{Q}$-algebra and both members of (7.58) coincide when evaluated at h_n, for all $n \geq 0$, they do coincide. The proof of (7.59) is similar and is based on the equality

$$\mathcal{D}_-(z)h_n = \sum_{j=0}^{n}\frac{h_{n-j}}{z^j} = \left(1 - \frac{1}{z}\frac{\partial}{\partial x_1}\right)^{-1}h_n = \exp\left(\sum_{i\geq 1}\frac{1}{iz^i}\frac{\partial}{\partial x_i}\right)h_n,$$

which implies (7.59), given that its right-hand side is the exponential of a derivation and hence a ring homomorphism. □

For sake of brevity set: $\Gamma(z) := \Gamma_\infty(z)\bigotimes_{\mathbb{Z}} 1_{\mathbb{Q}}$ and $\Gamma^*(z) := \Gamma^*_\infty(z)_{\mathbb{Z}}\bigotimes 1_{\mathbb{Q}}$.

Corollary 7.8.5. *The following two equalities hold:*

$$\Gamma(z)\ell^i\Delta_\lambda(H_\infty) = \ell^{i+1}z^{i+1}\exp(\sum_{i\geq 1} x_i z^i)\cdot\exp\left(-\sum_{i\geq 1}\frac{1}{iz^i}\frac{\partial}{\partial x_i}\right)\Delta_\lambda(H_\infty),$$

$$\Gamma^*(z)\ell^i\Delta_\lambda(H_\infty) = \ell^{i-1}z^{-i-1}\exp(-\sum_{i\geq 1} x_i z^i)\cdot\exp\left(\sum_{i\geq 1}\frac{1}{iz^i}\frac{\partial}{\partial x_i}\right)\Delta_\lambda(H_\infty).$$

Proof. In fact by Theorems 7.7.3 and 7.7.5 for $r = \infty$,

$$\Gamma(z)\ell^i\Delta_\lambda(H_\infty) = \frac{\ell^{i+1}z^{i+1}}{E_\infty(z)}\overline{\mathcal{D}}_-(z)\Delta_\lambda(H_\infty),$$

and

$$\Gamma^*(z)\ell^i\Delta_\lambda(H_\infty) = \ell^{i-1}z^{-i-1}E_\infty(z)\mathcal{D}_-(z)\Delta_\lambda(H_\infty).$$

The claim then follows by Definition (7.54) of the sequence $\mathbf{x} := (x_1, x_2\ldots)$, by (7.58) and (7.59). □

Let $R(z) : B(\ell) \to B(\ell)[z]$ be the rotation operator mapping any polynomial $f(\mathbf{x}, \ell)$ to $R(z)f(\mathbf{x}, \ell) = \ell z\cdot f(\mathbf{x}, \ell z)$. Let $R(z)^{-1}$ be its inverse:

$$R(z)^{-1}f(\mathbf{x}, \ell) = \ell^{-1}z^{-1}f(\mathbf{x}, \ell z^{-1}).$$

Then we have (re)proven the following:

Theorem 7.8.6 (Cf. [78, Theorem 5.1]).

$$\Gamma(z) = R(z)\exp(\sum_{i\geq 1} x_i z^i)\exp\left(-\sum_{i\geq 1}\frac{1}{iz^i}\frac{\partial}{\partial x_i}\right)$$

and

$$\Gamma^*(z) = R(z)^{-1}\exp(-\sum_{i\geq 1} x_i z^i)\exp\left(\sum_{i\geq 1}\frac{1}{iz^i}\frac{\partial}{\partial x_i}\right). \qquad \square$$

Corollary 7.8.7. *Let $f \in B$ a polynomial. Then f does correspond to a decomposable element of $F_0 := F_0^\infty$, via the boson–fermion correspondence, if and only if*

$$Res_z \exp(\sum_{i\geq 1} x_i z^i)\exp\left(-\sum_{i\geq 1}\frac{1}{iz^i}\frac{\partial}{\partial x_i}\right)f \otimes \exp(-\sum_{i\geq 1} x_i z^i)\exp\left(\sum_{i\geq 1}\frac{1}{iz^i}\frac{\partial}{\partial x_i}\right)f = 0.$$

Proof. In fact, by the boson–fermion correspondence, f can be identified with an element $\eta_f := \sum_{\lambda\in\mathcal{P}_{r,n}} a_\lambda \Phi_{0+\lambda}$. Now, by the same arguments of Theorem 6.1.7, η_f is decomposable if and only if

$$\sum_{j\in\mathbb{Z}}(u_j \wedge \eta_f) \otimes (u_j^* \lrcorner \eta_f) = 0,$$

i.e. if and only if

$$\text{Res}_z \sum_{i\in\mathbb{Z}}(u_i z^i \wedge \eta_f) \otimes (\sum_{j\in\mathbb{Z}} u_j^* z^{-j-1} \lrcorner \eta_f) = 0.$$

By the boson–fermion correspondence and the shape of vertex operators, this is the same as

$$\text{Res}_z \Gamma(z)f \otimes \Gamma^*(z)f = 0,$$

where $\Gamma(z)f = \ell \cdot z \cdot \exp(\sum_{i\geq 0} x_i z^i)\overline{\mathcal{D}}_-(z)f$ and $\Gamma^*(z)f = \ell^{-1}z^{-1}\exp(-\sum_{i\geq 0} x_i z^i)$ $\mathcal{D}_-(z)f$, which gives the result.

7.9 Notes and References

- When r goes to infinity, we may think $(u_0, u_{-1}, \ldots)$ as being a basis of solutions to the generic LRS of infinite order, with coefficients $(e_1, e_2, \ldots)$. Similarly, the inverse Laplace transforms $\mathcal{L}^{-1}(u_i)$, see Section 2.5, can be seen as a basis of solutions to the generic linear ODE of infinite order. When looking at u_i ($i \in \mathbb{Z}$) as a function of $(x_1, x_2, \ldots)$, according to (7.54), it solves in infinitely many ways the Hirota bilinear form of the KP equation. See Exercise 7.10.12. In particular a solution to a generic linear ODE of finite order can be thought of as an approximate solution to the Hirota bilinear PDE.

- A fact that we have passed completely over, but for which we refer the reader to [78, p. 52–53], is that the vertex operators computed in Section 7.8 allow an elegant compact expression of the representation in the bosonic Fock space $B_{\mathbb{C}} := B_\infty \otimes_{\mathbb{Z}} \mathbb{C}$ of the infinite-dimensional Lie algebra $gl(\infty)$ of infinite matrices with complex entries all zero but finitely many.
- We want to emphasize, once more, that to compute vertex operators, as in the last two chapters, the main tool we used was the formalism of Schubert calculus phrased as in [37] or [96, 97]. Our guiding principles were i) that $F_i^r \otimes \mathbb{Q}$ can be interpreted as the $\mathbb{Q}$-vector space generated by the generalized Wronskians (in the sense of [49, 50]; see also [40]), associated with a fundamental system of solutions to the generic linear ODE of order r, and ii) that the boson–fermion correspondence can be seen as a generalization of the classical result according which the derivative of the Wronskian of a fundamental system of solutions of a linear ODE is proportional to itself ([6, p. 195]).
- A generalization of the vertex operators $\Gamma(z), \Gamma^*(z)$ of Section 7.8 has been recently considered by Jing and Rozhkovskaya in [71], based on a truly intriguing generalization of the elementary and complete symmetric polynomials. These authors rely on arguments based on the Jacobi–Trudy formula and use, as in this book and in [31], the variables e_i and h_j, instead of the usual $(x_1, x_2, \ldots)$, like in [78]. The connection of the subject with Schubert calculus, notably Giambelli formula or Jacobi–Trudi formula, and symmetric functions was, on the other hand, already recognized in a more or less explicit way by many authors. We remember here, with no claim to be exhaustive, the references [140, p. 48], [3, 26, 119] and also [40, p. 82]. Vertex operators have also been considered in the realm of combinatorics, in connection with divided differences and Schubert polynomials, by Lascoux and Pragacz [100].
- Regarding the spin representation of the free charged fermion Clifford algebra, standard references are [32, 73, 78]. The connection of the boson–fermion correspondence with infinite exterior power is dealt with in many papers, also not strictly related to vertex operators, like in the appendix, *Odd and Ends*, by Kazarian and Lando, to the paper [84]. That presents a purely algebro-geometric proof of the Witten conjecture [150], first proved by Kontsevich [88] or in the very recent [21], to face the Bouchard–Mariño conjecture, and a new proof of the celebrated ELSV [23] formula (after Ekedahl, Lando, M. Shapiro and Vainshtein). Infinite exterior powers have been systematically used also in [9, 76, 107, 121].
- Computing in two different ways, the character of the ‘fermionic’ representation F of the Clifford algebra gives rise to the famous *Jacobi triple product* identity (see, e.g. [78, p. 186] of [73, p. 90]. As remarked in [31], the boson–fermion correspondence can be viewed as a ‘categorification’ of this identity.

- We have said very little about the representation of the oscillator Heisenberg algebra, and we spent no word about the bosonic/fermionic representation of the Virasoro algebra (central extension of the Lie algebra of the vector fields on the unit circle). This is treated in details in [78]. See [30, 72], for more on the representation of the Heisenberg algebra and its generalization. Another beautiful point of view which deserves to be aware of is the amazing relationship between the Heisenberg algebra and the geometry of the Hilbert scheme of points of a surface as developed by Grojnowski, Lehn and Nakajima [56, 102, 118]. This story is beautifully told in the comprehensive lecture notes [25].

7.10 Exercises

7.10.1. Using the fact that $u_{-j} = t^j u_0$ for all $j \geq 0$, prove that $B_r[[t]]$ is the completion of $\mathbb{U}_r \otimes_{\mathbb{Z}} B_r$ with respect to the topology in which $\{(t^m u_0)\}_{m>0}$ is a fundamental system of neighbourhoods of 0.

7.10.2. Let $\boldsymbol{\lambda} \in \mathcal{P}_r$. By substituting the expression

$$u_{-j+1+\lambda_j} = \mathrm{U}_0(u_{-j+1+\lambda_j})u_0 + \mathrm{U}_1(u_{-j+1+\lambda_j})u_{-1} + \cdots + \mathrm{U}_{r-1}(u_{-j+1+\lambda_j})u_{-r+1},$$

prescribed by the *universal Cauchy formula* (2.3.8), into

$$\Phi^r_{0+\boldsymbol{\lambda}} := u_{\lambda_1} \wedge \cdots \wedge u_{-r+1+\lambda_r} \wedge \Phi^r_{-r},$$

prove directly that $\Phi^r_{0+\boldsymbol{\lambda}} = \Delta_{\boldsymbol{\lambda}}(H_r)\Phi^r_0$, defines the *boson–fermion correspondence* $B_r \to B_r \otimes F^r_0$. (*Hint:* $\mathrm{U}_0(u_j) = h_j, \mathrm{U}_1(u_j) = h_{j+1} - e_1 h_j, \ldots$)

7.10.3. Consider the equation $E_r(z)\exp(\mathbf{x}_r(z)) = 1$, where $\mathbf{x}_r(z) = \sum_{i\geq 1} x_i z^i$. Prove that differentiating with respect to z gives

$$\partial_z E_r(z)\exp(\mathbf{x}_r(z)) + \partial_z \mathbf{x}_r(z) = 0.$$

Use such formula to deduce the recursions (7.9) and (7.10).

7.10.4. Prove formula (7.45) in two different ways: i) via a direct computation, looking at the first few terms of the contraction $\mathbb{X}^*_r(z) \lrcorner \Phi^r_{i+\boldsymbol{\lambda}}$ (hint: try first with $\boldsymbol{\lambda} = (0)$, the null partition) and ii) by studying the transpose action of D^j on $\mathbb{U}^*_r$ and obtaining the equality within a line.

7.10.5. Prove that

$$\mathbb{X}_r(z) = \frac{1}{E_r(z)} \sum_{j\geq 0} \left(\frac{t}{z}\right)^n = \frac{z}{E_r(z)} \cdot i_{z,t} \frac{1}{z-t}, \tag{7.61}$$

where $i_{z,t}(f(z,t))$ denotes the expansion of $f(t,z) \in \mathbb{Z}[t^{\pm 1}, z^{\pm 1}, (t-z)^{\pm 1}]$ in powers of t/z (Cf. [73, p. 16]). (*Hint:* multiply $\sum_{j\in\mathbb{Z}} u_j z^j$ by $E_r(z)$, and then use the fact that $\mathfrak{p}_r(D)u_{-j-r} = t^j$ and $\mathfrak{p}_r(D)u_{-r+1+j} = 0$ for all $j \geq 0$.)

7.10.6. Let r, n be positive integers. Show that for each integer $k \geq 0$,

$$\Delta_{(n,1^k)}(H_r) + \sum_{j=1}^{n-1}(-1)^j e_j \Delta_{(n-j,1^k)}(H_r) + (-1)^n e_{n+k} = 0, \tag{7.62}$$

where $(n, 1^k)$ denotes the partition $(n, \underbrace{1, \ldots, 1}_{k \text{ times}})$. If $n + k > r$, then $e_{n+k} = 0$, and formula (7.62) reduces to

$$\Delta_{(n,1^k)}(H_r) + \sum_{j=1}^{n-1}(-1)^j e_j \Delta_{(n-j,1^k)}(H_r) = 0.$$

Conclude that for all $r \geq 1$,

$$\sum_{n\geq 0} \Delta_{(1+n,1^{r-1})}(H_r)z^n = \frac{e_r}{E_r(z)}.$$

7.10.7. This exercise is not strictly related to the matter of this chapter but is of the same nature of 7.10.6. Let us work in $B_3 = \mathbb{Z}[e_1, e_2, e_3]$ with the series $u_0 = H_3(t) = \sum_{n\geq 0} h_n t^n$, such that $E_3(t)H_3(t) = 1$. To make notation shorter, let us set $h_{\boldsymbol{\lambda}} := \Delta_{\boldsymbol{\lambda}}(H_3)$ for all $\boldsymbol{\lambda} \in \mathcal{P}_3$.

i) For all $n \geq 0$, $h_{(n+1,n)} - e_1 h_{(n,n)} + e_3 h_{(n-1,n-1)} = 0$, putting $h_{(-1,-1)} = 0$ by convention;
ii) For all $n \geq 0$, $h_{(n+3,n+3)} - e_2 h_{(n+2,n+2)} + e_1 e_3 h_{(n+1,n+1)} + e_3^2 h_{(n,n)} = 0$;
iii) For all $n \geq 0$, $\displaystyle\sum_{n\geq 0} h_{(n,n)} z^n = \frac{1}{1 - e_2 z + e_1 e_3 z^2 - e_3^2 z^3}$;
iv)

$$\sum_{(\lambda_1 \geq \lambda_2 \geq 0)} h_{(\lambda_1,\lambda_2)} t_1^{\lambda_1} t_2^{\lambda_2} = \frac{1 - e_3 t_1^2 t_2}{E_3(t_1)(1 - e_2 t_1 t_2 + e_2 e_3 t_1^2 t_2^2 - e_3^2 t_1^3 t_2^3)}, \tag{7.63}$$

where t_1 and t_2 are two formal variables.

7.10.8. Formula (7.63) displayed at item iv) of Exercise 7.10.7 holds for $r = 2$ as well. In this case $e_3 = 0$, and the formula specializes to

$$\sum_{\boldsymbol{\lambda}\in\mathcal{P}_2} \Delta_{\boldsymbol{\lambda}}(H_2)\mathbf{t}^{\boldsymbol{\lambda}} = \frac{1}{E_2(t_1)(1 - e_2 t_1 t_2)}, \tag{7.64}$$

where we set $\mathbf{t}^{\boldsymbol{\lambda}} = t_1^{\lambda_1} t_2^{\lambda_2}$:

i) Prove formula (7.64) independently;
ii) Exploiting the expertise gained in 7.10.7 item iv), find a rational function $F(t_1,t_2,t_3) \in B_3(t_1,t_2,t_3)$ such that

$$\sum_{\boldsymbol{\lambda}\in\mathcal{P}_3} \Delta_{\boldsymbol{\lambda}}(H_3)\mathbf{t}^{\boldsymbol{\lambda}} = F(t_1,t_2,t_3),$$

where we have set $\mathbf{t}^{\boldsymbol{\lambda}} = t_1^{\lambda_1}t_2^{\lambda_2}t_3^{\lambda_3}$.

7.10.9. Let $\boldsymbol{\lambda}$ be a partition of length $0 \le k \le r$ and let $0 \le j \le k$. Prove that

$$t^j \wedge \Phi^r_{-1+\boldsymbol{\lambda}} = (-1)^{k-j}e_{k-j}u_{-k} \wedge \Phi^r_{-1+\boldsymbol{\lambda}} = (-1)^j\overline{D}_{-j}\Phi^r_{0+\boldsymbol{\lambda}}.$$

(*Hint:* use the fact that $\mathfrak{p}_r(D)u_{-j-r} = t^j$.)

7.10.10. Show that

$$\mathbb{X}_r(z) \wedge \Phi^r_{-1+\boldsymbol{\lambda}} = \frac{1}{E_r(z)}\left(u_0 + \frac{E_1(z)}{z}u_{-1} + \cdots + \frac{E_r(z)}{z^r}u_{-r}\right)$$

Use the formula to compute

$$\mathbb{X}_2(z) \wedge \Phi^2_{-1+(1,1)} \qquad \text{and} \qquad \mathbb{X}_3(z) \wedge \Phi^3_{-1+(1,1)}.$$

What is the difference between the two outputs?

7.10.11. For each $\boldsymbol{\lambda} \in \mathcal{P}_r$ and $1 \le j \le r$, let

$$\left[\hat{\partial}_i\,,\,\mathbb{X}_r(z)\wedge\right]\Phi^r_{i+\boldsymbol{\lambda}} := \hat{\partial}_i(\mathbb{X}_r(z) \wedge \Phi^r_{i+\boldsymbol{\lambda}}) - \mathbb{X}_r(z) \wedge \hat{\partial}_i\Phi^r_{i+\boldsymbol{\lambda}}$$

and

$$\left[\widehat{D^n}\,,\,\mathbb{X}_r(z)\wedge\right]\Phi^r_{i+\boldsymbol{\lambda}} := \widehat{D^n}(\mathbb{X}_r(z) \wedge \Phi^r_{i+\boldsymbol{\lambda}}) - \mathbb{X}_r(z) \wedge \widehat{D^n}\Phi^r_{i+\boldsymbol{\lambda}}.$$

Prove that

$$\begin{cases} \left[\hat{\partial}_i\,,\,\mathbb{X}_r(z)\wedge\right] = \dfrac{z^jE_{r-j}}{E_r(z)}\mathbb{X}_r(z)\wedge \\[2ex] \left[\widehat{D^n}\,,\,\mathbb{X}_r(z)\wedge\right] = \dfrac{1}{z^n}\mathbb{X}_r(z)\wedge \end{cases}.$$

Taking the limit for $r \to \infty$, one obtains formula (5.22a) of [78].

7.10.12. As in Section 7.8, let $B := \mathbb{Q}[h_1, h_2, \ldots]$, and for all $j \in \mathbb{Z}$, let, as usual, $u_j := \sum_{n\geq 0} h_{n+j} z^n \in B[[z]]$. Define

$$\frac{\partial u_j}{\partial x_i} := \sum_{n\geq 0} \frac{\partial h_{n+j}}{\partial x^i} z^n.$$

For an arbitrarily chosen positive integer n, show that u_j is a solution of the Hirota bilinear form (1.18) of the KP equation, by setting $x = x_n$, $y = x_{2n}$ and $t = x_{3n}$.

References

1. J. Abbott, A.M. Bigatti, G. Lagorio, CoCoA–5: a system for doing computations in commutative algebra (2015). Available via http://cocoa.dima.unige.it
2. R.P. Agarwal, *Difference Equations and Inequalities*, 2nd edn. (Marcel Dekker, New York, 2000)
3. A. Alexandrov, V. Kazakov, S. Leurent, Z. Tsuboi, A. Zabrodin, Classical tau-function for quantum spin chains. J. High Energy Phys. **64** (2013). doi:10.1007/JHEP09(2013)064. http://link.springer.com/article/10.1007%2FJHEP09%282013%29064
4. D. Anderson, Introduction to equivariant cohomology in algebraic geometry, in *Contributions to Algebraic Geometry. Impanga Lecture Notes*, ed. by P. Pragacz. EMS Series of Congress Reports, European Mathematical Society Publishing House, Zurich (2012), pp. 71–92. ISBN 978-3-03719-114-9. http://www.ems-ph.org/books/book.php?proj_nr=154. doi:10.4171/114
5. E. Arbarello, Sketches of KdV, in *Symposium in Honor of C. H. Clemens (Salt Lake City, UT, 2000)*. Contemporary Mathematics, vol. 312 (American Mathematical Society, Providence, 2002), pp. 9–69
6. V.I. Arnold, *Ordinary Differential Equations* (MIT, Cambridge, 1978). Translated from the Russian by Richard A. Silverman
7. M.F. Atiyah, I.G. MacDonald, *Introduction to Commutative Algebra* (Addison-Wesley, London, 1969)
8. A. Bertram, Quantum schubert calculus. Adv. Math. **128**(2), 289–305 (1997)
9. S. Bloch, A. Okounkov, The character of the infinite wedge representation. Adv. Math. **149**, 1–60 (2000)
10. R.E. Borcherds, Monstrous moonshine and monstrous Lie superalgebras. Invent. Math. **109**, 405–444 (1992)
11. R.E. Borcherds, What is a vertex algebra? (Expository). Unpublished Arbeitstagung talk (1997). Available via https://math.berkeley.edu/~reb/papers/bonn/bonn.pdf
12. A. Buch, Quantum cohomology of Grassmans. Compos. Math. **137**(2), 227–235 (2003)
13. R. Camporesi, Linear ordinary differential equations. Revisiting the impulsive response method using factorization. Int. J. Math. Educ. Sci. Technol. **42**(4), 497–514 (2011)
14. J. Cordovez, L. Gatto, T. Santiago, Newton binomial formulas in Schubert calculus. Rev. Mat. Complut. **22**(1), 129–152 (2009)
15. L. Costa, S. Marchesi, R.M. Miró-Roig, Tango bundles on Grassmannians. Math. Nachr. 1–12 (2015). http://onlinelibrary.wiley.com/doi/10.1002/mana.201500015/abstract. doi:10.1002/mana.201500015
16. S.C. Coutinho, *A Primer of Algebraic D-Modules*. London Mathematical Society Student Texts, vol. 33 (Cambridge University Press, Cambridge, 1995)

L. Gatto, P. Salehyan, *Hasse-Schmidt Derivations on Grassmann Algebras*, IMPA Monographs, DOI 10.1007/978-3-319-31842-4

17. P. Cull, M. Flahive, R. Robson, *Difference Equations (from Rabbits to Chaos)* (Springer, New York, 2005)
18. E. Date, M. Jimbo, M. Kashiwara, T. Miwa, Operators approach to the Kadomtsev-Petviashvili equation, Transformation groups for soliton equations III. J. Phys. Soc. Jpn. **50**, 3806–3812 (1981)
19. E. Date, M. Jimbo, M. Kashiwara, T. Miwa, Transformation groups for soliton equations. VI. KP hierarchies of orthogonal and symplectic type. J. Phys. Soc. Jpn. **50**, 3813–3818 (1981)
20. R. Donagi, On the geometry of Grassmannians. Duke Math. J. **44**(4), 795–835 (1977)
21. P. Dunin-Barkowski, M. Kazarian, N. Orantin, S. Shadrin, L. Spitz, Polynomiality of Hurwitz numbers, Bouchard-Mariño conjecture, and a new proof of the ELSV formula. Adv. Math. **279**, 67–103 (2015)
22. D. Eisenbud, *Commutative Algebra, with a View Toward Algebraic Geometry*. GTM, vol. 150 (Springer, New York, 1994)
23. T. Ekedahl, S. Lando, M. Shapiro, A. Vainshtein, Hurwitz numbers and intersections on moduli spaces of curves. Invent. Math. **146**(2), 297–327 (2001)
24. T. Ekedahl, D. Laksov, Splitting algebras, symmetric functions and Galois theory. J. Algebra Appl. **4**(1), 59–75 (2005)
25. G. Ellingsrud, L. Göttsche, *Hilbert Schemes of Points and Heisenberg Algebras*. School on Algebraic Geometry (Trieste, 1999). ICTP Lecture Notes Series, vol. 1 (The Abdus Salam International Centre for Theoretical Physics, Trieste, 2000), pp. 59–100
26. V. Enolski, J. Harnard, Schur function expansion of Kadomtsev-Petviashvili τ-functions associated with algebraic curves. Uspekhi Mat. Nauk **66**(4)(400), 137–178 (2011) (Russian). Translation in Russian Math. Surveys **66**(4), 767–807 (2011)
27. E. Esteves, Jets of singular foliations. Rend. Sem. Mat. Fis. Univ. Pol. Torino **71**(3–4), 401–420 (2013)
28. L. Falcidieno, A. Luvinson, A numerical approach to computing the companion matrix exponential. CSELT Tech. Rep. **4**, 69–71 (1975)
29. E. Frenkel, Vertex algebras and algebraic curves. Séminaire Bourbaki, vol. 1999/2000, Astérisque No. **276**, 299–339 (2002)
30. I. Frenkel, J. Lepowsky, A. Meurman, *Vertex Operator Algebras and the Monster*. Pure and Applied Mathematics, vol. 134 (Academic, Boston, 1988)
31. I. Frenkel, I. Penkov, V. Serganova, A categorification of the boson–fermion correspondence via representation theory of $\mathfrak{sl}(\infty)$ (2014). arXiv:1405.7553
32. E. Frenkel, D.B. Zvi, *Vertex Algebra and Algebraic Curves*. Mathematical Survey and Monographs, vol. 88, 2nd edn. (American Mathematical Society, Providence, 2004)
33. W. Fulton, *Young Diagrams*. London Mathematical Society Student Texts, vol. 35 (Cambridge University Press, Cambridge, 1997)
34. W. Fulton, *Intersection Theory*, 2nd edn. (Springer, Berlin, 1998)
35. W. Fulton, *Equivariant Cohomology in Algebraic Geometry*. Eilenberg Lectures, Columbia University, Notes by D. Anderson (2007). Available at http://w3.impa.br/~dave/eilenberg/
36. S. Galkin, V. Golyshev, H. Iritani, Gamma classes and quantum cohomology of fano manifolds: gamma conjectures (2014). arXiv:1404.6407v1
37. L. Gatto, Schubert calculus via Hasse–Schmidt derivations. Asian J. Math. **9**(3), 315–322 (2005)
38. L. Gatto, *Schubert Calculus: An Algebraic Introduction*. Publicações Matemáticas do IMPA, 25° Colóquio Brasileiro de Matemática (Instituto Nacional de Matemática Pura e Aplicada (IMPA), Rio de Janeiro-Brazil, 2005)
39. L. Gatto, The Wronskian and its derivatives. Atti Acc. Peloritana dei Pericolanti, Phys. Math. Nat. Sci. **89**(2), 14 pp. (2011)
40. L. Gatto, *Linear ODEs: An Algebraic Perspective*. XXII Escola de Algebra, Salvador-Brazil (2012) Publicações Matemáticas do IMPA, vol. 40 (2012). Available via http://www.impa.br/opencms/pt/biblioteca/pm/PM_40.pdf

41. L. Gatto, S. Greco, Algebraic curves and differential equations: an introduction, in *The Curves Seminar at Queen's*, vol. VIII (Kingston, ON, 1990/1991). Exp. B. Queen's Papers in Pure and Appl. Math. **88** (Kingston, 1991), 69 pp. Available via http://calvino.polito.it/~gatto/public/papers/ACI.pdf
42. L. Gatto, D. Laksov, From linear recurrence relations to linear ODEs with constant coefficients. J. Algebra Appl. **15**(6), 1650109 (23 pp.) (2016). http://www.worldscientific.com/doi/abs/10.1142/S0219498816501097?src=recsys. doi:10.1142/S0219498816501097
43. L. Gatto, P. Salehyan, Families of special weierstrass points. C. R. Math. Acad. Sci. Paris **347**(21–22), 1295–1298 (2009)
44. L. Gatto, P. Salehyan, The boson–fermion correspondence from linear ODEs. J. Algebra **415**, 162–183 (2014). http://www.sciencedirect.com/science/article/pii/S0021869314003263. doi:10.1016/j.jalgebra.2014.05.030
45. L. Gatto, P. Salehyan, Vertex operators arising from linear ODEs (2013). arXiv:1310.5132v1
46. L. Gatto, P. Salehyan, *On the Plücker equations defining the Grassmann Cone* (2016). arXiv:1603.00510v1
47. L. Gatto, T. Santiago, Schubert calculus on a Grassmann algebra. Can. Math. Bull. **52**(2), 200–212 (2009)
48. L. Gatto, T. Santiago, Equivariant schubert calculus. Ark. Mat. **48**, 41–55 (2010). http://link.springer.com/article/10.1007/s11512-009-0093-5. doi:10.1007/s11512-009-0093-5
49. L. Gatto, I. Scherbak, On one property of one solution of one equation or linear ODE's, Wronskians and schubert calculus. Moscow Math. J. to the seventy-fifth anniversary of V. I. Arnold **12**(2), 275–291 (2012)
50. L. Gatto, I. Scherbak, On generalized Wronskians, in contributions to algebraic geometry. Impanga lecture notes, ed. by P. Pragacz. EMS Series of Congress Reports (2012), pp. 257–296. ISBN 978-3-03719-114-9. http://www.ems-ph.org/books/book.php?proj_nr=154. doi:10.4171/114. Available via http://xxx.lanl.gov/pdf/1310.4683.pdf. arXiv:1310.4683
51. L. Gatto, I. Scherbak, Remarks on the Cayley-Hamilton Theorem. http://arxiv.org/pdf/1510.03022. arXiv:1510.030. http://arxiv.org/pdf/1510.03022, 22v1
52. M. Gekhtman, A. Kasman, On KP generators and the geometry of the HBDE. J. Geom. Phys. **56**(2), 282–309 (2006)
53. F. Gesztesy, R. Weikard, Elliptic algebro-geometric solutions of the KdV and AKNS hierarchies – an analytic approach. Bull. AMS **35**(4), 271–317 (1998)
54. V. Golyshev, L. Manivel, Quantum cohomology and the Satake isomorphism. http://arxiv.org/abs/1106.3120. arXiv:; http://arxiv.org/abs/1106.3120, 1106.3120.
55. P. Griffiths, J. Harris, *Principles of Algebraic Geometry* (Wiley Classics Library, New York, 1994)
56. I. Grojnowski, Instantons and affine algebras. I. The Hilbert scheme and vertex operators. Math. Res. Lett. **3**(2), 275–291 (1996)
57. M. Hall Jr., An isomorphism between linear recurring sequences and algebraic rings. Trans. Am. Math. Soc. **44**, 196–218 (1938)
58. J. Harris, *Algebraic Geometry*. GTM, vol. 133 (Springer, New York, 1992)
59. R. Hartshorne, *Algebraic Geometry*. GTM, vol. 52 (Springer, New York/Heidelberg, 1977)
60. H. Hasse, F.K. Schmidt, Noch eine Bergründung der Theorie der höheren Dif- ferentialquotienten in einem algebraischen Funktionenkörper einer Unbestimmten. J. Reine Angew. Math. **177**, 215–237 (1937)
61. R. Heluani, Florence Lectures on sheaves of Vertex Algebras (2013). Available via http://w3.impa.br/~heluani/files/florence-lectures.pdf
62. R. Hirota, Exact solutions of the Korteveg de Vries equation for multiple collision of solitons. Phys. Rev. Lett. **27**, 1192–1194 (1971)
63. R. Hirota, Direct methods of finding exact solutions of nonlinear evolution equations, in *Bäcklund Transformations, Inverse Scattering Methods, Solitons and Their Applications*. Lecture Notes in Mathematics, vol. 515 (Springer, New York, 1976), pp. 40–68
64. W. Hodge, D. Pedoe, *Methods of Algebraic Geometry*, vol. II, Book III (Cambridge University Press, Cambridge, 1947)

65. R. Hotta, K. Takeuchi, T. Tanisaki, *D-modules, Perverse Sheaves, and Representation Theory*. Progress in Mathematics, vol. 236 (Birkhäuser, Boston, 2008)
66. Y. Huang, C. Li, On equivariant quantum Schubert calculus for G/P. J. Algebra **441**, 21–56 (2015)
67. A. Iarrobino, Deforming complete intersection Artin algebras, in *Singularities*, Part 1 (Arcata, Calif. 1981). Proceedings of Symposia in Pure Mathematics, vol. 40 (American Mathematical Society, Providence, 1983), pp. 593–608
68. A. Iarrobino, The family G_T of graded artinian quotiens of $k[x, y]$ of given Hilbert function. Commun. Algebra **33**(8), 3863–3916 (2003)
69. A.J. Jerri, *Linear Difference Equations with Discrete Transform Methods* (Springer Science+Business Media, Dordrecht, 1996)
70. M. Jimbo, T. Miwa, Solitons and infinite dimensional lie algebras. Publ. RIMS, Kyoto Univ. **19**, 943–1001 (1983)
71. N. Jing, N. Rozhkovskaya, Vertex Operators Arising from Jacobi–Trudy Identities (2015). arXiv: 1411.4725v2
72. V.G. Kac, *Infinite Dimensional Lie Algebras*. Progress in Mathematics, vol. 44 (Birkhäuser, Boston, 1984)
73. V.G. Kac, *Vertex Algebras for Beginners*. University Lecture Series, vol. 10 (American Mathematical Society, Providence, 1996)
74. V.G. Kac, *Vertex Algebras*. Course Lecture Notes (MIT, Spring, 2003)
75. V.G. Kac, D.A. Kazhdan, J. Lepowsky, R.L. Wilson, Realization of the basic representations of the euclidean lie algebras. Adv. Math. **42**, 83–112 (1981)
76. V.G. Kac, D. Peterson, Lectures on the infinite wedge representation and the MKP hierarchy. Seminaire de Math. Superieures. Les Presses de L'Universite de Montreal **102**, 141–186 (1986)
77. V.G. Kac, A.K. Raina, *Highest Weight Representations of Infinite Dimensional Lie Algebras*. Advanced Studies in Mathematical Physics, vol. 2 (World Scientific, Singapore, 1987)
78. V.G. Kac, A.K. Raina, N. Rozhkovskaya, *Highest Weight Representations of Infinite Dimensional Lie Algebras*. Advanced Series in Mathematical Physics, vol. 29, 2nd edn. (World Scientific, Singapore, 2013)
79. V.G. Kac, M. Wakimoto, Exceptional hierarchies of soliton equations, in *Theta Functions, Bowdoin 1987*, ed. by L. Ehrenpreis, R.C. Gunning. Proceedings of Symposia in Pure Mathematics, vol. 49, Part 1, AMS, Providence, Rhode Island (1989), pp. 191–238
80. B.B. Kadomtsev, V.I. Petviashvili, On the stability of solitary waves in weakly dispersive media. Sov. Phys. Dokl. **15**, 539–541 (1970)
81. A. Kasman, Glimpses on solitons. J. Am. Math. Soc. **20**(4), 1079–1089 (2007)
82. A. Kasman, *Glimpses on Soliton Theory* (The Algebra and Geometry of Nonlinear PDEs). SML, vol. 54 (American Mathematical Society, Providence, 2010)
83. A. Kasman, K. Pedings, A. Reiszl, T. Shiota, Universality of rank 6 Plücker relations and Grassmann cone preserving maps. Proc. Am. Math. Soc. **136**(1), 77–87 (2008)
84. M.E. Kazarian, S.K. Lando, An algebro-geometric proof of Witten's conjecture. J. Am. Math. Soc. **20**(4), 1079–1089 (2007)
85. G. Kempf, D. Laksov, The determinantal formula of Schubert calculus. Acta Math. **32**, 153–162 (1974)
86. S.L. Kleiman, The transversality of a general translate. Comp. Math. **38**, 287–297 (1974)
87. S.L. Kleiman, D. Laksov, Schubert calculus. Am. Math. Mon. **79**, 1061–1082 (1972)
88. M. Kontsevich, Intersection theory on the moduli space of curves and the matrix Airy function. Commun. Math. Phys. **147**(1), 1–23 (1992)
89. D.J. Korteweg, G. de Vries, On the change of form of long waves advancing in a rectangular canal, and on a new type of long stationary waves. Philos. Mag. 5 **39**, 422–443 (1985)
90. I.M. Krichever, Integration of nonlinear equations by the methods of algebraic geometry. Funct. Anal. Appl. **11**(1), 12–26 (1977)

91. D. Laksov, Remarks on Giovanni Zeno Giambelli's work and life. Algebra and geometry (1860–1940): the Italian contribution (Cortona, 1992). Rend. Circ. Mat. Palermo **36**(Suppl. (2)), 207–218 (1994)
92. D. Laksov, Linear maps and linear recursion (Norwegian). Normat **55**(2), 59–77 (2007)
93. D. Laksov, Schubert calculus and equivariant cohomology of grassmannians. Adv. Math. **217**, 1869–1888 (2008)
94. D. Laksov, The formalism of Equivariant Schubert Calculus. Algebra Number Theory **3**(6), 711–727 (2009)
95. D. Laksov, A. Thorup, Weierstrass points and gap sequences for families of curves. Ark. Mat. **32**, 393–422 (1994)
96. D. Laksov, A. Thorup, A determinantal formula for the exterior powers of the polynomial ring. Indiana Univ. Math. J. **56**(2), 825–845 (2007)
97. D. Laksov, A. Thorup, Schubert calculus on grassmannians and exterior products. Indiana Univ. Math. J. **58**(1), 283–300 (2009)
98. D. Laksov, A. Thorup, Splitting algebras and Schubert calculus. Indiana Univ. Math. J. **61**(3), 1253–1312 (2012)
99. S. Lang, *Algebra*, 3rd edn. (Addison Wesley, Reading, 1993)
100. A. Lascoux, P. Pragacz, Orthogonal divided differences and Schubert polynomials, $\widetilde{P}$-functions, and vertex operators, dedicated to William Fulton on the occasion of his 60th birthday. Mich. Math. J. **48**, 417–441 (2000)
101. P.D. Lax, Integrals of nonlinear equations of evolution and solitary waves. Commun. Pure Appl. Math. **21**, 467–490 (1968)
102. M. Lehn, Chern classes of tautological sheaves on Hilbert schemes of points on surfaces. Invent. Math. **136**(1), 157–207 (1999)
103. J. Lepowsky, H. Li, *Introduction to Vertex Operator Algebras and Their Representations*. Progress in Mathematics, vol. 227 (Birkhäuser, Boston, 2004)
104. J. Lepowsky, R.L. Wilson, Construction of the affine Lie algebra $A_1^{(1)}$. Commun. Math. Phys. **62**, 43–53 (1978)
105. I.H. Leonard, The Matrix Exponential. SIAM Rev. **38**(3), 507–512 (1996)
106. C. Li, V. Ravikumar, Equivariant Pieri rules for isotropic Grassmannians. Math. Ann. (2015). doi:10.1007/s00208-015-1266-0
107. A. Licata, A. Savage, Vertex operators and the geometry of moduli spaces of framed torsion-free sheaves. Sel. Math. New Ser. **16**, 201–240 (2010). http://link.springer.com/article/10.1007/s00029-009-0015-1, doi:10.1007/s00029-. http://link.springer.com/article/10.1007/s00029-009-0015-1, 009-0015-1
108. E. Liz, A note on the matrix exponential. SIAM Rev. **40**(3), 700–702 (1998)
109. I.G. Macdonald, *Symmetric Functions and Hall Polynomials* (Clarendon Press, Oxford, 1979)
110. L. Manivel, *Symmetric Functions, Schubert Polynomials and Degeneracy Loci*, translated from the 1998 French original by John R. Swallow. SMF/AMS Texts and Monographs, vol. 6. Cours Spécialisés [Specialized Courses], vol. 3 (American Mathematical Society/Société Mathématique de France, Providence/Paris, 2001)
111. M. Marcus, *Finite Dimensional Multilinear Algebra*, Part II (Marcel Dekker, New York, 1975)
112. H. Matsumura, *Commutative Ring Theory* (Cambridge University Press, Cambridge, 1989)
113. A.V. Mikhalev, A.A. Nechaev, Linear recurring sequences over modules. Acta Appl. Math. **42**(2), 161–202 (1996)
114. J.W. Milnor, J.D. Stasheff, *Characteristic Classes*, Study 76 (Princeton University Press, Princeton, 1974)
115. S. Morier-Genoud, V. Ovsienko, R.E. Schwartz, S. Tabachnicov, Linear difference equations, frieze patterns and combinatorial gale transform. Forum Math. Sigma **2**, e22, 45 pp. (2014)
116. S. Mukai, Curves and Grassmannians, in *Algebraic Geometry and Related Topics (Inchon, 1992)*. Lecture Notes Algebraic Geometry, vol. I (Cambridge University Press, Cambridge, 1993), pp. 19–40

117. M. Mulase, Algebraic theory of the KP equations, in *Perspectives in Mathematical Physics*. Lecture Notes Mathematical Physics, vol. III (Cambridge University Press, Cambridge, 1994), pp. 151–217
118. H. Nakajima, *Lectures on Hilbert Schemes of Points on Surfaces*. University Lecture Series, vol. 18 (American Mathematical Society, Providence, 1999)
119. Y.A. Neretin, A remark on the boson-fermion correspondence. Coherent states, differential and quantum geometry (Białowieża, 1997). Rep. Math. Phys. **43**(1–2), 267–269 (1999)
120. H. Niederhausen, Catalan Traffic at the Beach. Electron. J. Comb. **9**(R33), 1–17 (2002)
121. A. Okounkov, The uses of random partitions, in *XIVth International Congress on Mathematical Physics* (World Scientific Publishing, Hackensack, 2005), pp. 379–403
122. P. Pragacz, La vita e l'opera di Józef Maria Hoene-Wroński. Atti Accad. Peloritana Pericolanti Cl. Sci. Fis. Mat. Natur. (Italian) **89**(1), C1C8901001, 19 pp. (2011)
123. A. Pressley, G. Segal, *Loop groups*. Oxford Mathematical Monographs, Oxford Science Publications (The Clarendon Press/Oxford University Press, New York, 1986)
124. E. Previato, Seventy years of spectral curves: 1923–1993, in *Integrable Systems and Quantum Groups* (Montecatini Terme, 1993). Lecture Notes in Mathematics, vol. 1620 (Springer, New York, 1996), pp. 419–481
125. E.J. Putzer, Avoiding the Jordan canonical form in the discussion of linear systems with constant coefficients. Am. Math. Mon. **73**, 2–7 (1996)
126. K. Ranestad, Private communication. Stockholm (2005)
127. P. Ribenboim, Dérivations d'ordre supérieur dans les catégories semi-additives. Cahiers de Toplogie et Géom. Differ. **11**, 437–466 (1970)
128. P. Ribenboim, Higher order Dérivations of rings I. Revue Roumaine Math. Pures Appl. **16**, 77–110 (1971)
129. P. Ribenboim, Higher order Dérivations of rings II. Revue Roumaine Math. Pures Appl. **16**, 245–272 (1971)
130. P. Ribenboim, Higher order Dérivations of modules. Portugaliae Math. **39**(1–4), 367–383 (1980)
131. J.S. Russel, Report on waves. Report of the fourteenth meeting of the British Association for the Advancement of Science, New York, September 1844 (London, 1845), pp. 311–390
132. T. Santiago, Catalan Traffic and integrals on the Grassmannian of lines. Discret. Math. **308**, 148–152 (2008)
133. M. Sato, Soliton equations as dynamical systems on infinite dimensional grassmann manifolds. RIMS Kokioroku **439**, 30–46 (1981)
134. M. Sato, The KP hierarchy and infinite-dimensional Grassmann manifolds, in *Theta Functions, Bowdoin 1987*, ed. by L. Ehrenpreis, R.C. Gunning. Proceedings of Symposia in Pure Mathematics, vol. 49, Part 1, AMS, Providence, Rhode Island (1989), pp. 51–66
135. S.A. Saymeh, On Hasse–Schmidt higher derivations. Osaka J. Math **23**, 503–508 (1986)
136. I. Scherbak, Gaudin's model and the generating function of the Wroński map. Geometry and topology of caustic–CAUSTICS '02. Banach Center Publication, vol. 62 (Polish Academy of Sciences, Warsaw, 2004), pp. 249–262
137. F.K. Schmidt, Die Wronskische Determinante in Beliebigen Diferenzierbaren Funktion körpern. Math. Z. **45**, 62–74 (1939)
138. B. Shapiro, Algebro-geometric aspects of Heine–Stieltjes theory. J. Lond. Math. Soc. (2) **83**(1), 36–56 (2011)
139. B. Shapiro, M. Shapiro, Linear ordinary differential equations and Schubert calculus, in *Proceedings of the Gökova Geometry–Topology Conference 2010* (International Press, Somerville, MA, 2011), pp. 79–87
140. G. Segal, G. Wilson, Loop groups and equations of KdV type. Inst. Hautes Stud. Sci. Publ. Math. **61**, 5–65 (1985)
141. J.H. Silverman, *The Arithmetic of Elliptic Curves*. GTM, vol. 106 (Springer, New York, 1986)
142. R.M. Skjelnes, Computing Hasse-Schmidt derivations and Weyl restriction over jets. J. Algebra **411**, 114–128 (2014)

143. C. Tomei, *Topics in Spectral Theory*. 30° Colóquio Brasileiro de Matemática (Instituto Nacional de Matemática Pura e Aplicada, Rio de Janeiro, 2015)
144. C. Towse, Generalized Wronskians and Weierstrass weights. Pac. J. Math. **193**(2), 501–508 (2000)
145. I. Vainsencher, *Classes Characteristicas em Geométria Algebrica*. 15° Colóquio Brasileiro de Matemática, Poços de Caldas-Brazil (1985)
146. P. Vojta, Jets via Hasse-Schmidt derivations, in *Diophantine Geometry*. CRM Series, vol. 4 (Ed. Norm., Pisa, 2007), pp. 335–361
147. M. Waddil, Using matrix techniques to establish properties of *k*-order linear recursive sequences, in *Applications of Fibonacci Numbers*, vol. 5, ed. by G.E. Bergum, A.N. Philippou, A.F. Horadam (Kluwer Academic Publ., Dordrecht, 1993), pp. 601–615
148. M. Wakimoto, *Infinite-Dimensional Lie Algebras*, Translated from the 1999 Japanese original by Kenji Iohara. Translations of Mathematical Monographs, vol. 195 (American Mathematical Society, Providence, 2001)
149. A. Weber, Equivariant chern classes and localization theorem. J. Singul. **5**, 153–176 (2012)
150. E. Witten, Two-dimensional gravity and intersection theory on moduli space, in *Surveys in Differential Geometry (Cambridge, MA, 1990)* (Lehigh Univ., Bethlehem, 1991), pp. 243–310
151. H. Yong-Nian, The explicit solution of homogeneous linear ordinary differential equations with constant coefficients. Appl. Math. Mech. (English Edition) **13**(12), 1115–1120 (1992)
152. D. Zagier, Introduction to modular forms, in *From Number Theory to Physics*, ed. by M. Waldschmidt, P. Moussa, J.-M. Luck, C. Itzykson (Springer, New York, 1995), pp. 238–291
153. N. Zierler, Linear recurring sequences. J. Soc. Ind. Appl. Math. **7**, 31–48 (1959)
154. N. Zierler, W.H. Mills, Products of linear recurring sequences. J. Algebra **27**, 147–157 (1973)

Index

L. Gatto, P. Salehyan, *Hasse-Schmidt Derivations on Grassmann Algebras*,
IMPA Monographs, DOI 10.1007/978-3-319-31842-4